SPICES, SEASONINGS, & FLAVORINGS

HOW TO ORDER THIS BOOK

BY PHONE: 800-233-9936 or 717-291-5609, 8AM–5PM Eastern Time

BY FAX: 717-295-4538

BY MAIL: Order Department
Technomic Publishing Company, Inc.
851 New Holland Avenue, Box 3535
Lancaster, PA 17604, U.S.A.

BY CREDIT CARD: American Express, VISA, MasterCard

BY WWW SITE: http://www.techpub.com

HANDBOOK OF

Spices, Seasonings, & Flavorings

Susheela Raghavan Uhl
Horizons Consulting

TECHNOMIC PUBLISHING CO., INC.
LANCASTER • BASEL

Handbook of Spices, Seasonings, and Flavorings
a **TECHNOMIC**® publication

Technomic Publishing Company, Inc.
851 New Holland Avenue, Box 3535
Lancaster, Pennsylvania 17604 U.S.A.

Printed in the United States of America
10 9 8 7 6 5 4 3 2 1

Main entry under title:
Handbook of Spices, Seasonings, and Flavorings

A Technomic Publishing Company book
Bibliography: p. 303
Includes index p. 309

Library of Congress Catalog Card No. 00-104018
ISBN No. 1-56676-931-0

To my parents . . . Pathmavathy Kumaran and Kattery Raghavan

I dedicate this book to my ma and cha who planted the seed of taste within me, and from which my thirst for knowledge of spices and flavors grew. It was this exposure that enabled me to truly appreciate and enjoy many diverse foods.

Contents

Preface

MY gastronomic heritage began while I was growing up in Malaysia. Watching my grandma grinding the soaked rice-lentil mixture for Sunday's breakfast, picking kari leaves for ma's aromatic crab curry and listening to my late cha's (father's) food adventures during meals, all created in me a passion for food, spices and cultures. For mom, cooking was a creative process—every day there had to be something new and different on the table. She never hurried her cooking and never settled for less than the best in her choice of spices. I observed and learned her pride in creating the ultimate flavor and absorbed her approach to freshness, flavor and healthy eating. For cha, food had no boundaries. It surpassed all cultures and religions. He taught me to explore and try all foods and flavors available in Malaysia, whether Chinese, Indian, Indonesian, Thai or Western. Mom's endless search for different tastes and a zest for cooking and cha's enthusiasm and appreciation of different foods gave me an appetite for adventurous eating and a curiosity about new flavors.

Their spirit ultimately influenced my career as a food developer. I have traveled to experience the floating markets of Thailand, to cook on the stone charcoal ovens of Kerala, India, to attend cooking classes in Oaxaca, Mexico, to dine in the country pubs of England and to taste the perfect chili in Texas, U.S. All of this, in search for authentic foods and flavors. I am not alone in my search. Nowadays, food professionals in the U.S. and around the world are continually looking for "new" and unique spice flavorings because of the growing global demand for authentic ethnic and fusion cuisines. People are also seeking natural foods and natural preservatives for healthier lifestyles and natural ways of preventing ailments. I wrote this book because I believe that food developers need to understand and gather a vast knowledge of spices, seasonings and flavorings to meet these new demands and to create and market successful products.

Today's food development is becoming "techno-culinary" by incorporating and connecting technology, and culinary skills, with cultural influences, food trends, nutrition and other disciplines. As such, a food developer needs technical knowledge, creative talent and an understanding of the cultural aspects of the consumer in order to develop successful products. This book incorporates technical information about spices, their varieties, properties and applications, with culinary concepts, food history, market trends and the cultural and flavor profiles of different ethnic populations and different regions of the globe. I designed this book as a tool for the many professionals who develop and market food. The product developer needs to use spices and flavorings with creativity in addition to technical know-how. The chef needs to have some technical information to balance creativity. The flavorist needs to understand the origins and varieties of spices and how they are prepared and used in ethnic cuisines to understand their differing flavor profiles. Nutritionists need to create flavorful menus for a diverse population, while marketers and sales professionals need cultural information and data on food trends. With the information in this book, they can work together to develop successful products.

First and foremost, I have tried to make this book a comprehensive guide to spices. Spices are the building blocks of flavors. They define and intensify flavors and are important tools for providing consistency and color. They create the desired taste, characterize cuisines and differentiate one recipe from another. Understanding spices in their fullest capacities is the cornerstone of successful product or seasoning development. Therefore, this book contains detailed descriptions of each spice, arranged alphabetically. While many reference books on spices include alphabetized descriptions, I believe that the similarity between this book and others ends there.

Consistent with my desire to create a truly comprehensive and global reference on spices, this book goes beyond a dry technical description of spices. It describes each spice's varieties, forms, and chemical components that typify its flavor and color. It includes functional and sensory descriptions and chemical and culinary information that will provoke curiosity and passion in the developer and assist in product development. This book also explains how each spice is used in cuisines around the world, lists global spice blends that contain the spice, describes each spice's folklore and traditional medicinal uses and provides translations of each spice's name in diverse languages.

In researching this book, I also became aware that there were no comprehensive guides to spices and seasoning blends that could help food developers create products based on the popular ethnic cuisines from around the world. Consequently, I have provided detailed descriptions of many varieties of each spice and have included many spices, seasonings and flavorings that are becoming increasingly popular, especially those used in Asian, Latin, Caribbean and African cuisines. In addition, this book goes beyond other spice books by describing other important ingredients found among the world's cuisines that provide flavors, textures, colors and nutritional value to foods. It describes how these flavorings are commonly used with spices to create characteristic ethnic spice blends and seasonings and how they can be used to create new, exciting or authentic flavors.

The last chapter is an important tool for creating ethnic and regional seasonings. It describes major ethnic and regional cuisines and their characteristic flavors and ingredients. Examples of seasoning blends that are significant in creating the typical flavors of these cuisines are described in detail. In addition, it provides many global and regional variations of a seasoning or spice blend, such as sofritos, adobos, curry blends, hot sauces and chile blends. This chapter is not intended as an in-depth study of each region's flavor profile, but it is written to provide a general understanding of some typical flavor profiles of each of these regions and a strong foundation for product development.

In writing this book, I have tried to create a complete modern book on spices, seasonings and flavorings. I have included traditional popular spices and flavorings, as well as those that are growing in popularity and are used to create authentic ethnic, fusion and regional American foods. It is designed to meet the challenges and demands of today's dynamic market. My ultimate aim, however, was to share with the reader some of the enthusiasm for food and sense of adventure that my parents gave to me.

A note on terminology in this book:

Throughout history, the various parts of plants have been cultivated and used for their aromatic, fragrant, pungent or other desirable qualities. This book uses the term "spice" to refer to all of the edible parts of a plant used for flavoring foods—including roots, stems, seeds, rhizomes and the leafy plant parts usually referred to as herbs in European and North American cuisines. There are several reasons for this usage. In the case of herbs, it avoids the shifting definitions of what an herb is, which have varied greatly over time. In addition, not all herbs are used in seasoning foods; many are not edible and do not function as a spice. Moreover, many traditional cultures today do not separate these leafy spices into a distinct herb category. This book attempts to discuss and define spices from a global perspective; therefore, a global approach to defining flavoring ingredients is most appropriate. Finally, as with roots, stems, seeds and flowers and other plant parts, the purpose of these leafy plant parts is to "spice up" food or beverage products, and their collective grouping is the most logical.

Acknowledgments

I wish to thank my daughter, Geeta, for encouraging me to write this book and showing patience for my absence during this period. My sincere gratitude to Bob Roach, who helped me immensely with the editing and organization and who was also my best critic throughout. I wish to thank my family members around the globe, who gave me moral support and help during this time.

I wish to express my thanks and gratitude to Eleanor Riemer, my editor and publisher, for her help with the final editing and organization and for sharing her excitement for writing this book. Finally, I want to thank many of my friends and family [Rachel Yohannes, Winsome Johnson, Karen Berman, Elizabeth Erman, Joe McLaren, Sree Raghavan, and many others] who had encouraged me to write this book or provided me with materials and information. Also, thanks go to friends, colleagues and family, particularly ma and Sathee, who provided help in translating the names of spices into many languages. In addition, my travels around the U.S. and the world, whether for work or for pleasure, have given me great insight and understanding of the many local and global flavors.

CHAPTER 1

Spices in History

TODAY'S search for unique and authentic spices is not new. In ancient times, spices were status symbols in Europe and throughout the Mediterranean for the wealthy who ate them. Spices had an enormous trade value, not only as flavorings for food, but as medicines, preservatives and perfumes.

A brief tour of the history of spices and modern trends will serve as a good introduction to the use of spices in today's global cuisines.

A "SPICY" TALE: A SHORT HISTORY OF THE SPICE TRADE

The history of spices is entwined with exploration, adventure, religious missions, commerce and conquest. Treasured like gold and precious stones, spices have had enormous commercial value in ancient and medieval times. Most spices and flavorings had origins in the tropics or subtropics. They were much sought after in the West, and the quest for spices tremendously changed the course of history.

The East is the birthplace of most popular spices and flavorings. India, Southeast Asia and China have given us anise, basil, cardamom, cinnamon, clove, garlic, ginger, mace, mustard, nutmeg, onions, pepper, star anise, tamarind and turmeric. Other spices, such as bay leaf, coriander, cumin, dill, fennel, fenugreek, rosemary, sage, sesame and thyme came from the Middle East, North Africa and other parts of the Mediterranean. The colder regions of Europe have provided us with juniper and horseradish, while the Americas gave us allspice, annatto, chile peppers, chocolate, epazote and sassafras.

Ancient civilizations, such as Asian Indians, Middle Easterners, Chinese, Aztecs and Incas, have used spices since time immemorial. As in modern times, these cultures spiced their foods to enhance them and to create different flavors. Spices were also used to preserve meats, to disguise tainted foods and disagreeable odors and even to create cosmetics and perfumes.

Early civilizations understood that spices had medicinal value and used them as antidotes for poisons, to help cure diseases and to prevent ailments. During medieval times, spices such as cinnamon, garlic and oregano were used as germicides to battle the spread of the plague.

People also believed that spices had magical properties, and they were used in religious functions and on ceremonial occasions.

EARLY USE OF SPICES IN THE AMERICAS

While stories of most spices begin in the East, a number of the more popular spices and flavorings in use today are native to the Western Hemisphere. Since the dawn of time, Native American Indians—Aztecs, Mayans and Incas—flavored their food and drinks with spices and offered them to their Gods in religious ceremonies. Chile peppers, sweet peppers, allspice, chocolate and vanilla originated in the New World before being introduced to Europe and Asia.

Chile peppers grew wild in the Andes and were used as early as 10,000 years ago. From South America, chile peppers were carried to Central America and the Caribbean. Archaeological excavations in Mexico reveal chile pepper remains dating back to 7000 B.C.

Anthropologists have been unable to define with certainty when chile peppers were first domesticated. It appears that Native Americans began domesticating chilies between 5200 and 3300 B.C. By the time the Spanish arrived in Mexico in the sixteenth century A.D., the Aztecs were growing dozens of pod types. Today, all domesticated cultivars are derived from five domesticated species of chile peppers, and none differ substantially from those domesticated by Native Americans.

In Pre-Columbian Americas, dried chile peppers were used in trade in what is now the Southwestern United States and regions of Mexico. Atole (a corn, cacao bean, sugar and chile pepper drink) and posole (a corn and chile pepper stew) were some of the foods flavored by chile peppers enjoyed by the Aztecs and Mayans.

Vanilla planifolia, a climbing, tropical orchid, grew wild in the hothouse jungles of Central America and northern South America. When the fruit pod of the vanilla orchid fell to the jungle floor before it was ripe, it would ferment and give off a marvelous aroma, which the Aztecs must have noticed. Called tlixochitl or "black pod" by the Aztecs, vanilla pods were harvested from wild climbing vines found in the jungles of southwest Mexico. Later, the Aztecs domesticated this exotic plant and cultivated its vines. They blended the smooth vanilla flavor with chocolatl and honey to create "royal" drinks reserved for the elite of society. By legend, the great Aztec emperor Montezuma presented a chocolate drink flavored with vanilla to the Spanish conquistador Cortez, and he served it in golden goblets. In addition to flavoring, vanilla was used in medicine as a nerve stimulant and was reputed to be an aphrodisiac.

In Mexico and Guatemala, Mayans, Toltecs and Aztecs took the seeds of cocoa pods, roasted them, crushed them into powder on stones and whisked the powder with boiling water to create tchacahoua (Mayan) or tchocoatl (Aztec). This drink, often mixed with chile pepper, honey or ground maize, was considered sacred food.

Allspice is the fruit of an evergreen-type tree that grew wild in southern Mexico, Central America and on several Caribbean islands, including Jamaica and Cuba. The Mayan Indians used allspice berries to help preserve or embalm the bodies of their leaders. The fruit of the unripe allspice berry looks like a large peppercorn, which was sought by early Spanish explorers. Thus, they called these berries pimiento or pepper, from which we get today the name pimento.

THE ASIAN SPICE EMPORIUM

Many of the spices that are popular today are indigenous to India, where they have been savored for thousands of years. The Harappa civilization, one of the first cultures of the Indus valley in northern India, ground saffron and other spices on stones around 3200 B.C.

One of the earliest written records regarding spices appears in the religious scriptures of the Aryan people of north India. The Vedas, written in Sanskrit between 1700 B.C. and 800 B.C., refer to mustard (baja), turmeric (haridra), long pepper (pippali) and sour citrus (jambira).

The Sanskrit language, itself, however, contains words for spices that reflect the well-established use of spices by the most ancient peoples in India. For example, the Sanskrit word for tamarind (chincha) has aboriginal origins. Harida or turmeric comes from the Munda, a pre-Aryan people who lived through much of North India. The Vedas, themselves, refer to a community called Nushadas, whose name translates literally into "turmeric eaters."

The Aryans looked down on some spice use. Vedic literature describes garlic, leeks, mushrooms and onions as native foods despised by the Aryans. Some scholars explain that this aversion arose from the common practice at the time of fertilizing these crops with a manure of human waste.

Later Vedic writings establish that early North Indians were engaged in a far-reaching spice trade. The Vedas report the Aryans using black pepper (maricha) imported from South India and asafoetida (hingu) from Afghanistan. In the Buddhist era (800 to 350 B.C.), we see the introduction in North India of ginger, cumin and cloves, which were generally imported from other parts of Asia.

The origins of ginger have been obscured by its wide domestication. It is native to Southeast Asia, but wild forms are found in India. Cumin appears in Vedic writing around 300 B.C. and appears to be native to the Middle East. The Sanskrit term for cumin, jeeraka, comes from Persian. Clove originated in the Moluccas Islands in Eastern Indonesia. It first appeared in the Ramayana, an Indian epic written between 350 B.C. and 1 A.D. Clove may have originally come to India through Malaysia because the Sanskrit word for clove, lavanga, appears to be derived from the Malay word for clove, bunga lavanga.

The Dravidians were the predominate civilization of South India. They used tamarind, black pepper, lemon, cardamom, cinnamon, cloves, turmeric and pomegranate to flavor their foods. Pepper plants, cardamom and cinnamon grew wild in the south of India, particularly in the states of Kerala and Karnataka. Mysore, in Karnataka, was known for its cardamom, and Kerala was known for its black pepper. While cinnamon also grew wild in South India, the best cinnamon came from Sri Lanka, off the coast of South India. In addition to flavoring foods, spices played a significant role in the religious and cultural lives of early Indian peoples. The colors yellow and orange were considered auspicious and festive because of their connection to the sun. Consequently, saffron and turmeric were used in religious ceremonies and in the important personal occasions in everyday life, such as childbirths, marriages and funerals.

It was a common practice of the Aryans in the north and the Dravidians in the south to mark the forehead with kumkum as a sign of religious respect and auspiciousness. The Aryans used saffron and the Dravidians used turmeric, made alkaline with slaked lime, to make kumkum. Arghya, consisting of water mixed with saffron, flowers and sandalwood powder, was presented to Deities in worship. Akshatas, or rice colored with saffron, was presented to God Vighneswara (presently called Ganesh) in the *Puniah Vachna* ceremony when praying for the removal of an obstacle in life. Saffron was also used to color other religious articles.

Turmeric had erotic significance for Indians and played an important role in wedding ceremonies of Hindus and Muslims. During the nuptial bath called *Nalangu,* the heads of the bride and groom were rubbed with sesame oil, and the exposed parts of their bodies were smeared with turmeric. In some communities, sweets made of nutmeg and saffron were also given to the newlyweds as aphrodisiacs, while perfumes of saffron, white sandalwood, cardamom, nutmeg and mace were poured on the sacrificial wedding fire. The exquisite golden complexion of Naga women of North India was reputedly obtained through their constant use of turmeric.

Turmeric was widely available and was considered auspicious, so it was also used in everyday life whenever good luck was desired. For example, garments dyed or marked on the corners with turmeric were considered lucky and possessed with protective powers.

Spices were also commonly used to cure disease and promote health in India. The sacred *Ayurvedic* texts, which were formulated before 1000 B.C. and dealt with matters of health and medicine, make frequent reference to the use of spices. For example, the *Ayurvedic* system of medicine suggested that cloves and cardamom wrapped in betel-nut leaves be chewed after meals to aid digestion. In about 500 B.C., the physician Susruta the Second described over 700 drugs derived from spices, including cinnamon, cardamom, ginger, turmeric and peppers. Vapors of white mustard were used to fumigate the rooms of surgery patients, and sesame poultice was applied to wounds as an antiseptic.

Spices were also used in China for thousands of years. Confucius, who lived from 551 to 479 B.C., mentions ginger in his analects. The use of cassia was noted in the *Eligies of Ch'u* in the fourth century B.C., and the name of the South China state of Kweilin, founded in 216 B.C., translates literally into "cassia forest." It is also reported that the Chinese officials ate cloves in the third century B.C. to sweeten their breaths when they addressed the Emperor.

THE FIRST SPICE TRADERS

Trade and travel have always been part of Indian culture. Some sources indicate that as early as 3000 B.C., Indian explorers and traders took sea trips from the Malabar Coast in South India to the Persian Gulf and the fertile valleys of the Tigris and Euphrates rivers. From at least as early as 600 B.C. until almost 1400 A.D., Hindu and Buddhist missionaries and traders from India colonized countries and converted peoples throughout Asia. During this time period, Indian culture was considered the height of civilization. Powerful Hindu/Buddhist kingdoms, influenced by India, arose in Sri Lanka and throughout Southeast Asia, including Malaysia, Thailand, Kampuchea and Indonesia. Hindu Brahmins were sought by the courts of regional leaders as teachers and bureaucrats.

These early Indian merchants and colonizers understood the value of spices and traded in spices with China and throughout the Malay Archipelago, Indonesia and the Spice Islands (or Moluccas). The Indian spice merchants brought back ginger from China, cinnamon from Sri Lanka, nutmeg from the Spice Islands and cumin, cloves and coriander from throughout Asia.

SPICE USE IN THE WEST

In the West, precious Eastern spices were collected and treated like jewels, given as gifts or used for ransom or for currency when purchasing cows, goats or sheep. From the Bible, we know that King Solomon counted spices among the valuables in his treasury. Ancient Egyptian rulers used spices such as sesame, fenugreek, cinnamon, anise, cardamom, saffron, caraway and mustard for embalming, as body ointments and as fumigants in their homes. Many of the spices described in ancient western texts and writings are not grown in the West but are native to India and other eastern countries. The great desire for these spices became a driving force in transcontinental trade between the East and the West.

Traders from the West seeking wealth in the spice trade came to India and other destinations in the Far East, such as China, and the Spice Islands (Indonesia) for at least 3000 years. Arabs, Greeks, Romans and other Europeans came to India's Malabar Coast, which they called the "spice emporium," for cloves, pepper, pippali, zedoary, nutmeg and turmeric. They carried their precious cargoes to Africa, the Mediterranean and Europe.

Spices came to Europe and the West from the Far East by land and sea. Spices were taken on long caravan trips from the Far East by the silk route that traveled from China through Northwest India, Afghanistan and Turkestan. They were also taken by the Incense Route that went through Southern Arabia to Egypt and other parts of the Middle East.

Arab and Phoenician traders were the first to bring eastern spices to the Middle East and Europe. From at least as early as 950 B.C., the Arabs were the dominant middlemen in the spice trade between India and the West. They braved rough sea trips to the Malabar Coast of Kerala, India, and brought back spices, such as black pepper, cinnamon, ginger and cardamom. They traveled through the Persian Gulf, using the Tigris and Euphrates valleys to Babylon or went around the coast of Arabia and up the Red Sea to the Middle East and Africa.

Over land, the Ishmealites, who were Arab merchants, took long caravan routes to India, Burma and the Spice Islands.

The Phoenicians were the dominant traders of the Mediterranean. Renowned as fearless seamen, they traded with the ancient people of Greece, Italy, Spain, Portugal, France and Africa. They brought fish sauce (called garum) and spice essences from Ethiopia and cassia and cinnamon from the Arabs. In turn, the Phoenicians traded them around the Mediterranean, with many of these spices being taken to North Africa, making it the focal point for spice trade between the Far East and the West. The Phoenicians introduced Asian spices, such as nutmeg, coriander, cumin and cloves, throughout the region. With the wealth obtained by the Phoenicians from this trade, they built their great colonial city of Carthage on the North African coast. They remained in control of the Mediterranean spice trade until Alexander the Great conquered Egypt and established Alexandria in 332 B.C.

GREEK AND ROMAN SPICE TRADERS

For centuries, Arab merchants sought to protect their spice trade by hiding their true sources. The Arabs told stories of a mythical land in Africa as the source of spices. They also spun tales of gods and creatures protecting the spices from harvest by human hands.

In the first century A.D., a Greek sailor discovered the secret of the monsoon winds to and from India that hastened the trip and broke the Arab monopoly. Early Tamil poems from that time tell about the *Yavanas* or Greeks who spoke a strange language and traveled in well-built ships. The Greeks were quickly followed by Romans who established trading posts and warehouses in South India. In his writings, Ptolomey from Alexandria listed 11 ports and 30 walled towns along the coast of India. A Roman warehouse was excavated in Tamil-Nadu at the town of Arikamedu. Artifacts discovered there date the ruins to the first or second century A.D. Coins from the Roman kings Augustus, Tiberius, Nero and Caligula have been found at 30 sights, mostly in South India, and they bear witness to the extensive trade that took place. The Romans paid gold, silver and wine for spices from India and, during the time of the Roman Empire, dominated the European trade for Asian spices.

During this period, the Romans sailed from Egypt to India to bring back spices such as black pepper and turmeric for food, wine, cosmetics and medicine. The Romans became the first Europeans to cook with spices and use them lavishly. Black pepper was the most popular and most expensive spice during this period. Cumin and coriander were used for preserving meats and sausages. Fish were preserved with salt and leafy spices such as dill, mint and savory and flavored with pepper, cumin and mint.

The Romans also carried spices overland using the Silk Road that passed from Xian in China, around the Himalayas in North India and across Persia and then by ship over the Caspian Sea to the Mediterranean. They brought back cumin, ginger, cloves, nutmeg and cassia, and Constantinople, the Roman Empire's eastern capital, became the "spice city."

The Romans also traded in spices with their colonies. The Romans brought to Northern Europe temperate spices such as garlic, parsley, dill, mint, sage, thyme and savory, as well as the exotic spices of Asia.

THE ARAB CONQUEST

The Arabs regained their monopoly on spices with their conquest of Alexandria in 641 A.D. and their subsequent expansion into Northern Africa and Southern Spain. With the growth of Islam, the Arabs again took control of the spice trade.

Arab influence also expanded beyond the lands they conquered because of the spread of Islam, which replaced Hinduism throughout Southeast Asia, including the Spice Islands, and influenced north and central India. The followers of the Prophet Mohammed traveled from Mecca, in Saudi Arabia, carrying Islam to the Far East and bringing back spices to North Africa, Turkestan and Spain. The Arabs continued to dominate the spice trade in Asia until the late fifteenth century.

In sub-Sahara Africa, Arab traders supplanted the Indians, who first brought spices to East Africa in the third century B.C. The Arabs established a clove-trading center on the island of Zanzibar.

Because the Arabs controlled trade routes in the Indian Ocean and throughout North Africa and the Middle East, the spice trade with Europe dramatically decreased during the seventh century A.D. Without access to Asian spices, Europeans grew temperate spices such as mint, fennel, lovage, rosemary, sage, dill, poppy and celery.

SPICE USE IN THE MIDDLE AGES

Trade with the Far East, especially with India and China, was reopened with the crusades in the eleventh century A.D. During this time, Genoa and Venice became important trading ports. In 1271 A.D., a Venetian trader named Marco Polo captured Europe's imagination with tales of exotic lands and exotic spices. Marco Polo had traveled with his father and uncle to China and eventually to many other countries of Asia and the Near East, including India, Indonesia, Turkey and Egypt.

Venice took control of the European trade in spices, buying the products of the Far East from Arab middlemen. Once again, Europeans enjoyed spices such as ginger and galangal (wild ginger) from China, cloves from the Spice Islands and cinnamon and pepper from the Malabar Coast of India. Venetians provided salt and a good deal of gold and silver in exchange.

As a consequence of this increased trade, the consumption of spices grew dramatically. In the Middle Ages, Arabic and Asian luxury goods became indispensable to the European upper classes. Asian spices, pepper in particular, became the most important luxury items. Indeed, the use of spices took on an almost ceremonial function. At dinner parties of the refined upper classes, spices were passed around on gold or silver trays from which guests helped themselves.

In cooking, spices were used in astonishing quantities by today's standards. Food was buried in pepper and other spices. Spices were also served in beverages, such as powerfully spiced wines. The more excessive a dinner host's use of spices, the higher was his guests' perceptions of his social rank.

As the Middle Ages drew to a close, the middle and upper classes expanded, and the European appetite for spices grew even larger. Pepper sauce became a staple of the middle class diet. Old overland transportation routes and numerous middlemen limited the supply of spices. Increased tariffs on this precious cargo also drove up its cost. As a consequence, the price of pepper increased 30-fold during the fifteenth century.

The ensuing crisis led to the age of exploration, conquest and the discovery of new trade routes by Europeans.

THE AGE OF EUROPEAN CONQUEST

Realizing the value of the spice trade, Europeans sought to discover new routes to Asia and to conquer the countries where spices grew. One by one, European nations took control over trade routes to Asia and the spice producing regions—first the Portuguese, next the Spanish, then the Dutch and lastly the English.

Around the end of the fifteenth century, the Portuguese, led by Vasco da Gama, were the first of these Europeans to reach Calicut, India. This ended the Arab and Venetian monopoly on spices. The Portuguese eventually took control of the Indian and Far East spice trade. They paid gold and silver to the local Indian rulers for spices.

The Portuguese established trading ports at Goa, India and Sri Lanka and moved further east to Malacca, Malaysia and the Spice Islands (the Moluccas) bringing back pepper, cloves, nutmeg and mace. Until the late sixteenth century, they dominated the spice trade to Europe.

Competing with the Portuguese for the lucrative spice trade were the Spanish. A Spanish explorer, Ferdinand Magellan, sailed to the Spice Islands in 1519 looking for spices. The Spanish also sought a quicker western trade route to the spices of Asia and its greatest prize, black pepper, called pimienta in Spanish. Christopher Columbus began his great exploration looking for pepper. What he found instead were the Americas and the chile pepper, used abundantly by the Native Americans. By 1529, Spanish colonizers learned that the Aztecs had developed dozens of pod-type chile peppers. They called the fiery new plant pimienta picante to reflect its stronger taste profile. Today, its name chile or chilli pepper is derived from the Nahuatl language, "chilli" meaning red.

The Spanish found that chile peppers were natural colonizers, readily transportable and remained viable for several years. As a consequence, chile peppers were brought back from the Americas and quickly spread to other parts of the world. Chile peppers were known in Spain in 1493, Italy in 1526, Germany in 1543 and the Balkans in 1569. As paprka (paprika), they revolutionized Hungarian cooking. The Spanish also brought back nuts, beans, allspice and other ingredients.

The Portuguese brought chile peppers from the New World to their colonies around the world, including Africa, Arabia and Asia, where they grew rapidly. Unlike the rare and exotic spices brought to Europe, which were expensive and unattainable except to the upper classes, chilc peppers grew easily in the tropical climates of Asia and were readily available to common people. Chile peppers were so widely used and grown in Asia that by the mid-sixteenth century, European colonizers in India were not sure whether the "Calcutta pepper" was native to India or came from the New World. They also brought corn, potatoes, beans and tomatoes from the New World and peanuts from Africa to the Spice Islands.

In the early seventeenth century, the Dutch took over from the Portuguese and became masters over Java, Malacca and southern India. The Dutch controlled the Spice Islands (which had a brief British rule from 1811–1816) until the Japanese came during World War II. During the period when the Dutch colonized South Africa, they introduced chile peppers from the Americas and brought in Malay slaves from the Spice Islands. The Malays introduced many spices such as aniseed, turmeric, cumin, cardamom, coriander, mustard seed, garlic, tamarind, ginger and fennel. These ingredients are now essential to Cape Malay cooking, which combines Malay cuisine with Dutch, English, Indian and indigenous African flavors.

In the eighteenth century, the British took control of the Indian and the East Indies spice trade. The British defeated the Moguls of North India, who had established themselves as rulers from the eighth to the sixteenth century. Then, spices such as mustard, poppy seed, sesame, coriander and cumin from the northern regions of India were available for export to Britain and its colonies. The British established Bombay and Calicut as the spice trade centers in India and Penang as the major eastern port in peninsular Malaya.

During the eighteenth century, the French introduced spices such as clove and nutmeg to Mauritius and the French territories in Vietnam and the Caribbean.

In the early nineteenth century, Chinese traders and merchants (mainly from south China), encouraged by a liberal immigration policy instituted by British colonial powers, traveled to peninsular Malaya and Singapore and married local Malay women. The descendants came to be called *Peranakan* or Straits Chinese (*Baba* for men and *Nonya* for women). Nonya cuisine evolved in these regions.

SPICES IN AMERICA

In the eighteenth century, the North Americans from Boston and other northern coastal towns sailed to the Far East, to places like Malabar in India and Sumatra in the East Indies, and brought back pepper, cloves, cassia, ginger and cinnamon. New York City, Baltimore and San Francisco became major ports for the spice trade to the U.S.

Today, the U.S. is the biggest spice importer and the largest consumer of spices in the world. As is shown in the chapter on trends, next, North Americans are now "exploring" for more seasoned products than ever. Thus begins the interest in authenticity of ethnic cuisines and the emergence of "new" North American flavors.

CHAPTER 2

Trends in the World of Spices Today

TRENDS IN FOODS AND SPICES

THROUGHOUT the ages, the opening of trade routes and changing immigration patterns have affected the way the world eats. Today, we are facing a new revolution in eating patterns and the way we use spices. North American palates are becoming more daring and adventurous. North Americans seek variety and something new. They want foods with more intense flavors and hotter or spicier profiles. North Americans also want foods that are fresh, light and healthy, having a perception of "natural," and that are convenient to prepare. At the same time, they also want to indulge.

As in the age of colonial adventure, we are seeking new routes to find foods that provide the tastes we demand. As a result, our interest in tastes and flavors from faraway places is increasing. Cuisines once considered unusual from other countries are becoming commonplace. The foods and ingredients of the world—Southeast Asia, India, Latin America and the Mediterranean—are more available than ever. Ingredients once considered "exotic" are infiltrating traditional North American foods through cross-cultural cooking and regional American fare, such as French-Indian, Thai-Italian, Pacific-Rim or Floribbean. New and diverse cooking styles and ingredients are not simply part of a passing fad. To the contrary, basic changes in who North Americans are and their lifestyles are driving these trends (Figure 1).

The United States is becoming an increasingly diverse country. The U.S. Census Bureau reports that since 1970, Asian and Hispanic populations have grown substantially.

Asian-Americans, who are the fastest growing ethnic group, increased by more than 30%, while Hispanic-Americans increased by about 20%. Latin Americans will soon be the largest minority group in the U.S., surpassing

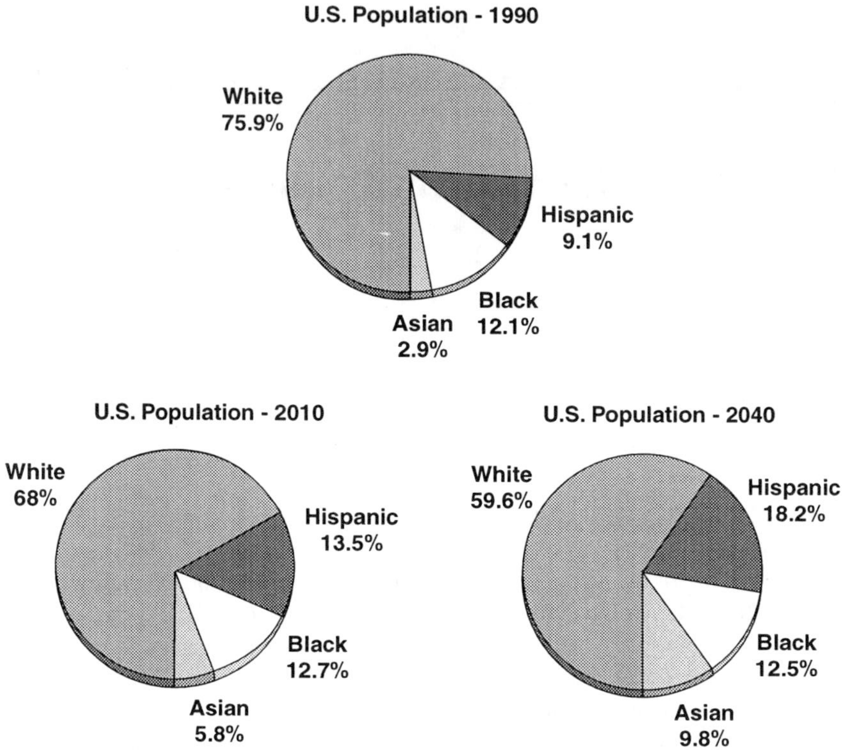

Figure 1 U.S. population (a) 1990, (b) 2010 and (c) 2040. Source: U.S. Census Bureau.

African-Americans in the early twenty-first century. In the mid 1990s, Hispanics, Asian-Americans and African-Americans made up 25% of the U.S. population [Figure 1(a)]. By 2010, these three groups will comprise one-third of the U.S. population. By the middle of the twenty-first century [Figures 1(b) and 1(c)], they will make up half of the U.S. population.

These statistics don't tell the whole story. Ethnic groups are becoming more diverse as well. There are not simply more Asian-Americans, but more Americans of Indian, Korean, Thai, Chinese and Vietnamese descent. Likewise, the growing Latin American population includes people of many different ancestries, including all of the regions in South America, Central America, Mexico and the Caribbean.

The increased presence of ethnic groups in our communities is increasing our exposure to many different cultures, foods and ingredients. Asian and Hispanic cuisines, which have complex flavor profiles, are becoming a greater part of our social fabric. These cuisines, which previously gave us familiar spices such as ginger, cilantro or cinnamon, are now introducing us to lemongrass, chipotle, epazote, and kari leaf and other new spices. Consumers are sampling these cuisines and their exciting new ingredients in a variety of ways. They are discovering flavorings such as fish sauces, flower essences, wrappers and fermented soybeans in restaurants that feature ethnic and fusion cuisines. Ethnic grocery stores and bodegas carry specialized items such as tamarind, kokum, banana leaf and galangal. Consumers can also find authentic ethnic ingredients, such as nigella, ajowan, star anise or black cumin, as well as prepared ethnic foods, in gourmet, health and natural food stores and even on the Internet.

Moreover, many Americans are traveling to Asia, the Mediterranean, the Caribbean and other destinations around the globe. Cooking schools in the U.S., as well as overseas, are helping Americans to learn about authentic ethnic ingredients and cooking styles. The Internet is also making global commerce, communication and the exchange of ideas an instantaneous affair.

What do these trends portend for the consumption of spices and their use in prepared foods? The answer is already available. Americans are buying an increasing volume and variety of Asian, Latin American, Caribbean and Mediterranean spices.

The American Spice Trade Association's 1998 Spice Statistics Report says that within the last 30 years, there has been a significant increase in the consumption of spices, especially hot spices such as ginger, mustard, white, black and red peppers. It reports that the hottest trend is our taste for hot spices that have shown an 80% increase in sales volume has increased 80% since the late 1970s.

The demand for spices will increase, not only in total volume, but also in variety. Thus, we can expect increased sales volumes for familiar spices, such as garlic, onion, allspice, cumin seed and mustard, while new demand for emerging spices, such as fennel seeds, anise, basil, guajillo and cardamom will grow as well.

There will also be an evolution in the nature of prepared foods. Prepared foods will be presented in Asian and Latin American styles. Smaller portions of entrees, with thinner cuts of meat that are marinated, seasoned or "sauced up" rather than being dry will be served. Entrees will also be perked up with a variety of seasoned side dishes and condiments. Asian and Latin concepts of one-dish/bowl meals, in which spices release multidimensional flavors, will become more popular because of their taste, convenience and economics.

A variety of preparation techniques for spices and other ingredients will bring a new dimension to foods. These include roasting, "tarkaring" and "tumising," which make spices become more fragrant, less bitter and more flavor intense.

The demand for "healthy" ingredients and natural ways of preventing illnesses or diseases will also contribute to the increasing use of spices in the U.S. For example, vegetarian Indian foods that have "spicier" profiles are becoming popular because they provide taste and are healthy. There will also be a growing consumer interest in using spices for their therapeutic properties.

Indians, Chinese, Native Americans and Middle Easterners traditionally consumed spices because they help provide good health as well as flavor.

All of these trends will arise from the demands of the new American consumer. As a result of increasing demographic diversity and greater global travel and communication, we will see two new types of consumers. One type is looking for familiar, traditional foods with new twists. These consumers want foods that are "safe" and comforting and use ingredients that are recognizable, but which give variety.

The second type of consumer wants excitement and adventure in meals. They are not afraid to try something totally different. Unique ingredients and bolder flavors appeal to them. These consumers are seeking new ingredients and new ways of preparing and serving meals. They are desiring spices that give depth and intensity. Ingredients and preparation techniques of Latin American, Asian, Caribbean and Mediterranean foods will inspire in them a greater desire for new flavors. Both types of consumers are looking for something different and flavorful, and that will create pleasure at the meal table.

How do we meet these consumers' needs for familiarity and tradition and at the same time create something new, tasty and exciting? First, we need to understand and effectively utilize authentic spices and other flavoring ingredients. Ingredients from different ethnic cuisines can create authentic flavors and a melting pot or a "salad bowl" of tastes that will satisfy the need for stronger flavors, health, variety, convenience and excitement.

Second, we must creatively combine "new" flavorings with familiar or traditional ingredients. By effectively connecting these ingredients, we can create variety and form new and flavorful products. Spices such as ginger, cilantro, lemongrass and saffron, which are identifiable with ethnic foods, can be great tools for "safely" giving authenticity and new flair to traditional foods. They can add a comforting new dimension to a traditional product or create a totally unique product.

Third, we can satisfy the niche markets by focusing on regional differences within the U.S. and developing products that reflect the flavors of the ethnic groups who live there.

UNDERSTANDING AND EFFECTIVELY MEETING THE GROWING DEMAND FOR AUTHENTICITY

Authentic ethnic flavors and their preparation techniques and presentation styles are becoming a regular part of our meals. Italian, Chinese and Mexican flavors have been popular for the last 15 to 20 years and are now considered mainstream. Consumers are now going further to explore the real, more authentic, regional flavor profiles of these cuisines. Consumers are seeking not just Mexican food, but the dishes of Oaxaca, Yucatan and Michoacán. They are looking beyond generic Chinese food and are asking for Szechwan, Cantonese, Hunan and Peking flavors. They want the Tuscan, Umbrian, Ligurian and Sicilian foods of Italy. They are also looking for the regional profiles of the newer ethnic cuisines of Thailand, India and Vietnam.

To create this authenticity, we need to look at the actual ingredients that give rise to these distinct flavorings (including their historical significance, varieties, forms, properties and uses), how they are prepared or cooked and when and how they are incorporated or presented in a meal.

AUTHENTIC ETHNIC INGREDIENTS

Certain spices have a mass appeal for many ethnic groups, such as garlic, black pepper, ginger or soy sauce. However, within each ethnic group or region, preferences for spiciness, sweetness or heat occur. Consequently, the types and amounts of spices desired vary as well. Regional ethnic cuisines use their characterizing, locally available spice varieties to bring out their authentic flavor profiles. Hence, we have Cajun, Thai, Tuscan or Cantonese.

In addition, there are many varieties of each spice. Different ethnic groups often prefer a specific variety. For example, with cinnamon, Mexicans use the more delicate canela, also called Ceylon cinnamon, while the Chinese and most Southeast Asians use the harsher and sweeter cassia. In the U.S., cassia has been traditionally used, but canela is becoming better known. In the past, substitutions have been used for an "exotic" ethnic spice that was unavailable. Today, consumers want the real thing. Therefore, we need to understand spices in all their varieties.

Let's look at some of the typical spices and flavorings used in some of the popular and emerging cuisines.

SPICES AND FLAVORINGS OF POPULAR AUTHENTIC ETHNIC CUISINES

Southeast Asian (Burmese, Cambodian, Filipino, Indonesian, Laotian, Malaysian, Singaporean, Thai, Vietnamese): black peppercorn, cardamom (green, brown), cayenne pepper, cinnamon, clove, coriander (leaf, seed and root), cumin, caraway, fennel, fenugreek, mace, nutmeg, cubeb pepper, saffron, turmeric, shallot, Chinese chive, mustard (seed, powder, oil), galangal, garlic, ginger, kaffir lime, basils (Thai, lemon, holy), kari leaf, lemongrass, spearmint, sesame (seed, oil), star anise, tamarind, bay leaf, zeodary, white pepper, bird pepper, candlenut, banana leaf, coconut milk, lily bud, curry blends, shrimp paste, soybean pastes, fish sauce, mushroom, palm sugar, pandan leaf, peanut, oyster sauce, adobo, annatto, soy sauce, laksa leaf, lime, kalamansi, jasmine essence, ginger flower.

South Asian (Afghan, Bangladeshi, Indian, Nepali, Pakistani, Sri Lankan, Tibetan): ajowan, anise, bay leaf, cardamom, cayenne pepper, cinnamon, clove, nutmeg, mace, caraway, coriander (seed, leaf), asafoetida, black cumin, cumin, kari leaf, fennel, fenugreek, onion, garlic, ginger, mustard (seed, oil), long pepper, nigella, tamarind, rose petal, saffron, black pepper, turmeric, celery, paprika, mints, curry blends, garam masala, pomegranate, screwpine leaf, kokum, almond, cashew nut, black sugar, amchur, ghee, coconut milk, sesame oil, dill seed.

East Asian (Chinese, Japanese, Korean): anise, black peppercorn, cayenne pepper, onion, shallot, garlic, ginger, ginseng, sansho, cinnamon, clove, fennel, sesame seed, shiso, soy sauce, star anise, szechwan pepper, horseradish, wasabi, parsley, sesame (seed, paste), licorice, coriander (seed, leaf), lotus (leaf, root and seed), scallion, chives, daikon, kudzu, five spice, hoisin, mirin, miso, soybean (whole, paste), dried shrimps, mushrooms, oyster sauce, goit chu jang, wine, orange peel, vinegar, bonito, ponzu, plum sauce.

Caribbean: allspice, annatto, black pepper, chive, coriander (seed, root), cilantro, garlic, ginger, clove, paprika, cumin, cardamom, nutmeg, mace, cinnamon, habanero, cayenne pepper, mustard (powder, pastes), onion, bay leaf, thyme, parsley, culantro, roselle, tamarind, turmeric, amchur, mango, papaya, coconut milk, cassareep, curry blends, garam masalas, jerk, colombo, adobo, sofrito, salted codfish.

Latin American (Mexican, Cuban, Puerto Rican, Dominican, Central American, South American): annatto, bay leaf, canela, black pepper, chepil, poblano, chipotle, aji, ancho, chiltepin, serrano, rocoto, guajillo, habanero, jalapenos, cilantro, parsley, coriander (seed, root), culantro, cumin, nutmeg, mace, clove, boldo leaf, epazote, garlic, tamarind, vanilla, chocolate, roselle, oregano, almond, moles, pepian, olive, recados, tomatillas, adobos, cassareep, aji dulce, naranga agria, lime juice, dende oil, ginger, bacalhau.

Mediterranean (Northern African, Middle Eastern, Saudi Arabian, Iranian, Turkish, Spanish, southern French, Italian, Greek, Cypriot): ajowan, anise, asafetida, bay leaf, garlic, ginger, pepper (black, white), caraway, cardamom, cayenne pepper, chervil, celery seed, cinnamon, clove, fennel, fenugreek, nutmeg, mace, coriander (seed, leaf), cumin, dill, juniper, long pepper, marigold, marjoram, meloukhia, mints, paprika, parsley, poppy seed, rosemary, rose petal, saffron, savory, sesame (seed, paste), sorrel, sumac, tarragon, sage, lemon verbena, lovage, lemon balm, oregano, sweet basil, myrtle, thyme, turmeric, vanilla, chocolate, capers, bell pepper, olive (fruit, oil), pomegranate, fig, harissa, berbere, zhoug, sofrito, tomato, ras-el-hanout, mushrooms, balsamic vinegar, caper, anchovy, pine nut, cheese, aioli, bacalao.

African: anise, clove, coriander (seed, leaf), cubeb pepper, cumin, egusi seeds, fenugreek, sesame paste, onions, garlic, ginger, grains of paradise/melegueta peppers, saffron, turmeric, licorice, mace, mint, nutmeg, piri-piri, seville orange, utazi leaf, tamarind, vanilla, ground nut, guedge, palm oil, berbere, curry powder, garam masala, cayenne pepper, allspice, rose petal.

North American: sassafras, red cayenne pepper, tabasco, jalapeno, pepper (black, white), allspice, cinnamon, bay leaf, oregano, sweet basil, cumin, cilantro, mustard, garlic, onion, chive, thyme, mints, nutmeg, rosemary, sage, chile powder, olives, cheeses, mushrooms.

AUTHENTIC ETHNIC FLAVORINGS

As the trend for authenticity and variety continues, "newer" spices such as ajowan seeds, nigella, turmeric leaf or shiso will emerge. Consumers are also discovering unique flavor properties of parts of plants other than seeds or leaves, such as stalks, stems, roots and flowers of spices. These are used in traditional spice blends of many global cultures, such as Latin American, Asian and Caribbean. In addition, there are many other ingredients that provide "spicing" or flavoring to foods, such as oyster sauce, pomegranate, olive, mushroom, candlenut and dried fish. These are being used with increasing frequency in the U.S. in pastes, purees or extracts and as natural and healthy ways to add flavor, color or texture to foods and beverages. Some of the emerging global spices and flavorings are examined in Table 1.

TABLE 1. Emerging Global Spices and Flavoring.

Seeds/fruits/bark	Ajowan, black cumin, canela, green cardamom, asafoetida, fennel seed, fenugreek, star anise, sumac, annatto, mustard seed
Leaves/stalks/stems	Thai basil, culantro, kari leaf, kaffir lime leaf, epazote, shiso, lemongrass, recao leaf
Chile peppers	Chipotle, aji, ancho, bird pepper, cayenne, guajillo, habanero, poblano, rocoto, cachucha
Fruits	Citrus, cranberry, guava, mango, persimmon, green papaya, kalamansi, pomegranate, olive, kokum
Vegetables	Mushrooms, tomatoes, squash, taro, yams
Nuts	Almond, candlenut, cashew, peanut, pine nut, pistachio, gingko biloba
Seeds	Pumpkin, sesame, sunflower, lotus
Food wrappers	Corn husk, lotus leaf, pandan leaf, banana leaf, nori, hoba leaf, hoja santa
Flowers	Jasmine, orange blossom, rose, squash blossom, violet, roselle, lavender, ginger flower
Roots	Coriander, ginseng, lotus, turmeric, galangal, wasabi, licorice, daikon
Fish and shrimp	Blacan, nam pla, nuoc mam, trasi, guedge, dashi (dried, fermented, smoked)
Legumes	Miso, oyster sauce, soybean pastes (fermented, pickled, salted), dals
Other	Balsamic vinegar, coconut milk, ghee, dende oil, gula melaka, chocolate, vanilla, coffee, tea

AUTHENTIC PREPARATION AND COOKING TECHNIQUES

Many spices are common to a number of ethnic cuisines, yet the same spice gives different flavors. Why? The way each spice is prepared and combined with other ingredients gives varying flavors. Dry toasting of coriander seeds, roasting of chili peppers, or "popping" of mustard seeds in hot oil produce distinct flavors and colors. Raw, bitter notes are removed, certain flavor notes are intensified and sometimes colors are modified. The timing in which the spices are added to the cooking process is also crucial in creating a well-balanced flavor profile.

In authentic ethnic cooking, corn husks and banana and pandan leaves are used to wrap chicken, fish, rice and meat, that are then steamed, smoked, grilled or barbecued. These wrappers provide unique flavors, moistness, color, texture and visual appeal to foods. Though commercially we may not be able to present food in these wrappers, we can simulate the flavors they create. For example, in a chicken dish, you can add a grilled flavor with pandan notes to simulate the flavor of chicken wrapped and grilled in pandan leaf. In addition, a pandan leaf for serving the chicken could be packaged as part of a meal kit that can be a presentation tool to create authenticity or excitement at the meal table.

Therefore, in order to meet the growing demand for authentic or new flavors, we will need to understand not only true ethnic ingredients, but also their preparation and cooking techniques.

PRESENTATION STYLES

Ingredients and preparation techniques affect flavors and the way foods are presented and served. (Globally, food is eaten from plates, bowls, banana leaves or "thalis" using different utensils: chopsticks, forks, knives, fingers or spoons.) To fully perceive the different sensations of flavors and textures, it is important to know not only how the food is seasoned, but also how it is served and eaten in traditional cultures.

Presentation of meals achieves visual appeal and balance. Ethnic cuisines such as Japanese, Thai or Caribbean create great eye appeal using contrasting colors, shapes or sizes of ingredients. Spices, fruits, vegetables, wrappers or flowers are sculpted or designed to create exciting appearances.

Serving containers are also an integral part of meal presentation. Various sized brass baltis (wok-like utensils) provide an authentic and familiar touch to many Indian and Pakistani cuisines. An authentic edge can be added to meal solutions by including authentic utensils in the package, such as chopsticks, chutney dip bowls, banana leaves, corn husks, comals, tagines or skewers.

In presenting authentic ethnic foods, side dishes play a prominent role. In traditional North American presentation, the focus is on a main course of meat or fish, with side dishes of rice or pasta and vegetables. In contrast, for many ethnic groups, the main dish is starch based, such as white rice, with side dishes of intensely seasoned chicken, fish or vegetables.

Curries, stews, pickled vegetables and other ethnic side dishes that are hot, sour, sweet or crunchy provide variety and enhance the flavor and texture of the main entree. Likewise, condiments are essential in many ethnic meals. Every ethnic group has its typical condiments to achieve each individual's desired taste. The Chinese have their sweet and sour sauces or plum sauces, Mexicans have salsas, Japanese have teriyakis, Tunisians have harrissa, French have rouille, Indians have pickles or chutneys, Indonesians have sambals, and Koreans have kimchis. As with side dishes, when condiments are eaten with a meal, totally new sensations are produced.

In many Asian cuisines, the concept of a "main entrée" and side dishes blend together in the form of a "one pot" or "one dish" meal such as stews, soups, sauced or souped noodles or fried rice, all of which are also served with condiments. Bowl meals provide varying tastes and textures simultaneously.

FUSION AND REGIONAL AMERICAN FLAVORINGS

While changing demographics are increasing the demand for authentic ethnic ingredients and foods in general, one of the most important trends in a diverse country such as the U.S. is the growth of fusion (or cross-cultural) and regional American foods. These new foods are emerging from the mixing of ingredients and preparation styles taken from the best of the traditional American and ethnic cuisines.

Fusion food is not something new. The transformation of cuisines through fusing flavors, ingredients and preparation techniques has happened around the world whenever two or more cultures have lived together. Mediterranean and Caribbean cuisines are two of the better-known types of fusion foods that combine flavors of many cultures and regions.

In the U.S. today, fusion flavors have evolved from two or more ethnic groups, such as Japanese and Cuban, Italian and Thai or French and Indian. These "new" cuisines have great appeal because they are the product of many cultures. Fusion foods also include newer versions of traditional American products with flavors derived from ethnic ingredients, such as ginger roast chicken, sofrito mashed potatoes or chipotle potato chips.

Within the category of fusion foods, there are also regional American fusion flavors, such as Creole, Tex-Mex, New England or Southern. These "traditional" fusion foods evolved when the cuisines of earlier immigrants to the U.S. were combined. Many of these regional foods are changing in flavors to meet the newer groups of immigrants and changing consumers' taste buds.

Regionalization of cuisines will continue in the U.S. More than half of the U.S population lives around large cities, such as New York, Miami, Chicago,

Los Angeles or Washington. The populations in these regions are becoming increasingly diverse and, thus, will give rise to unique fusion flavors. Newly arrived immigrants from Latin America, Asia or the Caribbean are producing the "newer" fusion and regional American fare, such as Floribbean, Nuevo York, Nuevo Latino or Pan Californian or Pan Asian.

Because fusion and regional cooking will increasingly affect our eating patterns, we will examine them in greater detail.

FUSION FLAVORS

Ethnic ingredients and their preparation styles blend well with North American cooking techniques, such as roasting, broiling or baking. Consumers are becoming more innovative and are creating exciting fusion foods using these varied ingredients. It is not uncommon nowadays to see these ethnic foods with traditional American ingredients. Ethnic groups have traditionally used wrappers, such as tortillas, pita breads and other flatbreads as meals or snacks. They are popular as fusion lunch entrees with salad ingredients and topping sauces, grilled chicken, chili con carne and other traditional American fillings.

Fusion sauces can be used as condiments to perk up traditional grilled, steamed or baked foods that are prepared with moderate seasonings. Alternatively, mainstream sauces, such as tartar sauce, steak sauce or ketchup, can be made into a "new" sauce with ethnic spices and flavorings.

To create great fusion products, flavors need not be compromised. Creative food product designers can develop foods for a wide market by "toning down" authentic ethnic ingredients or fusing them with more traditional American flavors. By adding, deleting or toning down chile peppers, fish sauces, fermented bean pastes and other flavor-intense ingredients, we can create appealing fusion foods or new flavor twists for traditional North American comfort foods.

For mainstream consumers seeking an ethnic flair or something new, lemongrass, adobo or cilantro can be added to chicken, rice or pizza. For the adventurous consumer, tandoori chicken, poblano pasta or sambal fish can be developed. Fusion foods can also be created to meet the needs of ethnic consumers. Ethnic comfort foods such as stir-fried noodles, pilafs, curries, rice and beans or pastas can also be mainstreamed to meet the tastes of second- and third-generation ethnic populations, who have adapted to a new way of living in the U.S.

REGIONAL AMERICAN FLAVORS

In the U.S., there is a long history of immigration. Often, various immigrant groups have been attracted to particular regions of the U.S. Italians moved to the Northeast, Japanese to the West Coast, Scandinavians settled in the Midwest and Cubans emigrated to Florida. As these various groups settled in a region, they mixed with the local population and influenced the local foods.

When food developers create products for the different regions of the U.S., they need to understand each region's specific preferences. This is also true when developing products for the ethnic market. For example, the market for Latin American foods varies around the U.S. In Miami, Hispanics are predominately Cuban, while Puerto Ricans and Dominicans dominate New York, with Mexicans the majority in Los Angeles. Therefore, when we create Hispanic-style foods for these niche markets, they must be directed to the tastes of those who live in these regions.

While the tastes of the new immigrants influence the foods of a region, their own tastes become less distinct. Bagels, chow mein, tacos, chilis and other "Americanized" ethnic foods emerged from the influence of the local population. These products were modified from their authentic profiles in order to suit regional consumers' palates. Thus, traditional tastes fade while new tastes grow.

New immigrant groups continue in the same path as those who arrived earlier. The interactions of these new ethnic groups with the existing population, the availability of a variety of ingredients and the adventurous spirit of cooks in the U.S. have given rise to some creative regional American cuisines. The flavors that are emerging from these regions reflect the cultural groups that are locally dominant.

In Miami, there is a large population of Cubans, Central Americans, Haitians and African-Americans from the South, and the cooking there reflects the influence of these groups. Yuca, black beans, habaneros or mango are combined with key lime, lima beans or potatoes to create a new Floribbean cuisine. In New York City, new fusion foods such as Italian-Thai, French-Indian and Japanese-Latino reflect the global diversity of its population.

For the future, the forces that have created the present demand for authentic ethnic and fusion flavors will continue to grow in the twenty-first century. The population trends that have changed the tastes of Americans today will continue to have an ever-increasing effect on the foods we eat tomorrow. The result will be a further evolution of unique regional flavors in North America.

In the future, the "new" foods will continually evolve that will have different or more sophisticated flavor profiles, colors and textures. In regional cooking, fusion themes will be taken to new levels of visual appeal and creativity, such as Malaysian-Tuscan, Chicago-Thai, Miami-Indian or Oaxacan-Japanese. Pasta sauces, ketchup, salsas or curries will develop clearly defined regional U.S. flavor profiles.

As consumers become more exposed to and knowledgeable about new and "exotic" ingredients, preparation techniques and presentation of meals, food and beverage designers need to create these new products.

CHAPTER 3

Forms, Functions and Applications of Spices

INTRODUCTION

SPICES are the building blocks of flavor in food applications. Food developers who wish to use these building blocks effectively to create successful products must understand spices completely. The word "spice" came from the Latin word "species," meaning specific kind. The name reflects the fact that all plant parts have been cultivated for their aromatic, fragrant, pungent or any other desirable properties including the seed (aniseed, caraway, coriander), leaf (cilantro, kari, bay, mint), berry (allspice, juniper, black pepper), bark (cinnamon), kernel (nutmeg), aril (mace), stem (chives), stalk (lemongrass), rhizome (ginger, turmeric, galangal), root (lovage, horseradish), flower (saffron), bulb (garlic, onion), fruit (star anise, cardamom, chile pepper) and flower bud (clove).

For people throughout the world, spices stimulate the appetite, add flavor and texture to food and create visual appeal in meals. Called rempah (Malaysian and Indonesian), beharat (Arabic), besamim (Hebrew), epices (French), kruen tet (Thai), masala (Hindi), specie (Italian), especerias (Spanish), sheng liu (Mandarin), specerjien (Dutch), krooder (Norwegian) or kimem (Ethiopian), spices have been savored and sought around the world from the earliest times because of their diverse functions.

Nowadays, food professionals continually search for "new" and unique spice flavorings because of the growing global demand for authentic ethnic and cross-cultural cuisines. Consumers are also seeking natural foods and natural preservatives for healthier lifestyles and natural ways of preventing ailments. So, spices are also being sought for their medicinal value, as antioxidants and as antimicrobials.

This chapter describes: the different forms in which spices are sold and their composition, the primary and secondary functions of spices in applications, the techniques for preparing spices, the methods for applying spices in product development, and the methods for assuring proper quality control in spices.

SPICE FORMS AND COMPOSITION

Spices are available in many forms: fresh, dried, whole, ground, crushed, pureed, as paste and as extractives. Each form has its respective qualities and drawbacks. The form chosen by the food product designer will depend on the specific application, processing parameters and shelf life.

FRESH WHOLE SPICES

Consumers prefer the taste of freshly made food. The "fresh" taste consumers seek from spices comes initially from their aroma. This aroma is due to the volatile component of the spice. It can be lost during harvesting, storing, processing or handling. For some spices, the fresh forms have different flavor profiles than the dried forms, examples include ginger, cilantro or basil. Fresh ginger has been found to be less pungent than dried ginger because fresh ginger has less shogaols (non-volatile constituents that cause pungency) than dry ginger.

Fresh ingredients, especially whole spices, when freshly ground give prepared foods a fresh taste. Fresh spices provide crunchy, crispy textures and colorful appeal. Fresh whole spices also become very aromatic when they are roasted or fried in oil, and their aroma transfers to the application. This is especially true of whole or cracked seeds and leaves, such as bay leaf, kari leaf or mustard seeds.

Whole spices provide aroma and, most importantly, texture and visual effect. Certain spices have a strong aroma when fresh, such as basil, garlic, onion and cilantro, due to their high volatile (essential) oils. The essential oils disappear quickly at high temperatures, especially if the spices are processed in an aqueous system, but they can also be lost at room temperatures or when the spices are cut or bruised.

While uneven distribution of whole spices in a product can be problematic, this effect is sometimes desired to achieve nuttiness or a sensation of "bite" into a whole spice, such as whole sesame seeds on a breadstick or ajowan seeds on Indian naan bread. In this regard, whole spices can become the major flavor characterizing a product. Also, whole spices, especially the leafy spices, provide great visual appeal as garnishes.

Flavor is intact in the whole spice and is more slowly released than with the ground spice, especially when subjected to preparation techniques such as frying or roasting, during which time, the whole spice slightly cracks open.

In a whole spice, the chemical components that provide the flavors vary in concentration throughout the spice. In chile peppers, the greatest concentrations of pungent compounds are found in the inner portions, such as the veins and seeds.

In many whole spices, cooking or processing changes the spices' chemical compounds and their proportions to varying degrees, often giving rise to different flavors. For example, smoking, grilling or drying certain chile peppers significantly changes their flavor and color. When jalapeno is smoked and dried, it changes its flavor and color completely, giving it a new identity, called chipotle.

Spices that do not have a strong aroma, such as bay leaf, chile pepper or sesame seed, develop intense flavor after roasting or boiling. Mustard seed, star anise and fagara (Szechwan pepper) are generally dry roasted to intensify their flavors for meat, fish and poultry dishes.

Many spices, such as lemongrass, spearmint, basil and chile peppers, are blended fresh and are used in making sauces and condiments with water, oil, wine or vinegar. The fresh pureed or paste forms have intense flavors and need to be mixed well before application in sauces, soups or gravies. Since the paste form usually contains oil, it can become rancid in a shorter period of time.

Consumers want to use "fresh" spices, but usually their flavors, colors and textures are lost during storage and prolonged processing. Preliminary preparations, such as grinding, roasting or flaking of whole spices, need to be done before adding them to processed foods.

Consistency is also more difficult to achieve in fresh spices because their origin, age and storage conditions cause flavor variations. Therefore, dry spices and spice extractives are, by necessity, the forms most often used to formulate foods or beverages. Fresh whole spices are not frequently found in processed foods, but are generally used in restaurants, in home cooking and in other smaller scale applications.

The goal for a food designer is to develop products that will have the "fresh" quality desired by consumers but that have spice-sensory attributes that can withstand processing, freezing and storage conditions.

DRIED SPICES

Spices are often used in their dried forms because they are not subject to seasonal availability, are easier to process, have longer shelf life and have lower cost. These dried forms are most frequently used for processed products or for wholesale usage. Dried spices come whole, finely or coarsely ground, cracked and as various-sized particulates. Spices are ground by milling them to various sized particulates. This grinding also generates rapid air movement and heat that dissipate some of the volatile oils and even change some natural flavor notes through oxidation.

Depending on its form, the same dried spice will deliver different flavor perceptions in the finished product. Ground spices have better dispersibility in food products than fresh whole spices. Some volatile oils are released through grinding, which partially breaks down the cellular matrix of the spice. In some spices, flavor is intensified through drying because of the elimination of most moisture. This leaves a greater concentration of the low volatile compounds that give stronger flavor but less aroma due to the loss of the volatile constituents. Dried spices can better withstand the higher temperatures and processing conditions than fresh spices.

Some dried spices can be used to characterize an application's flavor and texture. Garlic and onion, which come powdered, granulated, ground, minced, chopped and sliced and in various sized particles, characterize flavor and texture in garlic bread, onion bagels or chips.

Whether a dried spice is used ground, granulated, cracked or whole will depend on its use in specific applications. Many ground spices need to be "rehydrated" in order to develop their flavor, such as ground mustard that becomes pungent only when water is added. This addition of water triggers an enzyme reaction that releases the spice's aroma. Acidulents, oil or vinegar are also added to preserve the pungency or intense flavors of the spice in the finished product.

In processed foods, dried spices can be more economical to use than fresh spices. For example, dried leafy spices do not require the cutting, chopping or grinding preparation that the fresh forms do. Also, most dried spices retain a higher overall flavor concentration than fresh spices. For example, one pound of dried garlic has an equivalent flavor of five pounds of fresh garlic.

The sensory, physical and chemical characteristics of dried spices are determined by environment, climate, soil conditions, time of harvesting and post-harvest handling. The same type of spice can have different sensory characteristics depending on where it was grown and how it was harvested, stored and processed. For example, dried ginger from India has a subtle lemon-like flavor, dried ginger from southern China comes with slightly bitter notes and ginger from Jamaica has more pungent flavor. Similarly, ground black pepper, which comes from a dried berry called peppercorn, varies in flavor intensity depending on its origins. Black peppercorns from Tellicherry are highly aromatic (India), while Lampong (Sumatra) pepper is less aromatic with more pungency. The Malaysian and Brazilian peppercorns, in contrast, have milder aromas with stronger bites.

For most spices, the time period between harvesting and storage and between when the spice is ground and added to a food are crucial for obtaining its maximum potential.

The way a spice is treated or processed before being ground, and the conditions of storage before delivery to the food processor, create flavor and color differences. Spice flavor can be readily oxidized, and losses occur during milling and storage of spices.

Most spices such as cumin, coriander and cardamom give more aroma and flavor when freshly ground than when bought as a pre-ground spice. When spices are ground, the oils tend to volatilize, causing aroma losses. Anise, black pepper and allspice lose their aroma quickly as soon as they are ground. To better retain color, flavor and aroma, spices are sometimes milled using lower temperatures. While spices lose more aroma as they are ground more finely, the advantage is that finely ground spices blend better in finished products that require a smooth texture.

Ground spices should be stored in tightly closed containers and should not be exposed to light, high temperatures or humidity. Moisture and high temperatures will help mold growth that will cause spoilage. Generally, the moisture content of spice is 8% to 10%. High storage temperatures cause flavor loss, color changes and caking or hardening of the ground spice. Spices need to be stored at 50°F to 60°F (10°C to 15°C) with a relative humidity (RH) of 55% to 65%.

Dried spices can have some disadvantages. Some have poor flavor intensity, can cause discoloring in the finished product and can create an undesirable appearance in the product. For example, dried ground cayenne can cause irregular variations in flavor and color, sometimes creating "hot" spots in food products. Anticaking agents are added to ensure better flowability of dried spices. In applications with high moisture content, such as salad dressings or soups, where particulates are desired for visual and textural effects, there is a great risk in using dried spices, unless they are sterilized.

SPICE EXTRACTIVES

Flavor is a combination of taste and aroma. The sensations of sweet, piney, sour, bitter, spicy, sulfury, earthy and pungent are derived from an overall combination of aroma (due to volatile components) and taste (mainly due to nonvolatile components) in a spice.

Spice extractives, which are highly concentrated forms of spices, contain the volatile and non-volatile oils that give each spice its characteristic flavor. The volatile portions of spice extractives, also referred to as essential oils, typify the particular aroma of the spice. Most spices owe their distinctive "fresh" character to their essential oil content that generally ranges from 1% to 5% but even goes up to 15% in certain spices. The non-volatiles include fixed oils, gums, resins, antioxidants and hydrophilic compounds, and they contribute to the taste or "bite" of a spice.

Certain spices are prized for their bites and coloring, such as black pepper, chile peppers, ginger, saffron and turmeric. These properties are due to the non-volatile portions of spices.

Volatile oils contain several chemical components whose amount and proportion give rise to the spices' characteristic aromas. These can include one, two or several components. The major chemical components of essential oils are terpene compounds—monoterpenes, diterpenes and sesquiterpenes. Monoterpenes are the most volatile of these terpenes and give out strong aromas when spice tissues and cells are disintegrated through heating, crushing, slicing or cutting.

The taste of a spice such as sweet, spicy, sour or salty, is due to many different chemical components such as esters, phenols, acids, alcohols, chlorides, alkaloids or sugars. Sweetness is due to esters and sugars; sourness to organic acids (citric, malic, acetic or lactic); saltiness to cations, chlorides and citrates; astringency to phenols and tannins; bitterness to alkaloids (caffeine and glycosides); and pungency to the acid-amides, carbonyls, thio ethers and isothiocyanates.

The ratio of volatiles to non-volatiles varies among spices causing flavor similarities and differences within a genus and even within a variety. Within the genus *Allium* for example, there are differences in flavor among garlic, onions, chives, shallots and leeks, which differ in this ratio. They vary depending upon the species of spice, its source, environmental growing and harvesting conditions and storage and preparation methods. Even the distillation techniques can give rise to varying components—through loss of high boiling volatiles, with some components not being extracted or with some undergoing changes.

Non-volatiles in a spice also vary with variety, origins, environmental growth conditions, stage of maturity and post-harvest conditions. For example, the different chile peppers belonging to the *Capsicum* group, such as habaneros, cayennes, jalapenos or poblanos, all give distinct flavor perceptions, depending on the proportion of the different non-volatiles, the capsaicinoids.

Spice extractives come as natural liquids (which include essential oils, oleoresins and aquaresins) and dry encapsulated oils (spray-dried powders and dry solubles). Developed from fresh or coarsely ground spices, spice extractives are standardized for color, aroma, and, with some spices, for their antioxidant activity. They are more concentrated than dried or fresh spices and so are used at much lower levels. These extractives provide more consistency than dried spices in prepared foods.

ESSENTIAL (VOLATILE) OILS

Essential oils, such as oil of basil, oil of caraway or oil of black pepper, are produced by grinding, chopping or crushing the leaf, seed, stem, root or bark; then cold expressing, dry distilling or extracting through steam distillation (water, steam, steam and water) and recovering the distillate oil with a solvent. Sometimes the oil is distilled from a whole spice, such as the leaf or flower, or from a broken spice. Depending upon the method of extraction, the nature of the volatiles can differ with the same type of spice.

Essential oils are the major flavoring constituents of a spice. Each essential oil has many chemical components, sometimes even up to 15, but the characterizing aroma generally constitutes anywhere from 60% to 80% of the total oil (Table 2). The essential oils are composed of hydrocarbons or terpenes (e.g., α-terpinene, α-pinene, camphene, limonene, phellandrene, myrcene and sabinene), oxygenated derivatives of hydrocarbons (e.g., linalool, citronellol, geraniol, carveol, menthol, borneol, fenchone, tumerone and nerol), benzene compounds (alcohols, acids, phenols, esters and lactones) and nitrogen- or sulfur-containing compounds (indole, hydrogen sulfide, methyl propyl disulfide and sinapine hydrogen sulfate).

Terpenes usually contribute to the freshness of a spice and can be termed floral, earthy, piney, sweet or spicy. The oxygenated derivatives, which include alcohols, esters, acids, aldehydes and ketones, are the major contributors to the aromatic sensations of a spice. The compounds with benzene structure provide sweet, creamy and floral notes, while the sulfur- and nitrogen-containing compounds give the characteristic notes to onion, garlic, mustard, citrus and floral oils.

Essential oils are soluble in alcohol or ether and are only slightly soluble in water. They provide more potent aromatic effect than the ground spices. Essential oils lose their aroma with age.

Essential oils are very concentrated, about 75 to 100 times more concentrated than the fresh spice. They do not have the complete flavor profile of ground

TABLE 2. Examples of Characterizing Essential Oil Components in Some Popular Spices.

Spice	Components in Essential Oils
Allspice seed	Eugenol; 1,8-cineol; humulene, α-phellandrene
Basil, sweet	Linalool; 1,8-cineol; methyl chavicol, eugenol
Cardamom	1,8-cineole; linalool; limonene; α-terpineol acetate
Dill leaf	Carvone, limonene, dihydrocarvone, α-phellandrene
Epazote	Ascaridol, limonene, ρ-cymene, myrcene, α-pinene
Fennel	Anethole, fenchone, limonene, α-phellandrene
Ginger	Zingiberene, curcumene, farnescene, linalool, borneol
Juniper	α-pinene, β-pinene, thujene, sabinene, borneol
Kari leaf	Sabinene, α-pinene, β-caryophyllene
Lemongrass	Citral, myrcene, geranyl acetate, linalool
Marjoram	*cis*-sabinene, α-terpinene, terpinene 4-ol, linalool
Nutmeg	Sabinene, α-pinene, limonene, 1,8-cineol
Oregano	Terpinene 4-ol, α-terpinene, *cis*-sabinene
Pepper, black	Sabinene, β-pinene, α-pinene, limonene, 1,8-cineol
Rosemary	1,8-cineol, borneol, camphor, bornyl acetate
Star anise	Anethole, α-pinene, β-phellandrene, limonene
Turmeric	Turmerone, dihydrotumerone, sabinene, 1,8-cineol
Zeodary	Germacrone-4, furanodienone, curzerenone, camphor

spices, but they are used where a strong aromatic effect is desired. Essential oils are used at a very low level of 0.01 to 0.05% in the finished product. They can be irritating to the skin, toxic to the nervous system if taken internally (by themselves) and can cause allergic reactions and even miscarriages.

Sometimes, alternative oils extracted from a different part of the same spice plant or from another variety are used to enhance or adulterate the more expensive essential oils, but suppliers need to meet the quality specifications that are required from manufacturers for these essential oils.

OLEORESINS (NON-VOLATILES AND VOLATILES)

The non-volatile and volatile flavor components of spices, also referred to as oleoresins, are produced by grinding or crushing the spices, extracting with a solvent and then removing the solvent. Oleoresins have the full flavor, aroma and pungency of fresh or dried spices because they contain the high boiling volatiles and non-volatiles, including resins and gums that are native to spices.

The non-volatile components create the heat or pungency of black pepper, mustard, ginger and chile peppers. These components can be acid-amides, such as capsaicin in red pepper or piperine in black pepper, isothiocyanates in mustard, carbonyls such as gingerol in ginger and thioethers such as the diallyl sulfides in garlic or onion.

The different pungent or heat principles give different sensations—spicy hot, hot, sharp, biting or sulfury. The pungent sensation of onion or garlic is sulfury, while that of Jamaican ginger is spicy. Red pepper and white pepper do not contain much aroma because they have very little essential oils, whereas ginger, black pepper and mustard contribute aromatic sensations with their bites because of higher content of volatile oils. White pepper has a different bite sensation than black pepper because of their different proportions of piperine and chavicine contents.

Five types of capsacinoids have been isolated in chile peppers: capsaicin, hydrocapsaicin, homocapsaicin, dihydrocapsaicin and dihydrohomocapsaicin, each with its own characterizing "bite" sensation in the mouth. In any particular type of chile pepper, the levels of capsaicinoids vary, causing varying heat levels. Each type of capsacinoid also creates a different perception of heat. Habanero has an initial sharp and violent bite that quickly disappears, leaving behind an aromatic sensation, whereas the cayennes give an initial burn that lingers.

Similarly, the release of heat sensation in mustard is different from wasabi. In wasabi, heat is immediate and in front of mouth, while with mustard and horseradish, the release is delayed and comes at the back of the mouth, with a shooting sensation to the sinuses. Table 3 details some of the non-volatiles that contribute pungency to a spice.

Oleoresins come as viscous oils and thick pastes and are more difficult to handle than essential oils. Usually, oleoresins are mixed with a diluent such

TABLE 3. Hot and/or Pungent Non-Volatiles in Some Spices.

Spices	Pungent Components of Spices
Red pepper	Capsaicin, hydrocapsaicin, dihydrocapsaicin, homocapsaicin, dihydrohomocapsaicin
Black/white pepper	Piperine, chavicine
Sansho pepper	Sanshool
Mustard	Allyl isothiocyanate
Horseradish	Allyl isothiocyanate
Ginger	Gingerol, shogoal
Garlic	Diallylsulfide
Onion	Diallylsulfide

as propylene glycol, glycerol or other oils for better handling. An emulsifier is added to make it water-soluble, or gum is added to make it into an emulsion for use in beverages, sauces, soups, pickles and salad dressings.

Oleoresins are used at very low concentrations because they are highly concentrated. They have greater heat stability than essential oils. Oleoresins give more uniform flavor and color with less variability than their ground spice counterparts. They are typically used in high heat applications such as soups, salad dressings, processed meats and in dry mixes and spice blends.

Aquaresins are oleoresins that are dispersible in water and oil. They are convenient to use because of the ease with which they disperse into water-based foods such as soups, sauces, pickles or gravies.

OTHER EXTRACTIVES

Liquid soluble spices are blends of essential oils or oleoresins that are made for aqueous systems. Fat-based soluble spice is made from essential oil or oleoresin blended with vegetable oil and used for mayonnaise, sauce or soup. Dry soluble spices, usually used in dry blends, are prepared by dispersing standardized extractives on a carrier such as salt, dextrose or maltodextrin.

Encapsulated oils are prepared from essential oils and/or standardized oleoresins with gum arabic or modified starch as the encapsulent. These have five to 10 times the strength of dried ground spices. Spray-dried flavors are the traditional encapsulated products.

Spray-dried spice flavors or dry soluble spices are created to make liquid spices or extractives more convenient to handle and use in dry applications. They are a dispersion of up to 5% or more of spice oleoresin on a free-flowing carrier such as salt, dextrose, gum arabic, modified starch or maltodextrin. These encapsulated spice extractives are used for high temperature applications, such as baked or retorted products. The spice flavors are slowly released into the product at the appropriate processing temperatures.

Essential oils or oleoresins are encapsulated to keep the full flavor impact of spices over an extended shelf life. This process grinds and encapsulates the spices in a closed system so no volatile oils escape. They are encapsulated by creating an emulsion with modified starch, dextrose and maltodextrin or soluble gum (gum acacia) and spray dried under controlled temperature and humidity conditions. The spice extractives are entrapped in this matrix that protects the flavor from oxidation and high heat and thereby provides an extended shelf life.

Encapsulated oleoresins retain the fresh notes of spices better than the oleoresins. They have no particulates, are completely natural, and like essential oils or oleoresins, have a friendly ingredient label. They are water soluble and allow flavor to be liberated uniformly throughout the food. For application, a 1:1 or 1:2 ratio is used as a replacement for the non-capsulated extractives.

Encapsulation of an extractive renders it wettable and dispersible in water or oil and also decreases the dusting in production. The quantity of extractives that can be dispersed on the carrier varies with the type of carrier and the oleoresin. It is important to evenly disperse and blend the oleoresin onto the carrier.

There are new forms of encapsulation, such as coacervation, that show better heat stability and protect spice oils during high-heat cooking or extrusion.

Spice extractives are cost effective compared to fresh or dried spices because they are used at very low concentrations and provide similar or sometimes more acceptable flavor perceptions. One part oleoresin or aquaresin is equivalent to 20 to 40 parts of a ground spice. Also, the color, texture and flavor of dried and fresh spices are altered through heat and freezing, while extractives have some heat and freezer stability. Extractives are available throughout the year and have standardized flavor and color, whereas the dried or fresh spices fluctuate in availability and quality. Extractives are also free of microbes and other extraneous matter, so they do not cause microbial contamination in the finished product.

Spice extractives are labeled as natural flavors, natural flavorings or as spice extracts. They are typically used by food developers because of their consistency in flavor and aroma, instant flavor release, uniformity of color and stability in high-heat applications. By using natural extractives, the quality and consistency of products from development through production can be better controlled. Finished product quality, uniformity during mixing and consistency can be maintained from batch to batch production.

Proper usage levels in the finished product are very important to achieve the right flavor profile and to prevent bitterness. Usage level is generally 0.01 to 0.05% by weight in the final product but varies depending on the type of application. A more uniform dispersion of color and flavor is achieved with the liquid soluble extractives than with dried spices. This creates an acceptable appearance in finished products, unlike dried spices that sometimes mix unevenly in large quantities and leave pockets of flavor and color. However, oleoresins and essential oils need to be effectively dispersed on a soluble carrier, such as salt or sugar, before being added to the product. Otherwise, they will not blend well in the finished product. Over-optimum use of the liquid extractives will create "caking" of the finished product during storage.

Today, the trend is to capture varied flavor profiles from spices. For example, food processors are beginning to use chile peppers for flavor as well as pungency. Asians and Mexicans have traditionally used chile peppers for flavoring their dishes and have created unique flavors from chile peppers through different preparation techniques. The chile pepper's flavor is contained in its outer fleshy parts and is intensified when roasted or fried. Chile peppers vary in color, flavor and texture profiles when they are fresh, dry, smoked or grilled. As a result of the popularity of these ethnic cuisines, the flavor characteristics of chile peppers and many other spices are becoming a major portion of the profile sought.

Table 4 provides a summary of the advantages and disadvantages of various spice forms.

TABLE 4. A Summary of Advantages and Disadvantages of Different Spice Forms.

Spice Form	Advantages	Disadvantages
Fresh whole	Fresh flavor	Variability in flavor and color
	Release of flavor slowly at high temperatures	Short shelf life
		High microbes
	Label friendly	Unstable to heat
		Availability
Dried ground	Process friendly	Less aroma
	Longer shelf life	Hot spots—flavor and color
	Easy handling and weighing	Takes more storage space
	Higher taste intensity than fresh	Undesirable specks
		Variable flavor, color
		Availability
		Spice dust contamination in production
Extractives	Standardized flavor and color	Difficult to handle and weigh
	Uniform appearance	Aroma and taste usually not typical of natural spice
	Low usage	Loss of volatiles at high temperatures
	Less storage space	
	Available throughout the year	

THE FUNCTIONS OF SPICES

Edible spices serve many functions in food products. Their primary functions are to flavor food and to provide aroma, texture and color. Spices also provide secondary effects, such as preservative, nutritional and health functions (Table 5).

Spices are composed of fiber, carbohydrates, fat, sugar, protein, gum, ash, volatile (essential oils) and other non-volatile components. All of these components impart each spice's particular flavor, color, nutritional, health or preservative effects. The flavor components (volatile and non-volatile) are protected within a matrix of carbohydrate, protein, fiber and other cell components. When the spice is ground, cut or crushed, this cell matrix breaks down and releases the volatile components.

PRIMARY FUNCTION OF SPICES

Flavor, Aroma and Texture

The flavor, aroma, color or texture that a spice contributes to food or beverage determines its effectiveness in a recipe.

Every spice or flavoring contains predominating chemical components that create these sensual qualities. A spice's chemical compounds can contribute mild to strong flavors. The balance of these chemical compounds gives a spice its characteristic flavor profile.

Spices give characterizing tastes and aromas. They give six basic taste perceptions: sweet, salty, spicy, bitter, sour and hot. The other descriptive terms include pungent, umami (brothy, MSG or soy sauce like), cooling, floral,

TABLE 5. Primary and Secondary Functions of Selected Spices and Flavorings.

Taste	Basils, black pepper, cardamom, jalapeno, asafoetida lemongrass, star anise, vanilla, kokum
Aroma	Clove, ginger, kari leaf, mint, nutmeg, rosemary, cardamom, tarragon, cinnamon, mango, shrimp paste
Texture	Garlic, mustard, dals, onion, sassafras, sesame seed, shallot, candlenut, almond
Color	Annatto, cayenne, chocolate, paprika, parsley, turmeric, saffron, rose essence, marigold, basil
Antimicrobial	Cinnamon, clove, cumin, oregano, rosemary, sage, thyme, ginger, fenugreek, chile pepper
Antioxidant	Turmeric, rosemary, sage, clove, oregano, mace
Health	Chile pepper, cinnamon, fenugreek, ginger, turmeric, garlic, caraway, licorice, clove

earthy, woody or green. The taste sensations are generally experienced at different locations of the tongue: sweet is detected at the tip of the tongue, salty at the frontal sides of the tongue, sour at the posterior sides of the tongue, bitterness at the back of the tongue, and heat, depending on the type, at different areas of the tongue.

Most spices have more than a single flavor profile. For example, fennel has not only sweet notes but also bitter and fruity notes, tamarind has fruity and sour notes, while cardamom has sweet and woody notes.

A spice's textural qualities are derived from its specific physical characteristics, the form in which it is used in a recipe (e.g., whole or ground) and the techniques used in its preparation. Most textural characteristics are obtained through the preparation and cooking techniques of spices, which are discussed later in this section.

Let's look at the typical sensory characteristic obtained from each spice or ingredient (Table 6). (Note, however, that spice sensory profiles are frequently described from a mainstream, Western perspective. Western foods are generally seasoned in a blander and milder fashion. Therefore, Western descriptions are not necessarily consistent with the taste experiences of other cultures. For example, cumin, coriander or cloves are often described as spicy, but to Asians and many other cultures, who are accustomed to eating hot and spicy foods, they are not perceived as spicy. This book generally describes spice tastes from

TABLE 6. Typical Sensory Characteristics of Spices.

Sensory Characteristic	Spices and Other Flavorings
Sweet	Cardamom, anise, star anise, fennel, allspice, cinnamon
Sour	Sumac, caper, tamarind, sorrel, kokum, pomegranate
Bitter	Fenugreek, mace, clove, thyme, bay leaf, oregano, celery, ajowan
Spicy	Clove, cumin, coriander, canela, ginger, Thai basil
Hot	Chile peppers, mustard, fagara, black pepper, white pepper, wasabi
Pungent	Mustard, horseradish, wasabi, ginger, epazote, garlic, onion
Fruity	Fennel, coriander root, savory, tamarind, star anise
Floral Aromatic	Lemongrass, sweet basil, savory turmeric, ginger, chervil, black pepper
Woody	Cassia, cardamon, juniper, clove
Piney	Kari leaf, rosemary, thyme, bayleaf
Cooling	Peppermint, basil, anise, fennel
Earthy	Saffron, turmeric, black cumin
Herbaceous	Parsley, rosemary, tarragon, sage, epazote, dillweed
Sulfury	Onion, garlic, chives, asafetida
Nutty	Sesame, poppy and mustard seeds

a mainstream U.S. perspective but takes other cultural taste descriptions into account.)

Coloring Properties

Some spices, such as saffron, paprika, turmeric, parsley and annatto provide color as well as flavor to foods and beverages. Spices can meet consumer's demands for "natural" colorings (Table 7).

The components responsible for the coloring in spices are oil soluble or water soluble. Some typical coloring components in spices are crocin in saffron, carotenoids in paprika, capsanthin in chile pepper, bixin in annatto or curcumin in turmeric. The overall coloring given by a spice is sometimes a combined effect of two or more of its coloring components.

Saffron imparts a beautiful yellow color, which ranges from a deep red, yellowish orange to a reddish orange color, to paellas and pilafs. Its yellow color is primarily due to a terpene glycoside called crocin that is water soluble and not stable against acid or light. Crocin is hydrolyzed to crocetin that gives a

TABLE 7. Coloring Components of Selected Spices.

Spices	Coloring Component	Type of Color
Saffron	Crocin	Yellowish orange
	Crocetin	Dark red
	beta-Carotene	Reddish orange
Paprika	Carotenoids:	
	Capsanthin	Dark red
	Violaxanthin	Orange
	Cryptoxanthin	Red
	Capsorbin	Purplish red
	beta-Carotene	Reddish orange
	Lutein	Dark red
	Zeaxanthin	Yellow
Chile pepper	beta-Carotene	Reddish orange
	Cryptoxanthin	Red
	Capsanthin	Dark red
	Capsorbin	Purplish red
Turmeric	Curcumin	Orange yellow
Parsley	Chlorophyll	Green
	Lutein	Dark red
	Neoxanthin	Orange yellow
	Violaxanthin	Orange
Annatto	Bixin	Golden yellow
	Norbixin	Orange yellow
Safflower	Carthamin	Orange-red
	Saflor yellow	Yellow

darker shade of color. Though it is a permitted natural color in the U.S., its use is limited because of its cost. Safflower is frequently substituted for saffron in many countries, but it is not permitted in the U.S.

Turmeric, from the ginger family, is often called "Indian saffron." Its root is dried and ground, giving a yellow color with an orange tinge. It is used as a natural food coloring in salad dressings, pickles, mustards, soups and condiments. Its coloring is due to curcumin, a diketone, that accounts for 3% to 7% of this spice. The curcumin content varies depending upon its source, with Allepey (India) turmeric having a higher amount of curcumin than other varieties.

Turmeric's color varies from a bright yellow-orange to a reddish brown and is unstable to light and alkaline conditions. It can be used with high-heat products and in products with a pH of 2.5 to 6.5. Its color is yellow in an acid to a neutral pH, but reddish brown in an alkaline pH. It will break down when prepared in an iron utensil.

Paprika is produced from the mild to pungent dried red pepper. The U.S. generally uses the mild paprika that has a brilliant red color derived from many different carotenoids. Capsanthin accounts for about 35% of the total carotenoids, violoxanthin, 10%, cryptoxanthin and capsorbin, each 6% and other carotenoids, 2%. Capsanthin is oil soluble, stable against heat and has a strong red color. Paprika powder loses its color through oxidation, catalyzed by light and high temperatures. Its oleoresin, which has a reddish-orange shade, is more stable to light and heat and is used in snack products, spice blends, crackers and salad dressings.

Annatto is ground from annatto seeds. It exhibits a yellow to golden yellow shade and is used in cheddar cheese, bakery products and sometimes in combination with paprika and turmeric oleoresins. Its coloring is due to norbixin and bixin, which are soluble in oil and stable at pH 5 to 14. It is resistant to heat but is less resistant to light.

The coloring components of essential oils are extracted with ethanol or organic solvents, and the solvents are removed to give the coloring matter. Extracts of paprika, saffron, annatto and turmeric with emulsifiers (polysorbate 80) are used as natural colors to provide the bright yellow, red and orange hues in processed foods and beverages. They are available as oil-soluble (oleoresins) and water-dispersible extracts. Trace metals, oxidized fats and oils, intense light and exposure to oxygen will promote color losses in these extractives.

SECONDARY FUNCTIONS OF SPICES

Secondary effects of spices are becoming more important as the public's desire for natural or organic foods and natural ways of healing increases. Spices have been used traditionally to stimulate appetite, enhance digestion, relieve stress and increase energy. Spices can also aid nutrition when they are used in lieu of salt, fat or sugar to enhance taste in processed foods.

Spices may be used in food products as preservatives, which allows for a more "natural" or friendly label on processed foods.

Spices as Preservatives

Spices have long been known for their preservative qualities, as antimicrobials and as antioxidants. They have been used by many ancient cultures, Egyptians, Romans, Indians, Greeks, Chinese and Native Americans, to fumigate cities, embalm the royalty, preserve food and prevent diseases and infections.

Spices as Antimicrobials

In Europe, the Middle East and Asia, before the days of refrigeration, spices were used to preserve meat, fish, bread and vegetables. Spices were used alone or in combination with smoking, salting and pickling to inhibit food spoilage. The Romans preserved fish sauce with dill, mint and savory, and they preserved meats and sausages with cumin and coriander. The Greeks used garlic to prevent food spoilage, and in India, ginger, garlic, clove and turmeric were used to preserve meats and fish. In ancient Egypt, cinnamon, cumin and thyme were used in mummification. Spices are still used to preserve food in the villages of India, Africa, Indonesia and Thailand.

Spices have also been used for bactericidal and health reasons. During the Middle Ages, spices such as cinnamon, garlic and oregano were used to treat cholera and other infectious diseases. In the late nineteenth century, clove, mustard and cinnamon were shown to have antimicrobial activity. In the twentieth century, new research on spices, including ginger, garlic, fenugreek, turmeric and clove, as potential natural antimicrobials continues.

Spices have strong, moderate or slight inhibitory activity against specific bacteria (Table 8). Aldehydes, sulfur, terpenes and their derivatives, phenols and alcohols exhibit strong antimicrobial activity.

A combination of spices can be more effective as preservatives than one spice. Microorganisms differ in their susceptibility to specific spices. Gram-positive bacteria are more sensitive to spices than gram-negative bacteria. *B. subtilis* and *S. aureus* are more susceptible than *E. coli* bacteria. Certain spices can act as broad-spectrum antimicrobials, such as rosemary and sage, while others are very specific in their functions, such as allspice and coriander.

The essential oils of some spices have an inhibitory effect on bacteria and fungi in meats, sausages, pickles, breads and juices. Eugenol in clove, sage and cinnamon; allyl isothiocyanates in mustard; thymol in thyme, sage and oregano; and allicin in garlic are some of the components used as antimicrobials.

TABLE 8. Certain Spices and Their Antimicrobial Properties.

Spice	Effective Component	Microorganism[1]
Mustard	Allyl isothiocyanate	*Escherichia coli, Pseudomonas, Staphylococcus aureus*
Garlic	Allicin	*Salmonella typhii, Shigella dysenteriae,* molds, yeasts
Chile pepper	Capsaicin	molds, bacteria
Clove	Eugenol	*E. coli* 0157:H7, *S. aureus, Aspergillus,* yeast, *Acinetobacter*
Thyme	Thymol, isoborneol, carvacrol	*V. parahemolyticus, S. aureus, Aspergillus*
Ginger	Gingerone, gingerol	*E. coli, B. subtilis*
Sage	Borneol	*S. aureus, B. cereus*
Rosemary	Thymol, borneol	*S. aureus, B. cereus*

[1]Kenji and Mitsuo, 1998.

The more pungent spice non-volatile oils have also been shown to have strong antimicrobial properties, such as gingerol in ginger, piperine in black pepper, capsaicin in red peppers and diallyl sulfide in garlic.

Spices need to be used at high levels in order to be effective antimicrobials, but this typically causes flavor problems in food products. Thus, the level of spices typically added to foods is generally not enough to inhibit microorganisms completely but can inhibit spoilage to some extent. Spice extractives at the levels used are effective against certain bacteria. Thyme (thymol), anise (anetol) and cinnamon (eugenol) essential oils inhibit mold growth and aflatoxin production.

Spices as Antioxidants

Spices can be used in foods to help fight the toxins created by our modern world. Heat, radiation, UV light, tobacco smoke and alcohol initiate the formation and growth of the free radicals in the human body. Free radicals damage the human cells and limit their ability to fight off cancer, aging and memory loss. Many spices have components that act as antioxidants and that protect cells from free radicals (Table 9).

Some spices have more antioxidant properties than others depending on the food they are in. Combining spices with other spices or antioxidants such as tocopherols and ascorbic acid produces synergistic effects. The naturally occurring phenolic compounds (phenolic diterpenes, diphenolic diterpenes) in spices are effective against oxidative rancidity of fats and color deterioration of the carotenoid pigments.

Spices can prevent rancidity and extend shelf life by slowing the oxidation of fats and enzymes. Fats are broken down into peroxides (free radicals) with

TABLE 9. Spices as Antioxidants.

Spice	Chemical Component
Rosemary	Carnosol, carnosoic acid, rosmanol
Sage	Rosmanol, epirosmanol
Turmeric	Curcumin, 4-hydroxycinnmoyl methane
Clove	Eugenol
Oregano	Phenolic glucoside, caffeic acid, rosmarinic acid, protocatechuic acid
Mace and nutmeg	Myristphenone
Sesame oil	Sesaminol, δ-tocopherol, sesamol
Ginger	Shogoal, gingerol

exposure to air or oxygen and finally into aldehydes and alcohols that give a rancid taste. Spices can halt the oxidative process by blocking or "scavenging" the free radicals.

Today, with consumer demand for "natural" products, spices can be used commercially as natural antioxidants in foods. Sage, rosemary, oregano, thyme and marjoram are found to have stronger antioxidant properties than other spices. Rosemary and sage are currently used as natural antioxidants in foods, while other spices are being explored.

Rosemary and sage antioxidants are available as oil-soluble, water-dispersible or dry-soluble formulations and can be used in seasoning blends, salad dressing mixes, lard, sausage and instant potatoes. They are used as sprays, dips or surface coatings in comminuted poultry, seafood or meats before they are frozen to inhibit "warmed over" flavors that develop after cooking and reheating. For snack foods, they are added in the frying oil or atomized on the surface of snacks or put into dough. They are also added in glazes and injection marinades for meats and are extremely heat stable and withstand extrusion, spray drying or baking.

Rosemary and sage are the most effective of all spices in retaining the red color of processed meats by inhibiting the flavor and color degradation of fats and oils in them. The flavonoids and di- and tri-terpenes of rosemary and sage are responsible for their antioxidant properties. They exhibit antioxidant properties superior to BHA or BHT. Other effective spices are thyme, turmeric, oregano, ginger, clove, majoram, red pepper, mace, sesame and nutmeg.

Rosmanol, caffeic acid, myristphenone, curcurmin, eugenol, thymol and sesaminol in rosemary, clove, thyme, oregano, ginger, turmeric, nutmeg, sage and sesame oil are found to be strong antioxidants with meat, lard and soybean oil.

EMERGING SECONDARY FUNCTIONS OF SPICES

Spices as Medicines

The growing emphasis on healthy eating is drawing attention to spices as critical ingredients for creating tasty low-fat or low-salt foods. Consumers are also looking for natural cures and ways for improving their health. Consumers prefer eating a "natural" food product to taking medicine or drugs.

Spices can be used to create these health-promoting products. The active components in spices—phthalides, polyacetylenes, phenolic acids, flavanoids, coumarins, capsacinoids, triterpenoids, sterols and monoterpenes—are powerful tools for promoting physical and emotional wellness.

Spices such as celery, parsley, ginger, turmeric, fenugreek, mint, licorice, garlic, onion, mustard, horseradish and chile pepper can be used to stimulate production of enzymes that detoxify carcinogens, inhibit cholesterol synthesis, block estrogen, lower blood pressure, elevate immune activity and inhibit tumor growth.

Spices have been used by many traditional cultures, including Indian, Chinese, Latinos, African, Egyptian and Greek, to inhibit or relieve illnesses. Even today, rosemary, sage and basil fight against tumors; chile peppers inhibit blood clotting, stimulate digestion and circulation, induce perspiration and reduce pain; ginger prevents motion sickness and aids digestion and stomach ulcers; garlic lowers cholesterol and high triglycerides, prevents colds and flues and prevents tumor growth; turmeric inhibits tumors, heals wounds and acts as an antidepressant; licorice treats gastric and duodenal ulcers and relieves chronic fatigue, coughs and cold symptoms; peppermint combats indigestion, irritable bowel syndrome and inflammation of gums.

Aromatherapy, using essential oils, can relax or stimulate the body, create certain moods, relieve cold symptoms and respiratory problems and ease muscle pains. The vapors are inhaled to release neurochemicals in the brain through receptors in the mouth and nose that cause the desired effects. Cooking foods with spices is the oldest form of aromatherapy, since their aroma can stimulate gastric secretions that create appetites. Spices are also used as balms or massage oils and are applied on the skin, joints and muscles to relieve stress and pain. Examples of the health benefits of spices are almost endless.

In the U.S. and elsewhere, the growing trend of using foods as nutraceuticals that boost energy and improve health will promote spices that were historically used to cure ailments and prevent diseases.

Let's look at traditional healing methods to provide an understanding of how and why spices are used by many global cultures as medicines. This will also help in understanding the basis for the contrasting flavors presented by spices and the many taste sensations experienced in an Asian meal.

Traditional Medicine

Spices are used in many traditional cultures to boost energy, relieve stress, improve the nervous system, aid digestion, relieve cold symptoms and headaches and treat many diseases. A food creator can explore interesting food concepts that combine taste and "cure" by using authentic spices and methods of preparing and presenting them in meals.

Europeans have used oils from leafy spices or plants to provide physical, mental and spiritual benefits, such as to provide a calming effect, to soothe muscle aches or to cure many ailments such as colds and fevers. Similarly, Middle Easterners and Egyptians used spices for many therapeutic and cosmetic effects. The Greek system of medicine is similar to Indian and Chinese traditional medicine and is called *Unani tib*.

Indian cooking is based on the therapeutic principles of ancient *Ayurvedic* medicine, which is based on practical and spiritual concepts and uses abundant spices. *Ayurveda* combines two words, *ayu,* which means life, and *veda,* which means knowledge. This system of healing has been practiced in India for 5000 years. It is based on prevention and well-being and looks at the causes of disease and the ways they occur. It is based on the life force, or prana, that flows easily into every cell of the body and which is accomplished by eating the right foods, by deep breathing, and by following a healthy lifestyle. In the *Ayurvedic* meal, the ingredients are chosen not only for taste but also to assure physical and emotional harmony and well-being. *Ayurveda* emphasizes prevention of disease through the pursuit of mental, physical and emotional harmony.

Similar to the ancient Greek and Roman theories of medicine, spices are classified into five elements: earth, water, fire, air and ether. *Ayurveda* categorizes foods into six tastes or *rasas* (Table 10). Many foods and spices contain more than one taste. For example, fenugrek seed provides a bitter taste as well as astringency, and fennel seed has sweet and cooling tastes. Thus, there is more than one taste contributed by a spice (see Table 6). With regard to *rasas,* other spices can be explored to provide more variety. Understanding these *rasas* is essential to understanding and applying *Ayurvedic* medicine. They affect our digestion, disposition and health.

To maintain perfect balance and be well nourished, all six tastes or *rasas* should be part of every meal. This explains the complex spice combinations and depth of flavor in Indian foods. To harmonize the body, these tastes must be balanced in the meal according to each person's constitution or *doshas*. Illnesses and diseases occur when there is imbalance with foods and a person's constitution. There are three basic constitution types or *doshas* that came from the five elements of energy—earth, fire, air, space and water (Table 11). Generally, a person has influence from three doshas with one or two predominating doshas. The dosha or doshas that exhibit the stronger characteristics becomes or become the predominating dosha or doshas for that person.

TABLE 10. Foods and Contributing Tastes (*Rasas*).

Tastes (Rasas)	Example Foods
Sweet	Anise, fennel, cumin, coriander, sugar, butter, honey, jaggery, rice, legumes, fruits, milk, cardamom, coconut, most grains, ghee
Sour	Kokum, lemon, pomegranate, tamarind, tomatoes, lemon, grapefruit, fermented foods (yogurt, pickles, miso, soy sauce)
Salty	Salt, seaweeds, vegetables (high moisture)
Bitter	Fenugreek, turmeric, clove, cinnamon, endives, lettuce, purslane, bay leaf, ajwain
Spicy/pungent	Chile peppers, ginger, garlic, horseradish, mustard, onion, black pepper
Astringent	Asafoetida, teas, licorice, legumes, fenugreek

These *doshas* regulate the functions in our mind-body system. They control the basic functions of all organs and systems in our body. Certain spices and flavorings are used in our foods to balance our doshas and create harmony (Table 12). These *doshas* also have their own inherent natural tastes. The six tastes or *rasas* have a beneficial effect on the *doshas,* increasing one type and decreasing the others. *Ayurvedic* cooking calls on us to choose foods and spices with the tastes that balance our own *doshas* in order to maintain good health.

In *Ayurvedic* cooking, the art of using spices not only makes food taste better, but also increases the overall therapeutic value and negates any side effects in the body.

In Indian cooking, foods need to be well digested to be nourishing for the body. Foods can be hot, cold, moist, dry, heavy or light. Hot food will speed digestion, while cold food will slow it. Every meal should be well balanced between hot and cold foods to promote digestion and avoid illnesses. The way Indian dishes are served in a traditional meal illustrates these principles. In a typical vegetarian meal, rice and/or bread (chappati, naan or dosai) are placed on a thali or tray with an array of tiny silver bowls. They contain peppery sambar (lentil stew), spicy, sour rasam (spice broth), cool, minty raita (cucumber and tomato with yogurt), a couple of hot and spicy braised vegetables and a crunchy, astringent mango pickle. The "cold" raita, the sour broth and

TABLE 11. Doshas and Respective Personalities.

Doshas	Qualities
Vata	Quick thinking, creative, flexible, nervous
Pitta	Determined, strong willed, fiery, passionate
Kapha	Strong, with endurance and stamina, heavy, stable

TABLE 12. Spices that Balance and Reduce the *Doshas*.

Doshas	Natural Taste	Spices for Balance	Spices to Reduce
Vata	Bitter	Salty, sour, sweet	Pungent, bitter astringent
	Astringent pungent	cardamom, fennel, anise, tamarind	Ginger, asafetida chilies, fenugreek
Pitta	Sour salty pungent	Sweet, bitter, astringent Asafetida, ginger, cardamom, cinnamon, fenugreek	Pungent, sour, salty Chilies, salt, tamarind garlic
Kapha	Sweet, salty, sour	Pungent, bitter, astringent Turmeric, fenugreek, mint, mustard, clove	Sweet, salty, sour Salt, anise, fennel, tamarind, cardamom

the "hot" spicy vegetables balance each other, thus creating "harmony" in the body. The crunchy mango pickle balances the soft textured vegetables, while the hot lentil stew perks up the plain neutral bread or rice. Also, the order in which the different side dishes are eaten with rice creates a healthy balance in our body. Food developers can use these meal presentation concepts today to develop interesting and well-balanced meals.

In China, historically, there has been an integration of nutrition, medicine and foods. Even today, combinations of spices are made into tonics that nourish and strengthen the body and cure illnesses. Similar to the Indian *Ayurvedic* medicine, in Chinese traditional medicine, there are five different tastes—sweet, salty, bitter, sour and spicy hot. The proper balance of these different tastes with appropriate textures and colors creates good taste and health in Chinese cooking.

Each taste affects the *Qi* or (*chi*), an invisible, vital force which, similar to the prana in Ayurvedic practice of healing, circulates throughout the body along prescribed pathways. The proper balance of these five tastes is essential to creating harmony or good health. If *Qi* does not flow smoothly or "gets stuck," illness occurs. *Qi* can be "unstuck" by stimulants such as food, acupuncture or acupressure.

The different tastes as well as the movement of *Qi* are described by yin and yang. Yin and yang are very basic to Chinese culture, just as hot and cold are to Indian culture. They are similar to water/fire, female/male, moon/sun, etc. They describe *Qi*'s location, movement and functioning. There are yin (cold) producing foods (which are nourishing, cooler, softer, moister and alkaline) and yang (hot) producing foods (which are spicier, drier and acidic). Yin foods include mild spices (mint, parsley), seafood, melon, asparagus and steamed foods, and yang foods include chilies, ginger, meats and fried foods. Yin spices are sedative and slow metabolism, while yang spices are active and increase metabolism. The five tastes are also associated with major organs in the body. For example, sweet is associated with the spleen (yin organ) and stomach (yang organ); sour with liver (yin organ) and gall bladder (yang organ); and spicy with lungs (yin organ) and large intestine (yang organ). Balancing the yin and yang keeps the body in a state of equilibrium and good health and creates a wonderful release of flavors. Foods are prepared and cooked with spices based on this theory of well-being. This balancing is the basis of authentic Chinese cooking styles with contrasting flavors and textures.

Movement of *Qi* in our body is associated with yin and yang. For yin, the movement in the body is to contract and flow downwards and inwards; while for yang, the movement is to expand and flow upwards and outwards. One is not separate from the other, and in food combinations, yin follows yang and vice versa. Just as yin-yang creates movement, so do spices. With spices, the leaves and flowers produce an upward movement, while seeds, roots and fruits have a downward movement, both affecting the flow of *Qi*.

In Chinese cooking, foods must also be balanced between hot and cold. Hot foods are those generally rich in fats (peanuts, mutton or pork), spices such as chile peppers, sesame and ginger, and strong alcoholic drinks. Cold foods are generally fruits, leafy spices, bland vegetables (seaweed, watercress) and water. Neutral foods include rice, thus, it serves as a centerpiece in a meal. Contrasting ingredients or spices are added to a meal to manipulate its hotness or coldness. For example, chile peppers are added to seaweed to balance its effect on the body, or sugar is added to spicy pork to do likewise.

Cooking techniques are also classified under hot or cold. For example, grilling or deep-frying adds heat to foods while steaming or slow simmering cools it down.

The state of the person's health is also a factor. A person with a fever will not eat warm or hot ingredients. That person needs cooling ingredients to balance the system. Seasons also become important factors in balancing ingredients for health. During the cold season, spices that produce an upward and outward movement (yang spices) are more desirable for the flow of *chi*.

In addition to philosophies of balanced eating to promote well-being, traditional Asian cultures used spices for their healing properties. Unlike Westerners, spices were used daily in a meal and were taken in abundance to flavor dishes. Today, many of these spices are being examined and recognized by Western scientists for their positive therapeutic or pharmacological properties.

The healing and medicinal properties of spices in many traditional cultures are attributed to using the whole spice or a combination of spices that work synergistically. The medium in which it is applied with to be consumed also gives a synergistic effect. Spices were not only used in cooking but were also added to milk, tea, vinegar, salt, hot water or sugar to create healing effects. For example, chile peppers are added to milk and drank to reduce swellings and tumors, cinnamon is added to sugar and taken to prevent tumors and turmeric paste is added to milk and taken to reduce coughs and colds.

Modern Medicine

Today, the healing properties of spices are being studied using modern research methods (Table 13). The active components of spices classified as phytochemicals include sulfides, thiols, terpenes and their derivatives, phenols, glycosides, alcohols, aldehydes and esters. While the healing effects of many phytochemicals have been identified, questions still remain unanswered. Are phytochemicals by themselves the effective healing agents, or is the whole

TABLE 13. Reported Therapeutic Effects of Spices.

Spice	Chemical Component	Medicinal Value
Turmeric	Curcumin	Anti-inflamatory, antitumor (inhibits tumor initiation and promotion)
	Curcumene	Antitumor
Ginger	(6) Gingerol (6) Shogoal Gingeberane *Bis*-abolenenan Curcumene	Digestive aid—stomachache vomiting, indigestion, stomach ulcers
	(6) Gingerol	Antitumor
	Shogoal	Enhances gastrointestinal mobility Inhibits cholesteral synthesis
Fenugreek Seed	Trigonelline	Arrests cell growth and prevents hypoglycemic effect
	Diosgenin	Synthesis of steroid drugs and sex hormones
Seed & Leaf	Soluble dietary fiber (galactomannan) saponins, diosgenin, protein	Improved glucose tolerance (reduces plasma glucose levels), lowers cholesterol and triglyceride (decreases bile secretion)
Garlic	Allicin	Breaks down blood clots, prevents heart attacks, prevents gastric cancer
	Glutamyl peptides	Lowers blood pressure and blood cholesterol
	Allicin, diallyl	Inhibits platelet aggregation
	Sulfide, s-allyl Cysteine	Inhibits platelet aggregation
Licorice	Glycyrrhizin	Treats gastric and duodenal ulcers, prevents coughs and colds, treats chronic fatigue syndrome

*Data obtained from Mazza, G. and Oomah, B.D. eds. 2000. Herbs, Botanicals and Teas. Technomic Publishing Co., Inc., Lancaster, PA.

spice, as was traditionally used, required? Can the healing properties be strengthened or lessened through a synergistic effect with other spices or through a medium such as water, sugar, tea, milk, protein or starch?

Finally, it should be remembered that cultures that use spices as traditional cures consume them at much higher levels than Western cultures. For many cultures, "spiced up" foods are eaten at every meal on a daily basis. Moreover, "spice cures" are commonly taken through infusions in teas, milk, coffee, water and other beverages, which are not a common practice in the West. To meet the demands of today's consumer, spices and other flavorings can be explored as complementary medicine.

SPICE PREPARATION

Spices need to be ground, sliced, roasted, fried or boiled to release their characteristic flavors. Volatile oils are released from spices through grinding, cutting and heating. Spices release unique flavors when they are cooked. Several different flavor characteristics can be derived from the same spice by using different preparation methods. Around the world, many ethnic groups prepare spices to suit specific applications and to create totally different flavor profiles. Spices are dry roasted, roasted in oil, deep fried, simmered, pickled, barbecued or boiled in water to create a multitude of flavor releases. The flavor of spices is generally enhanced, intensified or simply becomes different. Cooking techniques are also discussed in Chapter 5, "Emerging Flavor Contributors."

Certain spices are more stable to heat and can be cooked at higher temperatures, while others have to be added at the end of cooking or just before serving. Certain pungent spices do not give any hot or sharp sensations until the spice is cut or roasted, which triggers enzymatic and other reactions that release these characteristic flavors. For example, whole onion needs to be chopped, ginger needs to be cut and ground mustard needs to be "wet" before their pungency can be experienced.

Spice volatiles are soluble in alcohol, so many cultures use wine or other alcohols to preserve the spice aroma. Because the volatiles are oil soluble, spices are often added to cooking oils before they are added to the finished product for better flavor retention. For example, in Asian and Indian cooking, spices are generally sautéed or roasted in oil before other ingredients are added to make chutneys or curries. This spice mixture gives more intense flavor to the finished product.

European and North American cultures tend to boil or simmer spices, such as bay leaf, coriander or cinnamon, to release their flavors. Those in other cultures, such as Asians or Middle Easterners, roast (dry or in oil) whole spices, such as cumin, coriander, fennel, cardamom, mustard seeds, clove, cinnamon sticks, nigella and ajwain, before adding them to sauces, tagines or stews. Even leafy spices are sometimes roasted in oil to give extra crunchiness. The pungent and aromatic sensations of many Indian foods are due to these prepared spices. Indians traditionally roast mustard seeds in oil to intensify their flavor, braise fresh kari leaf to remove its bitter green taste and fry asafetida to create a sweet flavor.

These preparation techniques tend to slightly crack open the whole spice, and when it is added to foods, it begins to slowly release its volatiles. Whole spices and ground spices are generally roasted for about one or two minutes, depending upon the spice, until they become slightly dark. Beyond their optimum cooking times, they become bitter and unacceptable.

Smoking, tumising or bagharing jalapeno, ground coriander or cumin seeds creates many desirable flavors. Tumising is a slow frying technique used with wet spice mixtures or spice pastes to create more aromatic and intense flavors. It is carried out at lower cooking temperatures with constant stirring. The cooking procedure is complete when the oil in which the spices are cooked in begins to separate or seep out of the tumised mixture or sauce. For example, in Malaysia, Myanmar, Singapore and Indonesia, chile peppers (dried or fresh) are blended with shallots and other ingredients and are then tumised in oil to create a fragrant "chili (cili) boh." This condiment is then used to flavor everything, from sandwiches to salads to vegetable curries to laksas to soups.

In India, spices are tarkared or bargared to obtain a broad spectrum of flavors. For example, when preparing South Indian dals and chickpea curries, fenugreek seeds are dry roasted and mustard seeds are cooked in hot oil until they "pop" (tarkared). Spice mixtures are also fried in oil (barghared) before other ingredients are added to create the unique, balanced flavors that characterize these vegetarian dishes.

In the U.S., with the increasing influence of Indian, Southeast Asian and Mexican cuisines, food preparation techniques of ground or whole spices are becoming an important, upcoming trend. Today, U.S. consumers seek fire-roasted chile peppers, oven-roasted garlic and grilled onions to provide enhanced flavor profiles. Fire roasting is done by holding the chilies over an open flame until the outer skin blisters and chars. The charred skin is then removed leaving a flavorful soft product. This product is then added to sauces, soups, spreads, stews or salad dressings to create an enhanced aroma and rich, smoky notes.

The order of addition of spices is important in creating authentic spice blends. Adding different spices at appropriate times or stages during cooking helps each spice to retain its individual flavor and balance the other spices. For example, when making a fish curry, mustard seeds are added to the heated oil first, followed by cumin, coriander, fennel and turmeric; then garlic and ginger are added, after which onions are added. When the oil separates, tomatoes are added with fenugreek seeds (already dry roasted) and saffron is added toward the end, to avoid bitterness. This is the method for developing a complexity of flavors in South Asian cuisine.

GLOBAL EQUIPMENT USED IN SPICE PREPARATION

In many cultures around the world, spices are prepared with certain basic tools. On a domestic level, mortars and pestles, woks or kwalis are the basic spice tools needed for spice preparation. Mortars and pestles are used for crushing, pounding and coarsely grinding whole dry spices, chile peppers or garlic and for coarsely blending spices for sauces or condiments. (In the section in Chapter 5, "Other Flavoring Ingredients," the effect of spice preparation and cooking techniques on flavors is discussed in further detail.)

Mortars and pestles have different names in many cultures—molcajete and tejolate in Mexican, krok and saak in Thai, batu lesung in Malaysian, idi kallu in Tamil, kootani in Hindi, wurk kacha in Amharic and almirez in Tagalog. They are made from volcanic or lava rock, stone, baked clay, wood, granite, marble or ceramic. They come in varying sizes and are essential tools in Southeast Asian, Indian and Mexican kitchens. In the U.S., a spice mill or coffee grinder serves the same purpose.

Other tools used in spice preparation include the rectangular, flat stone called metate in Mexican, batu giling in Malaysian, ammi in Tamil or chakki in Hindu. These are used to finely grind spices, such as chile peppers, onions or sesame seeds. Water is added to the grinding process to create a pasty form of ground spice. Therefore, some metates slope gently downward to allow the paste to be pushed forward, as it is worked, into a container placed under the front edge. The stone rolling pin or metlapil is used as a grinding tool. Mexicans use these tools to grind corn, chocolate and roasted chile peppers. Asian Indians grind onions, ginger, garlic, chile peppers, soaked rice and lentils. Today, a blender or food processor is used on a domestic or commercial scale.

The flat griddle, made of clay or cast iron, called comal in Mexican or tawa in Hindi is used to roast spices, tomatoes, garlic, corn and chilies. In Mexico, the clay-type comal is rubbed with slaked lime to prevent ingredients from sticking to it and creating hot spots in the spice being roasted.

The wok, kwali (Malaysian), tsao-guo (Mandarin) or kadai/karhai (Hindi/Urdu) is made of iron, brass, aluminum or cast iron and is used to roast whole spices, fry spices in oil or to sauté spice pastes.

SPICE APPLICATIONS

Spices provide savory, spicy, sweet, pungent, bitter or sour notes to foods and beverages. A vast spectrum of tastes and aromas can be created by combining spices and other flavorings. Asian Indians use various spices to create uniquely balanced curry blends, while the Chinese use them to contrast sweet, sour or pungent notes with vivid textures. Thais, Japanese and Caribbeans create great visual appeal with spices. Whether we are creating regional ethnic or fusion cuisines or merely adding ethnic zest to traditional American foods, spices form the principal basis of flavor, texture and visual appeal in finished products.

MARINADES, RUBS AND GLAZES

As consumers continue to seek tasty and healthy foods that are easy to prepare, unique seasonings introduced through marinades, rubs, glazes or as sprinkle-on-seasonings are becoming the hottest new trends. These marinades and dry seasonings can be used to create a variety of flavors and textures and to add convenience for consumers through easy-to-prepare or ready-to-eat products.

The perfect blends of dried spices or extractives that are balanced with acids, salt, sugar, starch and oil can be used to develop dry rubs, emulsions, topical seasonings, glazes and tumbling or injection marinades for foods and beverages.

Rubs and glazes provide flavor and texture to a product. Rubs in dry or paste forms can be externally added to chicken, meats or seafood. These contain spice particulates that give visual appeal. For example, blackened chicken or fish, which are popular Cajun dishes, have unique crunchy coatings from spice rubs. Unacceptable, charred flavors often develop when using spice marinades that contain high tomato solids or high D.E. (dextrose equivalent) maltodextrins. Thus, encapsulated spices are preferred for retaining flavors.

Dry or liquid glazes are surface applications containing starch, gums and spices, and they may or may not include spice particulates. Some glazes penetrate the product to provide flavor, while, at the same time, they stay on surfaces to create visual and textural appeal. In the processed meat industry, glazes are applied after the marinades; they are applied to meats in the tumblers after the meat absorbs the marinade.

The earliest marinade was a mixture of salt and spices with vinegar or fruit juices that was added to flavor meat and preserve it. Today, marinades typically contain coarsely or finely ground spices (with or without particulates), oil, vinegar or other acid sources, salt, sugar and alkaline phosphates. A marinade can be either a tumbling or an injection marinade. In a tumbling marinade, meat is placed in a tumbler, and marinade is added. The meat pieces are tumbled under vacuum until the marinade is absorbed by the product. Injection marinade is an internal soluble spice extractive with no particulates or insoluble spices. Flavor is delivered by injecting the spice solution into a whole bird or meat, such as rotisserie chicken, resulting in uniform flavor and color. To avoid color streaks on products, colorless spice solutions need to be used, such as decolorized capsicums, black pepper or turmeric. Sometimes, this injection is followed by a tumbling step.

SPICE BLENDS, SEASONINGS AND CONDIMENTS

Aromatic vindaloos, sour colombos, peppery sambals and spicy tagines owe their unique flavors to the spice blends contained in them. This is discussed in detail in Chapter 6, "Emerging Spice Blends and Seasonings." When creating or using ethnic spice blends, it is important to remember the great variety of spice blends that exist. Even for particular spice blends such as adobos, curry blends, ras-el-hanout or recados, flavor variations exist depending on regional and cultural preferences on the availability of ingredients.

For example, no chile blend, pasta sauce, salsa or curry blend is the same. Curry blends vary from region to region and even within a region (Table 14). Curry (or spice mixtures for sauces) originated in India and traveled to other

TABLE 14. Global Curry Blends and Their Characterizing Flavorings.

Global Region	Flavor	Characterizing Ingredients
Basic curry blend	Mild, medium or hot, aromatic	Turmeric, cumin, coriander, ground dried chile pepper or black pepper
India		
North India	Creamy, mild, nutty	Yogurt, almonds, bay leaf, paprika, mint
South India	Hot, fiery, coconutty	Coconut, tamarind, chile pepper, fennel, cayenne
East India	Sweet, medium hot, aromatic	Mustard, tamarind, kalonji, fenugreek seed
West India	Hot, sour	Vinegar, cayenne pepper, mint, saffron
Sri Lanka	Black, fiery	Coconut milk, toasted spices, cayenne pepper
Pakistan	Mild, creamy, nutty	Paprika, bay leaf
Southeast Asia	Aromatic, hot, pungent, coconutty	Coconut, star anise, lemongrass, galangal, coriander root, cayenne pepper, spearmint, tomato, fish sauce, shrimp paste, kari leaf, peanuts
East Asia	Mild, sweet, starchy	Soy sauce, turmeric, corn starch, caramel, fish sauce
Caribbean	Slightly sour, hot, fruity	Habaneros, allspice, turmeric, vinegar, black pepper, fruits
England	Sweet, fruity	Apples, raisins, cream, turmeric, sugar
Middle East	Spicy, nutty, intense	Black pepper, caraway, tomato, mint, olive, sumac, pistachio, sesame seed, nigella
South Africa	Mild, sweet, fruity	Turmeric, clove, mace, cinnamon, nutmeg, bay leaf

parts of the world, and ingredients were added and deleted to suit local tastes. Curry blends from India, Thailand, Japan, the Caribbean, England and Africa differ greatly because of what they contain. Even in India, curry blends vary in flavor depending on the geographic regions, who makes it and what its application is, whether for fish, lentils or lamb. Most curry blends have some of the basic spices that were originally present in the generic blend from India, such as turmeric, cumin, coriander and dried red chile pepper. But, other spices and flavorings may be added, depending upon preferences, such as lemongrass, cilantro, habaneros, soy sauce, coconut, yogurt, allspice or shrimp paste, that give them their distinct regional flavors.

The same principle holds true for adobos, dukkahs, recados, pestos or sofritos. The basic ingredients of adobo are garlic, oregano, black pepper and turmeric. Then, based on regional preferences, cumin, habanero, lime juice or soy sauce are added to the basic adobo.

SPICE QUALITY AND SPECIFICATIONS

Consumers need the highest quality spices in products. So, how do spice suppliers best achieve the quality and consistency they seek? When was the spice ground, what is its shelf life and how long has it been stored before shipment? When a spice supplier provides ground spice, we need to learn its specifications, origins, how long it was stored before grinding and if it was processed before grinding. When a spice extractive is provided, we need to learn its specifications, how it was extracted and how it was stored. These are some of the many quality control factors we need to ascertain from the spice supplier.

Spices have to meet certain requirements in order for them to be acceptable in foods. These are written in the form of specifications that are supplied by the manufacturer. Generally, the specifications contain the spice's form, its physical description and measurements (form, solubility, specific gravity, refractive index, moisture, granulation, etc.), its sensory description (color, aroma and taste) and its chemical measurements (volatile oil, ASTA units for color, HPLC for pungency value, piperine value, etc.).

Quality standards for spice specifications use analytical and sensory methods. Analytical methods are used for measuring physical and chemical properties in spices, such as moisture, granulation or volatile oils, piperine value in black pepper, HPLC method for measuring capsaicinoids in chile peppers, acid insoluble ash for sand and soil contamination or spectrophotometry for color values in paprika or turmeric. Using these methods, quality limits are set for essential oils, pungency or color values. Most times, pungency, color and other sensory values are correlated with organoleptic evaluations with trained sensory panelists.

Spice blenders need to develop their own spice specifications with spice vendors for better control of products. Specifications need to be based on a fresh sample of spice. For successful product development, the spice blender needs to understand spices in their entity, their forms (fresh, dried and extractives), their properties and their uses. The blender must have a knowledge of the spices' origins and varieties, cultural preferences, usage in different cuisines, their historical significance and their nutritional, preservative and medicinal properties. The understanding of folklore and names in other languages will also help to better formulate spice blends for diverse applications.

The blender also needs to be aware of the measurements and procedures used to monitor spice quality. Finally, recommended usage levels of the spice type and form should be obtained to create optimum flavors in the finished products.

Most spices come from the Far East—India, Southeast Asia and China—while many others come from the Middle East, Mexico, Central and South America, the Caribbean and Europe. The U.S. also supplies a significant proportion of spices, especially dehydrated onions, garlic, chile peppers, paprika, dillweed, sage, sweet basil, tarragon and marjoram.

When buying small amounts of spices, it is preferable to buy whole spices and then grind them before use. Grinding or crushing spices releases their volatile flavors. When grinding small amounts of spice, a mortar and pestle or coffee grinder can be used; for larger amounts, a spice mill is used. The fresher the whole or ground spice is, the more flavor it has.

When purchasing larger quantities of spices, spice specifications become important because the flavor and color of spices can vary due to the place of harvest, year of harvest, the length of time in storage, storage conditions, milling and separation processes and age since harvesting.

Spice specifications from the supplier, including volatile oil content, color and granulation, need to be examined. Also, in order to maximize flavor and color and to prevent caking in formulations, the spice supplier's storage requirements and capabilities must be known. Keep in mind that spices come from many regions of the globe, with varying processing procedures and quality standards. It is important to be aware of measures or controls taken in exporting countries. Cleanliness, insect and rodent infestation and microbial quality are important criteria for the spice processor and the food developer. Spice specifications are important forms of communication between the food developer and the supplier for ensuring that the correct spice is delivered within economic limits. When spice specifications are developed and maintained, product consistency is maintained over time.

INTERNATIONAL STANDARDS AND SPECIFICATIONS

Each country, whether exporting or importing, has its own quality specifications. The quality standards most used as international guidelines for spice importing are the American Spice Trade Association (ASTA) specifications, the U.S. Federal Specifications and the International Organization for Standardization (ISO) guidelines. The Association of Official Analytical Chemists (AOAC) defines many of the analytical test methods. Japan and many other countries use the ASTA specifications. India, Malaysia, Spain, Indonesia, Sri Lanka and Hungary have their own quality standards for exporting spices.

ASTA specifications are for unprocessed spices imported into the U.S., and they regulate the cleanliness of these spices. ASTA specifications place limits on extraneous matter (insects, insect excrement, stones, stems, sticks, molds, etc.) and set standards, sampling procedures and testing procedures. Imported spices not meeting these specifications are reconditioned at the port of entry, while local spices need to be reconditioned (using fumigants) before they are processed for use in a product.

The U.S. Federal Specifications establish quality standards for whole and ground spices. Specifications include total ash, acid-insoluble ash, volatile oil, moisture, color and granulation, and they are standardized for major spices. These specifications have maxima or minima values that differ considerably between importing countries. Standards for defective spices (stalks, stems or other contaminants) are also established.

The ISO has specifications for most spices, such as extraneous matter, moisture content, total ash, acid-insoluble ash and other chemical characteristics. Based on the ISO, the British and Canadians have established their own standards. Spices that are irradiated, reheated with ethylene oxide or contaminated with aflatoxin are prohibited under the Japanese Sanitation Law.

In India, quality standards are established for unprocessed and processed spices that especially regulate moisture content, volatile oil, total ash, acid-insoluble ash and starch and extraneous matter. Grades are also established with specifications. Many spices such as turmeric, black pepper, ginger and chile peppers have individual grade specifications that differentiate them from similar spices. Even curry powders are graded based upon the amount of spices and salt they contain. Similar specifications are established for spices in many other countries.

MAINTAINING SPICE QUALITY

Grinding and sterilizing procedures can decrease the volatile oils in spices. Excessive heat volatilizes and dissipates the essential oils in ground spices, and high humidity tends to cake ground spices. Exposure to light, humidity variations, air and certain metals can discolor many spices such as paprika, turmeric or the green leafy spices. Dry, ground chile peppers turn from a natural green or red color to an olive or dirty reddish-brown color when exposed to light. Flavor and aroma losses as well as insect and rodent infestation occur when spices are not stored in airtight containers.

Spices or spice extractives should be stored in tightly closed containers in cool, dark, dry conditions below 68°F and 60% humidity. Some spices need cooler refrigeration temperatures, such as 32°F to 45°F, to prevent mold infestation (capsicum peppers), color deterioration (paprika) and to avoid rancidity (in high fixed oil seeds, such as sesame seeds). Colder temperatures also help preserve volatile oil flavor and aroma, freshness and sanitary quality. Refrigeration slows microbial growth in ground or whole spices.

Light will fade the color and character of spices, and heat will dissipate the aromatic oils. Good storage conditions, monitoring and specifications are important in retaining quality attributes of spices. Proper packaging is essential for preventing oxidation and, thus, retaining color and flavor. To retain good aroma in spices, long-term storage is not recommended.

Whole spices tend to have about three to four times the shelf life of ground spices. The shelf life of ground spices is about three to four months at refrigeration temperature, and whole spices have about one to one and one-half years of shelf life. Ground spices for minimally processed foods such as salad dressings, condiments or "sprinkle on" seasonings should be well cleaned and sterilized.

High levels of moisture in ground or whole spices indicate mold and microbial growth. During storage, insects breed on spices in varying degrees, depending upon storage conditions, where they are harvested, transportation contamination and the extent of cleaning. Filth levels include foreign materials such as insect fragments (moths, mites, beetles), small stones, metal fragments and glass pieces. Insects and mold growth can change the color and, to some extent, the flavor of the spice.

Microbiological requirements for "clean" spices include counts for total bacteria, yeast, mold, coliforms and food pathogens such as *E. coli* and *Salmonella.* High microbial counts are caused by contamination during growing and post-harvesting handling. Spore-forming bacteria, such as the *Bacillus* species or aerobacter aerogenes found in the soil can be transferred to the spice during the drying process, especially with "under the ground" spices such as turmeric, ginger, galangal or garlic. The type and amount of molds and bacteria on a spice depends on the type of spice and the conditions under which it is harvested and dried. *Staphylocccus* and *Streptococcus* bacteria species predominate, but pathogenic bacteria tend not to exist on spices. Spices that show strong antimicrobial properties tend to have low counts of microbes.

Molds, such as *Aspergillus,* that produce toxin (aflatoxin and penicillium) are found on certain spices including red pepper, fenugreek and ginger, so there are specification limits for these toxins. Molds tend to multiply during the drying process and during storage.

The control of insects and microbes is important in receiving a quality spice. Spices need to be free of microbes so as to reduce the initial bacteria or mold content in processed foods. Spice extractives and sterilized spices tend to meet these objectives. Organic foods are becoming more popular, and for a food to be labeled as organic, irradiation is not permitted. Spices that not sterilized can be great sources for microbial contamination and growth.

Several methods that are used to control insects and microbes to meet the specifications of the buyer are discussed to briefly provide an understanding and awareness of the procedures involved.

Fumigation

Spices are most commonly fumigated with methyl bromide, formaldehyde, carbon disulfide, chloropicrin or phosphine (aluminium phosphate) to control insect growth and breeding during storage. Some of these chemicals impart undesirable odors and flavors. Also, they are not totally effective when applied to whole spices.

Controlled Atmosphere Storage

Insects do not survive in atmospheres containing less than 2% oxygen. So, oxygen can be substituted by carbon dioxide or nitrogen in storage areas. Also, low-temperature storage conditions can be used because most insects do not survive in low temperatures. Relative humidity is controlled to prevent mold growth and aflatoxin production.

Sterilization

Microbes can be decreased by sterilizing with ethylene oxide gas, a mixture of ethylene oxide and carbon dioxide (90:10), propylene oxide, steam and irradiation.

Ethylene oxide is a widely used cold sterilization method. It has minimal effects on flavors and effectively reduces microbial loads. The gas reacts with the protein portion of the living insects and microbes and changes its structure. The efficiency of this method on microbes is further affected by the temperatures (25–30°C is needed). The problem with ethylene oxide is its toxicity to humans at concentrations typically used for sterilization. It also reacts with nutrients. Its residue decreases during storage. Vitamin C and vitamin B1 are decomposed, and the colors of some spices are changed. Ethylene oxide reacts with chlorides in foods to form chlorohydrins that are toxic and remain in processed foods. It irritates the skin and produces edema. Its residual limit in the U.S. is 50 ppm.

Steam sterilization using dry steam (saturated or supersaturated steam) reduces microbes on spices and is commonly used in Japan. The volatile oils are decreased, the color of certain spices is altered and spices develop unacceptable sensory attributes using this high-temperature method.

Irradiation can sterilize spices through gamma radiation generated from cobalt-60 and cesium-137 and sometimes through electron and X-ray radiation from an electron accelerator. Many countries use irradiation to sterilize spices, but this method is banned in Japan. There is a minimum irradiation dosage based on the type of spice, its microbial load and types of previous treatment on the spice. Recommended irradiation doses of 7 to 10 kGy can reduce microbes. In certain spices, color and flavors are changed when irradiated at a dose of 7.5 kGy. The terpene oils decrease at the same dosage. Coliform bacteria and fungi are reduced at a dose of 6.5 kGy, aerobic spores are reduced at 4 kGy, and thermophiles are reduced at 10 kGy. Disadvantages of irradiation include cost and consumer perception.

CHAPTER 4

A to Z Spices

AJOWAN

SOMETIMES mislabeled as lovage seeds, ajowan is a popular spice in Indian, Pakistani, Iranian and Ethiopian cooking. It is an essential ingredient in a Bengali seasoning called panchphoron. *Omam* water, an infusion of ajowan seeds, has been used since ancient times in India for stomach pains, colic, diarrhea and other bowel disorders.

Scientific Name(s): Trachyspermum ammi, T. copticum or *Carum copticum*. Family: Apiaceae (parsley family).

Common Names: ajwain, carom, wild parsley. It is also called netch azmud (Amharic), taleb el koub (Arabic), jowan (Bengali), bishop's weed (English), nanava (Farsi), ajowan (German, Italian), ajwain/carom (Hindi), ayamodakam (Malayalam), assamodum (Singhalese) and omam (Tamil).

Origin and Varieties: ajowan is indigenous to southern India but is also cultivated in Europe, Egypt, Pakistan, Iran and Afghanistan.

Form: it is a small, caraway-like seed that is used whole or ground.

Properties: ajowan is a close relative of caraway, dill and cumin. It has large curved and ridged oval, celery-like seeds that are light brown to purplish red in color. Ajowan seed looks like celery seed and when bruised has a taste similar to thyme but stronger. It has piney, phenol-like and slight lemony notes. When crushed or ground, it has a more intense flavor. It can be bitter and slightly spicy. Its leaves, stems and roots are aromatic.

Chemical Components: ajowan contains 2.5% to 5% essential oil, mainly thymol (35% to 60%), along with α-pinene, ρ-cymene, limonene and α-terpinene.

Ajowan contains calcium and iron.

How Prepared and Consumed: ajowan is commonly used by North Indians, Pakistanis, North Africans and Iranians. It is used whole or ground and has a natural affinity with starchy foods, such as root vegetables, legumes, breads, snacks and green beans. Ajowan makes starch and meats easier to digest and is added to legumes to prevent flatulence. It goes well with cumin, ghee, garlic, ginger and turmeric.

In North India, ajowan seeds are fried in ghee with other spices, and this aromatic mixture is added to cooked legumes and vegetables. The seeds are also used in a fried bread called paratha, added to snacks (pakora) and pastries and served with nuts. In Bengal, it is used as part of a seasoning called panchphoran, which goes well with fish and vegetable curries. To enhance its flavor, ajowan is roasted or fried in oil until it becomes light brown. It then provides a more intense aroma to fish curries, lentil stews and potatoes.

In Ethiopia, ajowan is an integral part of a spice blend called berbere, which is used in meat stews and vegetables.

Spice Blends: berbere, chat masala, panchphoran and pakora filling blend.

Therapeutic Uses and Folklore: ajowan is highly valued in India as a stomach medicine and an antiseptic. It is combined with salt and hot water to control flatulence and indigestion. Ajowan was also a traditional remedy for cholera and fainting spells.

Ajowan is an ingredient in mouthwashes and toothpastes because of its antiseptic properties.

ALLSPICE

The English derived the name allspice because the flavors of several spices, such as cloves, cinnamon, nutmeg and black pepper are combined in it. Native to the Caribbean and the Americas, Spanish explorers named it pimienta or pepper because the allspice berry resembles a large black peppercorn. Pimienta was eventually anglicized to pimento. Also called Jamaican pepper, allspice is not related to the peppercorn. It was first imported into Europe in 1601 as a substitute for cardamom.

Scientific Name(s): Pimenta dioica, formerly *P. officinalis*. Family: Myrtaceae (myrtle family).

Common Names: Jamaican pepper, pimento, clove pepper. Also, pepe di Giamaica, baharat, lb'harhelo (Arabic), piment (French, German, Dutch), pimento (English, Italian), toute-epice/poivre de Jamaique (French), nelkenpfeffer (German), pilpel hodi (Hebrew) orusupaisu (Japanese), kappalmulagu (Malayalam), kryddpeppar (Swedish), pimienta gorda and pimienta de Jamaica (Spanish), kattukaruva (Tamil) and yeni bahar (Turkish).

Origin and Varieties: there are many types of allspice, each with varying tastes. It is indigenous to the Caribbean Islands, specifically Jamaica, South America (Brazil, Leeward Isle), Central America (Guatemala, Honduras) and

Mexico. Allspice is also grown in India and Réunion. The U.S. buys mainly from Central America and Mexico.

Form: the berry/seed of the pimiento tree is picked green/unripe and then dried until it turns dark reddish brown in color. It is globular and has a rough textured surface. It is slightly larger than the black peppercorn. The Mexican type is the largest and darkest in color. Jamaican allspice berries are smaller. Allspice is used whole or ground. The aromatic leaves and bark can also be used to provide an allspice-type flavor to foods, especially to smoked meats and beverages.

Properties: allspice has a warm, pungent taste and the aroma of cloves with sweeter, floral background notes. Its flavor has a hint of cinnamon, mace and nutmeg with peppery overtones. Jamaican allspice is the most aromatic. The Mexican variety is less sweet and more mellow than the Jamaican or Central American types. Allspice berries lose their aroma upon ripening, so they are collected in an unripe stage and are dried in the sun until they turn dark reddish brown in color. The leaf has a different, woodier aroma and a less intense, coarse, flavor. The bark has coarse woodier notes than the leaf.

Chemical Components: the allspice berry contains 1.5% to 5% essential oil, which is colorless to reddish yellow. The Jamaican type has up to 5% essential oil, Guatemalan—3%, Mexican—1.4% to 3% and Honduran—1.3% to 4%. Jamaican allspice has a minimum of 65% phenols, mainly eugenol (68–78%), methyl eugenol (2.9–13%), 1,8-cineol, α-phellandrene, humulene, terpinolene and caryophyllene. The fixed oil content is about 6%. The Mexican variety has a high myrcene content (5%). Allspice has over 8% quercitannic acid that provides its astringency.

The Jamaican variety produces the most leaf oil—its fresh leaf contains 0.35% to 1.25% essential oil (dried leaf has higher oil, from 0.7% to 2.9%). The essential oil contains 80–90% phenols, mainly eugenol, α-pinene, caryophyllene, limonene, 1,8-cineole and good amounts of tannin. The bark contains small amounts of eugenol and higher levels of tannin.

The oleoresin is brownish green to dark green in color. About 2-1/2 lbs. of essential oil will replace 100 lbs. of freshly ground spice, while 5 lbs. of oleoresin will replace 100 lbs. of freshly ground spice.

Dried allspice has calcium, potassium, sodium, manganese and beta-carotene.

How Prepared and Consumed: the Aztecs and Mayans flavored their chocolate drink with allspice seeds. Caribs and other indigenous Americans used it for preserving fish and meat. This practice was learned by the Spanish who also used allspice to preserve meats. During the seventeenth century, pirates in the Caribbean smoked and barbecued meat with allspice, which they called boucan.

Nowadays, allspice seeds are typically used whole as part of a spice blend for pickling and marinating fish. In the U.S., allspice is ground for use in seasonings and sauces, and its extracted oils are used in sausages. Allspice is also used in ketchup, jams, pumpkin pies, gravies, roasts and ham. It goes well with smoked pork, beef and fish and with habaneros, cumin, onions, tamarind, cinnamon and cloves. Allspice leaf is used in baked goods, chewing gum, candy, ice cream, fruit soups, teas and liqueurs.

Allspice is the most important spice in Caribbean cooking and is used in curries, stews, barbecues and sweet potatoes. Jamaica's popular jerk seasoning has ground allspice as its main ingredient. It is rubbed over pork, chicken or fish that are then cooked over a fire. It gives a smoky and spicy flavor to the barbecued product. Allspice leaves are sometimes stuffed into the meat that is then barbecued over allspice wood or bark to give it the typical flavor of "jerk."

In Oaxaca, Mexico, allspice is used in certain moles. In India, it becomes part of some curry blends. Scandinavians use it to preserve herring. The English use ground allspice in cakes, puddings, mincemeats, pickled vegetables, sausages and cured meats. In North Africa, allspice is used in Ethiopian berberes and Moroccan ras-el-hanout spice blends. Middle Easterners flavor stews, kibbeh (ground lamb with cracked wheat) and pilafs with ground allspice.

Spice Blends: jerk seasoning, berbere, ras-el-hanout, quatre-epices, fish pickling blend, ketchup blend and Jamaican curry blend.

Therapeutic Uses and Folklore: the Aztecs and Mayans used allspice to embalm bodies because of its preservative qualities. It was also considered an aphrodisiac.

Allspice has been used to promote digestion and remove gases from the upper intestinal tract. It is used as a mild anaesthetic for aching gums and teeth and as a mild pain reliever for muscles and joints.

Allspice has bactericidal, fungicidal and antioxidant properties.

ANISE/ANISEED

Anise/aniseed is a popular spice used throughout the world. It was called anysum by early Arabs, anison by Greeks and then anise by the English. Ancient Assyrians used anise as a medicine, Greeks found it to be a digestive aid, and the Romans used anise to soothe sore throats. It was used as an aphrodisiac and as a charm against nightmares.

Associated with the taste of licorice, the Portuguese called anise erva doce, the "sweet herb," and the Indonesians called it jintan manis, the "sweet seed." Asian Indians often confuse anise with fennel because its flavor and name, saunf, are similar to fennel. In China, anise is commonly used with fennel and star anise to add a savory, sweet flavor to barbecues.

Scientific Name(s): Pimpinella anisum. Family: Umbelliferae (carrot family).

Common Names: sweet cumin, aniseed and common anise. It is also called yansoon (Arabic), anijs (Dutch), anisun (Farsi), anis vert (French), anis (German, Swedish, Russian), anison (Hebrew), patli saunf (Hindi), anise (Italian), anisu (Japanese), jintan manis (Malaysian, Indonesian), huei shiang (Mandarin), erva doce (Portuguese), anis (Tagalog), shombu (Tamil, Malayalam), anis (Spanish) and cay vi (Vietnamese).

Origin and Varieties: it is indigenous to Greece, Egypt, Crete, Turkey and Lebanon. Anise is also grown in Mexico, Chile, Argentina, Syria, Spain, Italy, India, Pakistan, China, Russia, Japan and Germany.

Form: anise is a dried ripe fruit or seed. It is small, oval, greenish gray to yellow brown, with a ridged or ribbed surface. Anise is sold whole, cracked or ground. When it is ground into powder, anise quickly loses its flavor.

Properties: anise seed has a sweet licorice-like taste and is warm, fruity and camphoraceous. Anise's flavor is similar to fennel and star anise, but it is more delicate. Its leaves are also aromatic.

Chemical Components: depending on its source, anise seed has 1-1/2% to 6% essential oil, mainly trans-anethole (80–90%), methyl chavicol (10–15%), iso-anethole (2%) and anisaldehyde (less than 2%). It has 8% to 20% fixed oil. The leaf has a much lower level of essential oil.

About 2-1/2 lbs. essential oil is equivalent to 100 lbs. freshly ground spice, and 8-1/4 lbs. oleoresin will replace 100 lbs. freshly ground spice.

Anise contains iron, potassium, phosphorus and calcium.

How Prepared and Consumed: the early Romans used anise to flavor a special cake called mustaceum that was served as a dessert to aid digestion. They also mixed anise with vinegar and honey and used the mixture as a tonic to soothe sore throats. This spice tends to be used in sweet foods in Europe, while in Asia, anise is combined with pungent, spicy ingredients for savory applications. Anise goes well with fruits, sugar, fennel, wine and cinnamon. Anise leaves and stalk can garnish fruit salads and are sometimes added to fish soups and cream sauces. They are roasted or sautéed in oil with other spices to enhance stewed vegetables, roasted meats, curries and tomato sauces.

The Portuguese, Germans, Scandinavians, French and Italians use anise to flavor cakes, sweet rolls, cookies, sweets, applesauces, rye bread, churek, pancakes, cheeses, relishes, marinated meat and fish, beef stew, salad dressings, sausages and luncheon meats.

Europeans flavor many liqueurs and spirits with anise, such as anisette, arrack, sambuca, pernod, pastis and even juice drinks and teas.

Middle Easterners use anise in sweet and savory dishes, and it is the fundamental ingredient in their local spirits, ouzo and raki. Syrians use it in a beverage called miglee and in their popular fig jams.

Spice Blends: curry blends, hoisin, tomato sauce blends, sausage blends and betel leaf mixture.

Therapeutic Uses and Folklore: traditionally, Europeans used anise to treat epilepsy and to ward off evil. The Aztecs drank tea made from its flowers and leaves to relieve coughing. Anise can dispel gas and aid digestion, improve appetite and alleviate cramps, nausea and colic in infants. Anise is commonly used in lozenges and cough syrups because it is a mild expectorant. It also soothes insect bites and is chewed to induce sleep. In India, anise seeds are served after meals to aid digestion and sweeten breath.

Anise shows antifungal activity.

ANNATTO

Annatto is called urucul by the Tupi-Guarani Indians of the Amazon region, achiote in the Nahuatl language of the Aztecs in Mexico, annatto by the Caribs and achuete by Filipinos. Today, annatto seeds are an important coloring and flavoring spice for Latin Americans, Native Americans, Spanish, Asian Indians and Filipinos.

The Caribs, Mayans and other Native Americans dyed their bodies with annatto oil to protect against the sun, thus giving rise to the term "redskin" by the early European settlers in the Americas. Aztecs used annatto seeds to intensify the color of their chocolate drink. This practice of coloring traveled to Europe, where annatto is now used to give a deep yellow color to butter and cheese. Often called "saffron" by Puerto Ricans, this spice was introduced to India by the Portuguese and to the Philippines by the Spanish.

Scientific Name(s): Bixa orellana. Family: Bixaceae.

Origin and Varieties: Annatto is indigenous to the Caribbean, Mexico, Central America and South America. It is now cultivated in Brazil, Peru, Guatemala, India and western Africa.

Common Names: achiote (Nahuatl, Spanish), bija (Mexican), bijol (Spanish), lipstick tree (English), roucou (French, Caribbean), annatto (German), anato (Hebrew), urucum (Portuguese-Brazil), achuete (Tagalog) and hot dieu do (Vietnamese).

Form: annatto is a small dark-red seed in a prickly, heart-shaped fruit. It is sold as a paste, as oil (extracted from seeds) or ground (from the whole seed).

Properties: annatto is deep golden yellow to orange red in color. It has a delicate, mild peppery flavor with flowery and earthy undertones.

Chemical Components: bixin, an oil-soluble apocarotenoid, is the main coloring pigment, with norbixin, other carotenoids and apocarotenoids making up 7% of the dry seed.

How Prepared and Consumed: annatto is used as a cooking oil and a gentle flavoring and coloring agent in Colombia, Ecuador, Venezuela, Mexico, Puerto Rico, Spain, Central America, India and the Philippines. It is very common in Puerto Rican and other Caribbean cooking and is used with rice, polenta, beans, chicken and fish stews, soups, pork, okra, yuca and tomatoes. In Jamaica, the popular salt cod and ackee dish, fried chicken and pork are colored with annatto.

Annatto combines well with cumin, garlic, oregano and coriander. Typically, it is used to color or flavor food in the form of aceite (oil) or manteca de achiote (annatto lard) in Latin American and Filipino dishes. The seeds are fried in oil or lard that becomes a golden orange in color. Then, the seeds are discarded, and this colored oil is used to fry vegetables, rice, chicken or meats. It is also blended with other ingredients to create the Puerto Rican sofrito. In the Yucatan and in Guatemala, the whole annatto seed is ground into a paste and is used with other spices as a rub and to provide a deeper flavor to barbecued pork, poultry and fish dishes.

In Europe, annatto extract is used to color butter, margarine, ice cream, sausages and many cheeses including red Cheddar, Muenster, Livarot and Leicester. In the U.S., annatto is used in relishes, snacks, beverages, gravies, seasonings, baked goods and margarine. The Filipinos and Vietnamese color fried or grilled pork and fish with annatto.

Spice Blends: sazon, recados, achiotes, sofritos and adobos.

Therapeutic Uses and Folklore: the ancient Mayans and Caribs used annatto to paint their faces and bodies in religious ceremonies and in preparation for wars. To them, it represented courage and strength.

ASAFOETIDA/ASAFETIDA

The name asafoetida is derived from the Persian word aza, meaning resin, and the Latin word *foetida,* meaning bad smelling. Asafoetida was known to early Persians as "the food of the Gods" and to the Romans who used it to flavor sauces and wines as Persian sylphium. Europeans equated its smell to truffles and called it devil's dung. In ancient India and Iran, asafoetida was used as a condiment and as a medicine. Today, asafoetida is commonly used in the vegetarian cooking of South India.

Scientific Name(s): Ferula asa-foetida L (hing type); *F. foetida* (hingra type). Family: Umbelliferae or Apiaceae (parsley family).

Common Names: asafetide, stinking gum and devils dung. It is also called abu kabeer (Arabic), retshina fena (Farsi), assa foetida/ferula persique (French), asant/stinkasant (German), hiltit (Hebrew), hing (Hindi), assafetida (Italian), ah wei (Mandarin), asafetida (Russian), asafetida (Spanish), dyvelstrack (Swedish), perungayam (Tamil, Malayalam) and anjadana (Urdu).

Origin and Varieties: asafoetida is indigenous to Iran, India, Pakistan and Afghanistan. It is also found in Russia and China. There are many varieties, but the two most commonly sold varieties are called in Indian, *hing* (water soluble) and *hingra* (oil soluble). The former is more popular because of its aroma. Each type shows more sweetness or bitterness depending on its country of origin.

Form: it is a congealed, dark brown to black resin-like gum obtained from the juice of the rhizome of the ferula or giant fennel plant. After drying, it becomes a darker brown, resin-like mass. It is sold as different grades of resin, dried granules, chunks or powders.

Properties: the resin is a pale brown color that darkens after drying. It is acrid with a strong garlic-like odor and bitter, unpleasant back notes. If it is used sparingly and fried in oil, pleasant shallot and garlic-like notes develop. The powder is less intense and can be added without frying. The commercial paste and powdered forms are resin mixed with diluents such as rice flour, cereal flour or gypsum and are, thus, less intense in taste. The hing type is white or pale in color, while the hingra type is darker, almost black, in color.

Chemical Components: its repugnant strong smell is due to sulfur compounds (that disappear during cooking) and ferulic esters. The true gum resin consists of essential oil (10–17%), gum (25–48%), resin (40–60%) and ashes (1–10%). The essential oil has an abundance of sulfur compounds, mainly 50% of 2-butyl-1-propenyl disulfide, methylthiopropyl disulfide and 2-butyl-3-methylthioallyl disulfide with some terpenes (α-pinene, phellandrene).

How Prepared and Consumed: Iranians and Asian Indians use it abundantly in vegetarian dishes. Asian Indians fry the resin in oil for a few minutes to disperse it well before it is mixed with other ingredients. The unacceptable smell disappears when it is cooked. At very low or "pinch" levels, asafoetida enhances many dishes such as fish curries, brined or pickled fish, spiced legumes, vegetables, chewra (Indian snack), relish and even Worcestershire sauce.

South Indians use it in sambar podi, a spice blend added to legume dishes to enhance their flavor and to prevent flatulence. Jains, a religious group in India, do not eat root vegetables or root spices such as garlic, onion, ginger or turmeric, for fear of killing living organisms. Therefore, they rely on asafoetida as an alternative flavoring. The Brahmins, who will not eat garlic because they consider it an aphrodisiac, also use asafoetida as a substitute flavoring.

It pairs well with nuts, grains, legumes, vinegar and grilled or roasted meat. Iranians rub asafoetida on warmed plates before placing meat on them. Afghans and Persians also eat the stem and the leaves as vegetables, the odor disappearing once they are boiled.

Spice Blends: sambar podi, dal podi, chat masala and chewda blend.

Therapeutic Uses and Folklore: Romans used asafoetida to aid digestion and as an aphrodisiac. In India and Iran, it is used to treat hysteria and as an antispasmodic, anticoagulant and sedative. Asafoetida is also used to reduce flatulence and to treat nervous disorders. In India, singers take asafoetida before singing because it supposedly mellows their voices and produces a sensation of warmth.

Asafoetida has antibiotic and antimicrobial properties.

BASIL

Basil is best known for its wonderful "royal" fragrance. The name basil is derived from the Greek word *basileus,* meaning "king." The French call basil the "herb royale." In ancient times, basil was considered to have magical properties. The Greeks gave basil the name basilisk because it was reputed to provide protection from a half-lizard, half-dragon monster of the same name.

Holy basil is used in India mainly for religious purposes in Hindu temple ceremonies. This basil, called tulsi, is woven into garland to grace the Hindu God, Lord Vishnu. Tulasi is named after Vishnu's wife, the Goddess Tulasi, who took the form of this herb when she came down to earth.

Today, sweet basil is the most popular type used in North American, Italian and other European and Mediterranean cuisines. Thai basil (also called anise basil), lemon basil and holy basil are popular in Southeast Asian and Indian cooking.

Scientific Name(s): Sweet basil: *Ocimum basilicum,* holy basil: *Ocimum sanctum,* lemon basil: *Ocimum basilicum citriodorum,* cinnamon basil: *Ocimum basilicum cinnamon,* curly basil: *Ocimum basilicum crispum,* and Japanese basil or shiso/jiso: *Perilla frutescens.* Family: Lamiaceae (mint family).

Common Names: sweet basil or great basil—also called besobila (Amharic), reehan (Arabic), basilicum (Dutch), basilie/basilic commun (French), basilikum (German), reihan (Hebrew), barbar (Hindi), basilico (Italian), bajiru/meboki (Japanese), pachcha (Malayalam), lo le (Mandarin), manjericao (Portuguese), bazilik (Russian), albahaca/ albega (Spanish), basilika/basilkort (Swedish) and tirunitru (Tamil).

Anise basil or true Thai basil—bai horapha (Thai), daun selaseh, (Malaysian, Indonesian) and rau que (Vietnamese).

Lemon basil—daun kemangi (Malaysian, Indonesian), bai maenglak (Thai).

Holy basil or sacred basil—laun (Burmese), tulsi (Hindi), Sivatulasi (Malayalam), tulasi (Tamil) and bai krapao (Thai).

Japanese basil—shiso/jiso (Japanese) and perilla.

Origin and Varieties: basils are indigenous to Europe and India. They are cultivated in Iran, Africa, Seychelles, Southeast Asia, Greece, Italy, France, Egypt, Hungary, Morocco, southern Europe and the U.S. (California). Many types of basil exist. They vary in size, color and flavor intensity based on their origins and climatic and soil conditions. Sweet basil is the most commonly used type in the U.S., but there are many other varieties, including holy, lemon, Thai, dark opal, Cuban, West African, cinnamon, East Indian, purple ruffle, minty Egyptian and many more. Even with sweet basil, flavor variations occur, depending on its country of origin.

Form: basil comes fresh, dried or as a paste in oil. Fresh basil is used whole, chopped or pureed. Dried basil is used as ground and as particulates of varying sizes. The dried form is less aromatic than the fresh form.

Sweet basil has bright green leaves. Thai basil is similar in size to sweet basil, with purplish stems and veins. Holy basil is smaller and narrower, with a dark green to almost reddish purple tinge. Lemon basil is paler in color than Thai basil. Japanese basil (shiso) is light to dark green.

Properties: the European-type sweet basil has a sweet and floral, anise-like aroma with cooling clove-like undertones. Its taste is delicate and fresh with slight minty notes. Its delicate aroma decreases with cooking. Thai basil has a sweet anise-like aroma and licorice-like notes with a spiciness that sweet basil does not have. It has strong, phenolic notes, with a lingering aftertaste. Holy basil has a strong anise-like, slightly musky and lemony taste with a camphoraceous aroma. Lemon basil has a slightly spicy, lemony taste with a distinct fruity aroma. Cinnamon basil has overtones of cinnamon, while curly basil has sharper and harsher notes than sweet basil.

Japanese basil or shiso has a flavor that is a cross between mint and basil. Eaten by the Chinese in the past, it fell out of favor, and when it was introduced to Japan, it became popular in fresh or pickled forms. It is sold in Japan as a green-type (ao-jiso) and a red-type (aka-jiso) shiso. The green shiso is used as a garnish, in sushi rolls, in salads or fried with tempura batter. The red type, which has larger leaves, is used to pickle plums, with the resulting product called umeboshi.

Chemical Components: the differences in flavor among basils are due to their different chemical components, especially methyl chavicol (or estragol), linalool, citral, methyl cinnamate, eugenol or 1,8-cineole. The dominant aroma component in sweet basil is linalool, in holy basil is eugenol and in Thai (anise-type) basil is methyl chavicol. Sweet basil has about 0.5–1.1% essential oil, mainly linalool (40%) and methyl chavicol (25%), with the remainder consisting of eugenol, 1,8-cineole and geraniol. In anise or Thai basil, 85% of the oil is methyl chavicol (which oxidizes when exposed to light and air), less than 1% is linalool, and the rest consists of camphor, borneol, eugenol and 1,8-cineole.

Basil oleoresin is dark green and viscous, and 0.75 lbs. are equivalent to 100 lbs. of freshly ground basil.

Fresh basil contains folic acid, potassium, magnesium and calcium, while dried basil contains vitamin A, potassium, vitamin C, sodium, phosphorus, calcium, iron, folic acid and manganese.

How Prepared and Consumed: basil is widely used in Italian, Southeast Asian and Mediterranean foods. It is mainly used as a garnish—whole, chopped or minced. Basil turns black in an acid medium and loses its aroma easily when heat is applied. Because sweet basil's delicate aroma is easily destroyed during cooking, it is frequently added whole or chopped to cold or warm dishes just before serving. Basil combines well with tomatoes, nuts, olive oil, olives, ginger, pungent sharp cheeses, coriander leaf, garlic, galangal, lemongrass and white and black pepper.

The popular Italian pesto that is served over pasta is made from sweet basil mixed with olive oil, garlic, Parmesan cheese and pine nuts. Italians also use it with capers and tomato sauces, as pizza toppings, in stews, sausages, pasta salads and beans. The French use sweet basil to create their version of a pesto called pistou, which is served in soups, omelettes, sauces or stews. Sweet basil goes well with rosemary, parsley, tomatoes, mozzarella, cilantro and sage. Lemon basil goes well with fish and lobster dishes.

Basils are also commonly used in Thai and Vietnamese foods, but these basils have more pungent notes. Thai basil is used to add pungency to stir fries, salads and curries. These local basils go well with coriander root, galangal, turmeric, garlic, coconut milk and chicken and pork dishes. Asian Indians use holy basil mostly in teas or as a garnish on meats and vegetables. It is not generally used in cooked foods. The Vietnamese, who top every dish with fresh herbs, use fresh basil in soups, condiments, stews and freshly made salad rolls.

Spice Blends: pestos, pistous, green Thai curry blend, Malaysian kurma blend, bouquet garni and pizza sauce blends.

Therapeutic Uses and Folklore: while the Greeks and French called basil the herb fit for a king, in many other parts of Europe, it symbolized hatred and hostility. For Romans, it was a symbol of love and fertility. The Italians wore a sprig of basil as a sign of interest in marriage. And, it was used as an antidote to venom or bites.

Basil is revered by Hindus in India and is grown near every temple. In India, holy basil is planted around many homes to ensure happiness. Basil leaf is often given with gifts as a sign of hospitality. In some Hindu weddings, parents of a bride give her away by presenting a basil leaf, sometimes made of solid gold, to the groom.

In Asia, basil is used for most stomach disorders, cramps, diarrhea, headaches, whooping cough and head colds. It is also used as a diuretic. In China, it was used as an antidote to fish poisons. The Italians believe that basil keeps insects away.

BAY/LAUREL LEAF

Greek legend says that the gods turned a beautiful nymph named Daphne into an evergreen laurel tree when she was fleeing from Apollo's (Greek god of prophecy, medicine and poetry) love. Today, the Greek name for bay leaf is daphnee. In ancient Greece, the winners of the Olympic games were decorated with laurel wreaths and these leaves became an immortal symbol of victory and courage. When Greek physicians completed their studies, they were crowned with laurel branches called the *baca lauris,* which later became baccalaureates. The Romans, who also used these wreaths, made bay/laurel leaf a popular spice in their cooking. The word bay is derived from the Latin *baca* meaning berry.

Scientific Name(s): Laurus nobilis. Family: Lauraceae (laurel family).

Common Names: true laurel, sweet bay/sweet laurel (English), warak al ghar (Arabic), laurier (Dutch), barg bu (Farsi), laurel (Tagalog) laurier (French), lorbeer (German), daphnee (Greek), aley dafna (Hebrew), tejpatta (Hindi), alloro/lauro (Italian), gekkeiju (Japanese), yueh kuei (Mandarin), loureio/louro (Portuguese), lavr (Russian), hojas de laural (Spanish), lager (Swedish), bai kra wan (Thai) and dafne yapregi (Turkish).

Origin and Varieties: native to the Mediterranean and Asia, bay leaf is grown in Mexico, Guatemala, Turkey, France, Greece, Russia and North America. There are many "other" leaves that closely resemble the laurel leaf in flavor and are thus called bay leaves in other regions of the world. The West Indian bay leaf (from the allspice family), the Indian bay leaf, the California bay leaf, the Indonesian bay leaf and boldo leaves are not true bay leaves, even though their flavors may be similar or sometimes stronger than bay leaf. The U.S. buys bay leaves mainly from Turkey and Greece.

Form: bay leaf is a thick, leathery, aromatic leaf with a bright green, glossy upper surface and a pale green color beneath. If the leaf has a brownish green color and no aroma, then it is overaged. It is elliptical in shape, tapering to a point at the base and tip. Bay leaf is used whole or crushed, fresh or dried. The dried leaf is also ground into powder.

Properties: bay leaf has a strong, spicy and pungent flavor with cooling undertones. Its flavor is described as piney, nutmeg and clove-like with slight camphor-like notes. It has a slightly bitter aftertaste. The crushed leaf releases a strong aroma and has a grassy, sweet flavor. When the whole dried leaf is cooked in foods, its aroma is slowly released and becomes more intense. When it is used in ground form, bay leaf releases its aroma quickly and loses it quickly the longer the food is cooked. When the dried leaf is fried in heated oil, it develops a more intense flavor.

Chemical Components: bay leaf has 0.8% to 3% essential oil, mainly 1,8-cineole (35%), with others including methyl eugenol (4%), α-pinene (12%), β-pinene (6%), linalool (11%), α-terpineol (6%), limonene (4%), α-terpinyl acetate (10%), sabinene (5%) and eugenol (2%).

The dried fruit has higher essential oil (0.6% to 10%), mostly 1,8-cineole, α-terpineol, α- and β-pinene, cinnamic acid and methyl ester.

The leaf contains calcium, iron, potassium, beta-carotene and vitamin C.

How Prepared and Consumed: bay leaf is indispensable to Mediterranean cooking, especially in Turkish, Greek and Armenian dishes. It is used in soups, stews, soup stocks, pickles and in meat and fish marinades. The French use it in their popular bouquet garni, bouillabaisse, bourride and bean soups. Turks use it to flavor grilled fish, fish casseroles and kebabs. Moroccans add it to their chicken tagines and pickled fish. Bay leaves can be cooked a long time before losing their aroma. Steaming tends to bring out more of its flavor into the dish. Too high a level of ground bay leaf can create a bitter flavor. The whole leaf is usually removed from the cooked food before serving.

The fruits of bay leaf are sometimes used to season wild game or potatoes in European regional cooking.

Spice Blends: bouquet garni, curry blends, rice and bean blends, bean blends, bouillabaisse blend.

Therapeutic Uses and Folklore: ancient Greeks considered bay leaf holy. For Italians, it signifies good luck and protection. Bay leaf helps relieve pain in joints, chest, womb and stomach. It also eases cramps and earaches. Bay leaf aids digestion by stimulating gastric functions.

"OTHER" BAY LEAVES

There are many other so-called bay leaves that are not related to true bay leaves, such as the Indian bay leaf, West Indian bay leaf (leaf of allspice) and the Indonesian bay leaf. The boldo leaf, which tastes similar to but stronger than bay leaf, is a distant relative of the bay leaf. The California bay leaf is milder and less pungent than the "true" bay leaf from Turkey or Greece.

Indian Bay Leaf

Indian bay leaf is not related to the true laurel leaf, but it is a cinnamon relative, and it is popularly used in North Indian meat curries and vegetables.

Scientific Name(s): Cinnamomum tamala. Family: Lauraceae.

Origin and Varieties: Indian bay leaf is the leaf of a tree that is closely related to cinnamon and is cultivated in North India.

Common Names: this is also called thitchabo (Burmese), tejpat (Hindi), tejpatra (Punjabi), talishapattiri (Tamil), indisches Lorbeerblatt (German), kanelilaakeri (Finnish) and laurier des Indes (French).

Form: whole, crushed or ground.

Properties: Indian bay leaves are tough and leathery, very aromatic and have a flavor somewhat like cinnamon or cloves.

How Prepared and Consumed: Indian bay leaf is popular in North Indian meat and lentil curries. Indians fry it in oil with other spices and curry powder before adding meat, lamb or chicken. Indian bay leaf is commonly used in Mogul-type biryanis and kormas. It is an important ingredient in the garam masalas of North India. In the U.S., these same dishes can be prepared using the Mediterranean bay or laurel leaf.

Spice Blends: garam masala, korma blend, biryani blend and dal curry blend.

West Indian Bay Leaf

Scientific Name(s): Pimienta racemosa Miller or *P. acris* Kostel. Family: Myrtaceae.

Origin and Varieties: it is cultivated in the Caribbean islands, and it has a harsh, penetrating, clove-like flavor.

Chemical Components: it has an essential oil of 1% to 2%, mainly phenols (50–60%). The essential oils are eugenol (38–75%), myrcene (14–32%), chavicol (11–21%), linalool (92–93%), limonene (1.5%), 1,8-cineole (0.2–2.0%) and 3-octanone (1%).

How Prepared and Consumed: West Indian bay leaf is used to flavor rice and beans, stewed beans, meats, curries, relishes and beverages.

Indonesian Bay Leaf

This spice is called daun salam in Indonesian, meaning "peace leaf." It is a popular ingredient in Indonesian cooking. Indonesian bay leaf is from the cassia family.

Scientific Name(s): Eugenia polyantha. Family: Myrtaceae.

Origin and Varieties: Indonesia.

Common Names: daun salam (Indonesian) and Indonesisches lorbeerblatt (German).

Form: fresh or dried.

Properties: the leaves are small and brownish when dried. It has an aromatic and sour taste.

How Prepared and Consumed: Indonesian bay leaf is commonly used with meats in the Indonesian islands of Bali, Java and Sumatra. It is initially cooked or stir-fried to release its flavor into the dish.

Boldo Leaves

Scientific Name(s): Peumus boldus Molina. Family: Monimiaceae. This is a close relative of the bay leaf family.

Origin and Varieties: Chile, Bolivia, Peru and other regions of South America.

Common Names: boldoblatter (German) and boldo (French, Spanish, Italian).

Form: the leaves are used fresh or dried.

Properties: boldo leaf has a strong flavor that is slightly bitter and unpleasant. Its aroma is camphoraceous and cinnamon like.

Chemical Components: its essential oil is about 2%, with terpene derivatives such as ascaridol (40%), ρ-cymene, 1,8-cineole, eugenol, cinnamic aldehyde and linalool. Boldo has alkaloids such as boldine, sparteine and isocorydine.

How Prepared and Consumed: boldo leaf enhances meaty flavors and mushroom notes. In South America it is popular with meats and fish. It also goes well with pickled vegetables, soups and stews.

CAPER

Caper is derived from the Latin word *capra,* which means "goat," a name that reflects its strong smell. Thought to originate from the Near East or Middle East, it has been used by Arabs for medicinal purposes.

Scientific Name(s): Capparis spinosa. Family: Caparidaceae.

Origin and Varieties: caper is cultivated in the Mediterranean regions of Spain, France, Italy, Greece and Algeria.

Common Names: caper is also called kubar (Arabic), kappertjes alcaparra (Dutch), capres (French), kaper (German, Norwegian), tzala (Hebrew), kiari/kabra (Hindi), cappero (Italian), alcaparra (Portuguese), kabarra (Punjabi), kapersy (Russian), alcaparra (Spanish), kapris (Swedish) and kabar (Urdu).

Form: caper is the green, dried bud of an unopened flower. It is graded based on its size—the smaller, the higher the grade. Usually, it is cured with vinegar or vinegar and salt.

Properties: caper has a sharp fermented bitter taste, and its characteristic taste is developed when placed in vinegar or brine. Pickled capers have an acrid, tart and pungent taste with a lemony tang.

Chemical Components: caper contains mainly water (85%), bitter glycosides (such as rutin and glucocapparin), pentosans, rutic acid, pectin and saponin. Methyl glucosinates upon enzymatic action give methyl isothiocyanate, which gives it its pungency. It has a high sodium content.

How Prepared and Consumed: capers have been pickled by Southern Europeans for over 2000 years. Today, they are consumed abundantly in the Mediterranean regions of Sicily, Apulia (in Italy), France, Spain and Greece. Sicilians use capers in game, tomato sauce, pizzas, chicken, caponata (a salad that includes eggplant and tomatoes), tartar sauce and fish. Apulians in Italy use caper in meatballs, string beans and other boiled vegetables. The Spanish crush it, combine it with almonds, garlic and parsley and then serve over fried fish. In Provence, France, caper is combined with olives, potatoes, eggplants or cheese and is then used as a salty sharp spread.

Caper pairs well with fish, olives, chicken, basil, mustard, black pepper, garlic, oregano and tarragon. Because heat easily destroys its aroma, it is added to cold dishes of fish, meat and vegetables. In the U.S. and Europe, it is served as a garnish for cold roasts and salads, as a spread and in pickles and relishes. Capers are also used to add tartness to the curried dishes of northern India.

Spice Blends: tapenade, pickling blend, caponata blend and pizza sauce blend.

Therapeutic Uses and Folklore: capers have been used to aid digestion, prevent diarrhea and increase appetite. In India, they were used as a traditional treatment against scurvy.

CARAWAY

Caraway was used at least 5000 years ago during Mesopotamian times. Its name is derived from the Arabic word *karawya*. It is called "foreign" cumin by Asian cultures, because it is perceived to be a relative of cumin due of its similar flavor profile. In ancient times, Romans seasoned sausages with caraway seeds. Hungarian herdsmen used them to flavor goulash. Midwives in Europe have used caraway to stimulate production of breast milk in nursing mothers. Today, caraway seeds characterize South German and Austrian rye breads and cooking.

Scientific Name(s): Carum carvi. Family: Umbelliferae or Apiaceae (parsley family).

Origin and Varieties: caraway is native to North Africa, the Mediterranean, Germany and Holland. It is also grown in North India, Britain, Canada, Russia, Poland and North America.

Common Names: Roman cumin, German cumin and Persian cumin. It is also called karawya (Arabic), karwij/komijn (Dutch), carvi (French), kummel (German), kravya (Hebrew), shia jeera (Hindi), alcaravia (Portuguese), karuwai (Japanese), shima jirakam (Malayalam), shiang tsai (Mandarin), tmin (Russian), alcaravea (Spanish), kummin (Swedish) and shimai shembu (Tamil).

Form: caraway is a dried, ripe fruit that is used whole or ground. Its leaf is feathery and used as a garnish, while its root has a crispy texture, like parsnips.

Properties: caraway seed is curved with tapered ends and is dark brown with light-brown ridges. It has a sweet, warm, slightly dill and anise-like flavor. Its aftertaste is somewhat sharp, bitter and soapy. If cooked too long, a bitter taste develops. The leaves have a milder taste than the seeds.

Chemical Components: caraway seeds have about 4% to 7.5% essential oil, which is pale yellow in color. It has mainly 50–85% *d*-carvone and (20–30%) limonene. Other essential oils in caraway are carveol, dihydrocarveol, α- and β-pinene and sabinene. Its fixed oil content is 15%. The oleoresin is greenish yellow, and 5 lbs. are equivalent to about 100 lbs. of freshly ground spice.

Caraway contains calcium, potassium, magnesium, phosphorus and beta-carotene.

How Prepared and Consumed: caraway pairs well with garlic, vinegar, breads, pork, vegetables and fruits. It dominates the flavor in foods. It adds a savory sweetness to cakes, crackers, cheeses, applesauce, onion bread, potato soups and cabbage. In Europe, it is a popular spice used in meats, vegetables, condiments, ice cream, marinades and pickling solutions. Germans use caraway abundantly in their cooking to flavor breads, meats, vegetables, sauces, sauerbraten and sauerkraut. It is an indispensable ingredient in rye bread and roast pork. In Dutch foods, it is a popular flavoring for cheese. The British serve tea and seedcake with cinnamon and caraway seeds. Central Europeans coat caraway with sugar or use caraway candies to sweeten their breath after meals. Caraway is an ingredient in a European liqueur called kummel. Italians boil chestnuts with caraway before roasting them. Alsatians serve a bowl of caraway seeds with Muenster cheese.

Tunisians combine caraway with dried chile peppers, garlic and other spices to make their famous fiery condiment called harissa. Yemenites use caraway with coriander and other spices in a blend called zhug. The Nigerians savor a crispy caraway twist called chin chin. In North India, caraway is used to season many dishes such as dahi wada, a dumpling that also contains lentil flour, chile peppers and almonds and which is eaten as a snack.

Caraway leaves are used in soups and salads, while the root is stewed or boiled and eaten in salads.

Spice Blends: pickle blends, harissa, zhug, sauerkraut blend and dahi wada blend.

Therapeutic Uses and Folklore: ancient Egyptians placed caraway on their dead to keep bad spirits away. The ancient Arabs, Greeks and Romans used caraway not only as a flavoring, but also as a medicine. Caraway had magical powers of protecting prized possessions, and it was also used in love potions to keep a lover from wandering away.

In Europe, caraway was chewed after a starchy meal to help digestion. Simmered with milk and honey, caraway was a remedy for colic. It has also been used as a sedative, to relieve uterine and intestinal cramps, and to help calm a queasy stomach after taking medicines that cause nausea.

CARDAMOM/CARDAMON

Known as the "queen of spices" in India, green cardamom is the world's third most expensive spice, after saffron and vanilla. The name cardamom comes from the Greek word, *kardamom,* and was used by Greeks, Romans and Chinese over 2000 years ago in foods, medicines and perfumes. The Vikings bought cardamom from traders who obtained it through Constantinople and introduced it to Scandinavia. To this day, it is an essential ingredient in Danish pastry. Cleopatra filled her chambers with the sweet smell of cardamom smoke before Mark Anthony's visit to Egypt.

It is a popular ingredient in Indian, Iranian and Arabian cooking. An essential ingredient in Arab coffees, the Bedouin Arabs place cardamom pods in the spouts of coffee machines so that when coffee is poured, it takes the flavor of the cardamom.

There are two types of cardamom, true or lesser and false or greater cardamom. There are many grades of true or green cardamom, depending upon their origins. The cheaper false brown cardamom substitutes are from Nepal, Thailand and Bengal. The bigger brown or black cardamom is a false cardamom that has a very different flavor profile.

Scientific Name(s): True or "lesser" cardamom: *Elettaria cardamomum* (Malabar, Sri Lanka); false or "greater" cardamom: *Amomum* or *Afromomum subulatum* (North India, West Africa, Cambodia); *A. korarima/kewrerima* (Ethiopian type); *A. compactum* (Java cardamom); *A. globosum* (Chinese round cardamom); and *A. melegueta* (grains of paradise). Family: Zingiberaceae.

Origin and Varieties: true (or lesser) cardamom is native to India (Malabar, Mysore) and Sri Lanka. It is also grown in Thailand, Sumatra and Guatemala. The U.S. buys this cardamom from Guatemala, India and Sri Lanka. False (or greater) cardamom is cultivated in Bengal (India), Madagascar, Nepal, Laos, Cambodia and West Africa. Grains of paradise (discussed later), a relative of cardamom, is relatively unused today except in parts of Africa. Also called guinea pepper or Guinea grain, it is common in West Africa (Nigeria, Benin) and North Africa. The Yorubas of Nigeria call it *atare* and use it in many of their food products.

Common Names: true or lesser cardamom, also called korerima (Amharic), hab el-hal (Arabic), kardemom (Dutch), hel (Farsi), cardomome (French), kardamom (German), kardamomum (Greek), hel (Hebrew), chotti elaichi (Hindi), kapulaga (Indonesian), cardamomo (Italian, Spanish, Portuguese), hel (Hebrew), karudamon (Japanese), elathari (Malayalam), buah pelaga (Malaysian, Indonesian), sha jen (Mandarin), kardamon (Russian), enasal (Singhalese), kardemumma (Swedish), elakai (Tamil) and luk krawan (Thai).

False or greater cardamom, also called Indian cardamon, Nepal cardamom, cadamome noir (French), Nepal cardamom (German), kali, motti or badi elaichi (Hindi), cardamomo nero (Italian), perelam (Malayalam), cardamomo negro (Spanish) and periya yalam (Tamil).

Form: cardamom is the dried, firm unripened fruit packed with many seeds. It is used as pods (whole or crushed), as whole seeds (decorticated cardamom) or as ground seeds. The pods are handpicked and are generally sticky to the touch.

Properties: true cardamom pods are small and dark green, light green or white, depending how the pods have been processed. They are dark green when the pods are oven dried, light green when the pods are air dried and white or buff colored when they are bleached. The seeds in all of these pods are reddish brown to black and are highly aromatic. The green pods have a delicate clean, sweet and spicy floral flavor with a lemony scent. They have overtones of mint, eucalyptus and black pepper. The white pods have very little flavor. The ground seeds have an intense flavor when just removed from the pods, but they lose their flavor quickly. Toasting the whole pod intensifies cardamom's sweet flavor.

False cardamom pods are larger sized, dark brown to black in color (when dried) and coarsely ribbed. They give a smokey, camphorous aroma and a harsher, more pungent flavor. The seeds are darker and have a menthol-like taste.

Grains of paradise (discussed later) have a pungent and peppery flavor that is in between the flavor range of black pepper and dried chile pepper.

Chemical Components: true cardamom has 2% to 10% volatile oil depending on its regional origins and storage conditions. The essential oils include 1,8-cineole (25–45%), α-terpineol acetate (28–34%) with linayl acetate (1–8%), sabinene (2%), limonene (2–12%) and linalool (1–4%). The oil is spicy, sweet, citrusy and musty, while the oleoresin is dark green, pungent, cool and burning.

False cardamom has more than 70% of 1,8-cineole. Its camphor taste is due to bornyl acetate. About 4 lbs. of the oleoresin are equivalent to 100 lbs. of freshly ground spice.

Cardamom contains calcium, magnesium, potassium and manganese.

How Prepared and Consumed: true cardamom enhances sweet and savory dishes. It pairs well with fruits, almonds, saffron, butter, parsley, cilantro, clove and cinnamon. It is a very important flavoring in Saudi Arabian foods and beverages. In Saudi Arabia, dark-roasted coffee mixed with ground cardamom called gahira/gahwa is served to guests as a symbol of hospitality. It is added to fiery Arabic spice blends such as baharat and jhoug and to rice-meat dishes called kabsah. North Africans use it in ras-el-hanout and berbere seasonings.

Green and large brown cardamoms are commonly used in Asian Indian cooking. In India, the green pods are used whole, or its seeds are ground and added during cooking or sprinkled on savory dishes to give a sweet aroma. Green cardamom is an essential ingredient in Indian sweets, puddings, yogurt (called shrikand) and ice creams. Green cardamom gives the distinct flavor to India's famous creamy kormas. It is also an essential ingredient in Southeast Asian curries, pilafs and desserts and other foods that have an Indian influence. Whole green cardamom is an important ingredient in chai masala, a spiced, hot tea beverage that is consumed all over India and that is now emerging as a popular drink in the U.S.

The false cardamom pods are coarsely crushed for use in curries, vindaloos, pilafs, biryanis, snacks and teas of North India. The whole or coarsely crushed cardamoms are braised before they are added to in North Indian and Pakistani curries and pilafs to intensify their flavors. They are usually removed before serving. Their flavor is too strong, so generally they are not added to sweet dishes, unlike green cardamom.

Cardamom is not commonly used in Europe except in Scandinavia, where ground cardamom is used instead of cinnamon in Danish pastries, cookies, cakes, breads (kaffekage, julekake), glogg (hot spiced wine punch), apple pie and meatballs.

Grains of paradise are used in the spice mixture ras-el-hanout and other savory dishes of North and West Africa.

Spice Blends: garam masala, curry powder, korma blend, berbere, baharat, zhoug, ras-el-hanout, kabsah blend and kesari blend.

Therapeutic Uses and Folklore: Indians use cardamom to sweeten and clean their breaths after meals. In India, many chew on it after dinner to help settle their stomachs and aid in digestion. Cardamom also prevents nausea and vomiting. It soothes colicky babies, induces sweating and cools the body during summer months. Arabs traditionally used cardamom as an aphrodisiac. In Scandinavia, it is used to mask the smell of alcohol, fish and garlic.

CELERY

The Greeks historically used celery seed as a medicine and as a sign of victory. The Romans first used it as a seasoning, and later, it became a delicacy for Italians and French. It was only in the nineteenth century that North Americans began to use celery seed, mainly in pickling solutions. Europeans commonly eat the leaves and roots of celery plants as vegetables. Chinese and Southeast Asians use the leaf to flavor many of their dishes.

Scientific Name(s): Apium graveolens. Family: Umbelliferae.

Origin and Varieties: celery seed is native to eastern and southern Europe and the Mediterranean. It is cultivated in India, France, Britain, Japan, China, Hungary and the U.S. The seeds are cultivated from the original wild celery variety. Another cultivated celery variety is eaten for its stalk, leaves and seeds. Celeriac or celery root, yet another celery variety, is savored for its root.

Common Names: garden celery and wild celery. Many global names call the seed, leaf and stalk by similar names. Celery seed is also called karafes (Arabic), chiluri (Bengali), selderij (Dutch), celeri (French), sellerie (German), seleri (Hebrew), ajmud (Hindi), serori (Japanese), sedano (Italian), qing chai (Mandarin), aipo (Portuguese), seldjerey (Russian), apio (Spanish), selderi (Malaysian, Indonesian), selleri (Swedish and Norwegian), kinchay (Tagalog), kheun chai (Thai) and ajmod (Urdu).

Celeriac: celeriac, also called turnip root or knob celery, and celery stalk are used as vegetables.

Form: celery seeds are tiny globular seeds that are sold whole, slightly crushed or ground. The leaves, which are light green, are used whole (fresh or dried), flaked or ground. The stalks or stems and root are sold fresh.

Properties: The dried seed is dark brown with light ridges. It has a harsh, penetrating, spicy aroma and a warm bitter taste, that leaves a burning sensation. The seeds have a stronger and more intense flavor than the leaf, stem or root. The French type of celery seed is herbal and sweet with a citrus bouquet, while the Indian type is more herbal with a slight lemon-like aroma.

Celeriac root, leaf and stalk have a strong celery, herbaceous and parsley-like taste.

Chemical Components: the seeds have 2% to 3% essential oil, which is yellow to greenish brown. It consists primarily of terpenes, mainly limonene (68%), with sesquiterpenes such as 8% β-selinene, 8% *n*-butylidene phthalide and myrcene. The characteristic aroma is due to the phthalides. The fixed oil is 16%.

The seed has calcium, vitamin A, vitamin C, iron, magnesium, sodium, potassium and phosphorus.

Celery stem has 37% limonene, 19% *cis*-β-ocimene, 12% *n*-butyl phthalide, including myrcene, γ-terpinene and *trans*-β-ocimene.

Celery root has 18% γ-terpinene and *trans*-β-ocimene, 16.7% 3-methyl-4-

ethyl-hexane and 13.2% ρ-cymene, including limonene, *cis*-β-ocimene and α-terpineol.

Celery leaf has about 0.1% volatile oil, mainly 33.6% myrcene, 26.3% limonene, 14.2% *cis*-β-ocimene, 6.2% *n*-butyl phthalide and 3.7% β-selinene.

The seed oleoresin is green, with 4.75 lbs. equivalent to 100 lbs. of freshly ground celery seed.

How Prepared and Consumed: celery is a popular spice in European and North American foods and beverages. The seeds are used in fresh tomato juices, chicken soups, pickles, salad dressings, cole slaw, breads and meats. Scandinavians and Eastern Europeans add celery seeds and leaves to sauces, soups, stews and salads. The ground form or extractives are used in salami, bologna, frankfurters, knockwurst, sausages, corned beef and Bloody Mary drinks.

Cooking tends to reduce its bitterness and enhance its sweetness. North Indians and Bengalis add the seeds to curries, pickles and chutneys. Celery pairs well with chicken, turmeric, sage, cumin, soy sauce, ginger and vinegar.

Celery stalks and roots are not used as spices but are discussed because they add flavor to many foods and beverages. Celery stalks are braised to give distinct flavorings and crunchy textures. Leaves of celery are chopped and used as a garnish for soups and sauces, while stalks of celery are cut and used to flavor soups, stuffings and casseroles. Celery root is eaten raw in salads or is cooked and served as a vegetable. In North America, the leaves are added to Creole gumbos and soups. In East Asia and Southeast Asia, it flavors soups, stir-fries and sauces and is used as a garnish for Chinese-style rice dishes.

Celeriac root is eaten like the turnip—raw, blanched, pureed, stir-fried or boiled. It provides a clean celery-like flavor and a crunchy texture.

Spice Blends: bouquet garni, gumbo blends, curry blends, stuffing blends, pickling blends and tomato juice blends.

Therapeutic Uses and Folklore: Romans used celery seeds in herbal tonics as an aphrodisiac, while the Greeks used it in love potions. Celery was traditionally used as a sedative for nervousness or to promote sleep. It reduces swelling and was used to treat gout and arthritis. In India, it was taken as a remedy for rheumatism.

CHERVIL

To Europeans, chervil is a symbol of new life. Romans used these leaves more than 2000 years ago because of their pleasant aroma. The Romans introduced it to France and the rest of Europe. Chervil is now widely used by central and western Europeans. One of the most common leafy spices used in French cooking, it has become part of the popular local spice blends, bouquet garni and fines herbs.

Scientific Name(s): Anthriscus cerefolium. Family: Umbelliferae (carrot family).

Origin and Varieties: chervil is native to Russia, Europe and Northwestern Asia. It is now cultivated around the Mediterranean regions of Greece, France, Spain, Italy, in Britain and in the U.S.

Common Names: garden chervil, gourmet parsley, French parsley. It is also called maqdunis afranji (Arabic), san lo po (Chinese), kervel (Dutch), cerfeuil (French), kerbel (German), tam'ha (Hebrew), cerfoglio (Italian), chabiru (Japanese), cerefolho (Portuguese), kervel' (Russian), perifollo (Spanish) and korvel (Swedish).

Form: chervil has small and delicate, feathery light-green leaves. It is used fresh or dried, whole, chopped, crushed or as a paste in oil.

Properties: it has a sweet, aromatic, anise-like flavor with slight hints of pepper and parsley. The dried herb is less aromatic than the fresh herb.

Chemical Components: it has 0.03% essential oil, mainly *iso*-anethole.

Dried chervil contains potassium, calcium, iron, sodium, magnesium and phosphorus.

How Prepared and Consumed: in French cooking, chervil can be an alternative herb to parsley. It is widely used in many other European cuisines, where it is chopped fine and used as garnish over eggs, soups, meats, tartar sauce, potato salads and fish dishes. It is also pureed with nuts, olive oil and sun-dried tomatoes for a pesto-like sauce.

Chervil is usually added at the end of cooking to retain its delicate flavor. It pairs well with cheese, peppermint, dill, stuffings, creamy soups, béarnaise sauce, mustards, potatoes and pastries.

Spice Blends: fines herbes, bouquet garni, béarnaise sauce blend and pesto sauce blend.

Therapeutic Uses and Folklore: for Greeks, chervil symbolized sincerity. It was used to perk up memory, lower blood pressure and help menstrual cramps. In ancient times, the Romans used it as a cure for hiccups.

CHILE PEPPERS

The "mother" chile pepper is thought to have originated in the Andean region of Bolivia, Peru and Ecuador almost 10,000 years ago. Since 7000 B.C., chile peppers have been part of the diets of Mayans and Aztecs in Central Mexico and the Yucatan. They were widely cultivated when the Spanish first landed in the Americas.

Chilies or chile peppers add flavor as well as heat to foods. They tend to enhance and provide a background note for spices and other flavorings. Mexicans have mastered the knowledge of use of different types of chilies to achieve that characteristic flavor, aroma, mouth feel, color and bite in their foods. The sophisticated use of different chilies that began with the ancient Mayans and Aztecs continues around the world today. They were taken to India by the Portuguese and to Southeast Asia by the Arabs, Indians and Portuguese. Chilies dominate the flavors of many cultures—South Indians, Sri Lankans, Southeast Asians, Latin Americans and the Caribbean islanders.

Scientific Name(s): Within genus *Capsicum* (*C.*), there are five species: *C. annum, C. frutescens, C. chinense, C. pubescens,* and *C. baccatum.* Family: Solanaceae.

Globally, there are more than 3000 known varieties of chile peppers that differ in shape, color, size, flavor and degree of pungency. There are at least 150 known types in Mexico and about 79 in Thailand. Most chile peppers belong to the *C. annum* species, which are the principal chilies of Asia and other regions outside of Latin America and the Caribbean. New varieties of chile peppers are continually evolving.

Origin and Varieties: chile peppers are indigenous to South and Central America, Mexico and the Caribbean but are now cultivated in India, the U.S., Mexico, China, Africa, Japan and Southeast Asia. Pods of different species have characteristic shapes, sizes, colors and pungency, and there are many different varieties within each species. Some of the more popular chile peppers are listed by species in Table 15.

TABLE 15. Chile Pepper Species and Varieties.

Species	Varieties
C. annum	Ancho, Banana wax, cabe merah/hijau, cascabel, cayennes, chilacate, chile de arbol, chile guero, chile pasado, chilhuacle negro, chipotle, Cubanelle, exotics, Fresno, guajillo, Guntar red, Hungarian, jalapeno, Kashmiri, lal mirch, lombok, mirasol, mulato, New Mexican, pasilla, poblano, pungent bells/chile dulce, Sante Fe, serranos, tabasco, tabia Bali, tipico, xcatik, shi feng jiao, cherry-sweet/hot, chi yang jiao, kochu, pepperoncini, prik chee fa
C. annum (variety aviculare)	Bird pepper, chile padi, chile rawit, ata wewe, chiltepins, piquin, tabia kerinji, devil peppers, ululte, parado, costeno, uvilla garande, max-ic, pili-pili, prik khee noo suan
C. chinense	Chinchu uchu, habanero, Scotch bonnet, Bahama Mama, Uxmal, Congo pepper, Bonney pepper, datil, ata rodo, ixni-pec, rocotillo, rosas-uchu, piment bouc, cachucha, seasoning pepper, mak phet kinou
C. pubescens	Rocot uchu, rocoto, locoto, chile manzano, chile peron, canario, chile caballo
C. baccatum	Aji/kellu uchu, puca uchu, cusqueno, aji amarillo, Escabeche, cuerno de oro
C. frutescens	Tabasco, malagueta pepper

These five species are crossed with each other to produce many new varieties. Even within each variety, such as jalapeno or cayenne, there are several cultivars with different pungencies, flavors, textures and colors. Variations occur because of differences in breeding, environmental growing conditions, stage of maturity when picked, and post-harvest conditions such as storage and processing. Some of the different cultivars are listed in Table 16.

TABLE 16. Chile Pepper Varieties and Cultivars.

Variety	Description and Cultivars
Cayennes	Cayennes were first referred to as milchilli in 1542. They appear to have originated in French Guiana, South America, and were named after the Cayenne River. Today, they are not cultivated in Latin America but are grown widely throughout Asia, where they were introduced by the Portuguese. In the U.S., they are grown in Louisiana and New Mexico. Cultivars include Hot Portugal, Ring of Fire, Hades Hot and Exotics. The Exotics are cayenne types and include santaka, Hontaka, Thai chilies and Chinese kwangsi. The dried forms of cayennes are called cili kering, lal mirchi or Ginnie peppers.
New Mexican	These peppers were originally called Anaheim. They are mild to hot. Cultivars include Big Jim, Sandia, Fresno, Nu-Mex No. 6-4, Nu-Mex sunrise, Nu-Mex sunset, Nu-Mex Eclipse, Sandia, Fresno, TAM Mild Chile, long green/red chile, prik chee fa, la chiao, chimayo, chile verde, chile Colorado and California long green or red. In Mexico, the dried green New Mexican chiles are called pasados, and red ones are called chile Colorado.
Jalapeno	Jalapeno peppers originated in Jalapa, Mexico. The name means "looking at the sun." Some cultivars are San Andres, TAM mild jalapeno-1, 76104 Jumbo Jal, Morita, Tipico, Peludo and Espinalteco. In Mexico, the short fat version is called chile gordo. In Oaxaca, the smoke or dried jalapeno (or chipotle) is called chile huachinango, and the pickled jalapeno is called morita.
Serrano	Serrano peppers originated in Puebla, Mexico. Serrano means "from the mountains." When fresh green, they are called chile verde, and when dried, they are called balin/chico, tipico and largo. Commercial cultivars are Tampiqueno 74, VeraCruz S69, Altamira, Panuco and Hidalgo.
Ajis	This is the general term for chile peppers that were cultivated as early as 2500 B.C. in Peru and were an integral part of the Inca diet. Ajis come in diverse shapes, sizes and colors, including aji amarillo (or kellu uchu as the Quechua Indians call it) and cuerno de oro in Costa Rica. They are commonly used in South America (Peru, Bolivia, Colombia) and Central America.
Rocotos	Rocotos has its origins in Bolivia, was domesticated in 6000 B.C. and was commonly used by the Incas. Called rocotos by Peruvians for fruit-shaped chilies, it is also called caballo in Guatemala, locoto in South America and chile manzano, canario or chile peron in Mexico.

TABLE 16. *(continued).*

Variety	Description and Cultivars
Poblano	Poblano means "people chile," and its name was derived from Puebla, Mexico, where it was first cultivated. The dried versions are ancho and mulato. Mulato is popular in Mexico and is known as chile negro, Verdano, Esmeralda or morita. In Baja, Mexico, fresh and dried forms are called ancho or pasilla.
Chiltepin	Chiltepin peppers are also known as piquins, pequins and tepins. They were part of the historic migration of peppers from the Andes. Because birds were responsible for their migration, chiltepins were called bird peppers. They are also called chile mosquito, chile de pajaro or chile bravo (macho) in Mexico, maxic in Mayan, huarahuao (in Haiti), ululte in Yucatan and parado in Oaxaca and bird pepper, chili padi and chile rawit in Southeast Asia.
Habanero	Its name comes from the word "Havana like" and is thought to have originated in Cuba. They are now used abundantly in the Caribbean, Mexico, Brazil and Belize. They are also called Scotch bonnet (smaller size), Congo or Bahama Mama, Bonney pepper, pimient, Jamaican hots in the Caribbean and Uxmal and INIA in Yucatan.

Common Names for Chile Pepper (in General): berbere (Amharic), filfil (Arabic), spaanse peper (Dutch), chile (English), sili/siling (Filipino), piment (French), roter pfeffer (German), pilpel harif (Hebrew), mirch (Hindi), cabe/tabia/lombok (Indonesian), peperone (Italian), togarashi (Japanese), mak phet (Laotian), cili/ladah (Malaysian), la jiao (Mandarin), pimiento and pimentao picante/piripiri (Portuguese), miri (Singhalese), chile/aji (Spanish), chilipeppar (Swedish), mulagu (Tamil/Malayalam), prik (Thai), marach (Urdu), ot (Vietnamese) and ata (Yoruba, West Africa).

Form: the fruit is used ripe or unripe, fresh (whole, sliced, pureed), smoked or dried (whole, crushed or ground). The fresh and dried forms can be made into a paste or pureed (as cili boh in Malaysia/Indonesia).

Whole fresh or dried chile peppers can be roasted to develop more intense flavors. They are called by different names when the chile pepper is smoked or dried. For example, when green or red jalapeno is smoked or dried, it becomes a dark red chipotle, and when green poblano is dried, it becomes a mahogany ancho or reddish black mulato. Examples of fresh and dried chile peppers are listed in Table 17.

Dried chile peppers are available commercially as ground or crushed red peppers, ground chile pepper and chile powder. Ground or crushed red pepper usually consists of blended red peppers from any variety of hot, dried, red chile peppers. Heat rather than flavor or color is expected from these products. Ground chile pepper in the U.S. is ground New Mexican, cayenne or ancho chile peppers. It is used in chili powder with other spices. Chili powder

TABLE 17. Chile Peppers—Fresh and Dried Versions.

Fresh/chiles frescos	Dried and/or Smoked/chiles secos
Jalapeno	Chipotle, morita, tipico, huachinango
Serrano	Balin, tipico, largo
Poblano	Ancho, mulato
Cayenne	Sanaam, tien tsin, cili kering
New Mexican	Chile pasado, chile Colorado, lal mirchi
Chilaca	Pasilla (chile negro)
Cili padi	Cili padi kering
Chiltepin, piquillo	Cascabel, chiltepin, piquin, coban
Aji amarillo	Cusqueno

has varying degrees of pungency and different shades of color ranging from bright red to deep mahogany. In Asia and many other countries, chili or chilli powder means only ground chile pepper, usually cayenne. In the U.S., chili powder is a blend of ground, dried cayenne/New Mexican chile peppers, cumin, oregano, paprika and garlic powder. This blend forms the basis of the flavor for chili con carne or chili from Texas.

Properties: chile peppers have many shapes, sizes, colors, flavors and heat. They can be hot, sweet, fruity, earthy, smoky and floral. They are variable in their heat levels. Fruits harvested from the same species but from different regions can differ greatly in heat. Chiles can enhance and round up other ingredients in a dish. General physical descriptions of several of the more popular chile peppers are listed in Table 18.

Chemical Components: Chile peppers contain 0.2% to 2% capsaicinoids, (vanillylamides of monocarboxyl acids) which are responsible for the pungency or bite in capsicums. Heat varies widely among the different chile peppers depending on their chain length. Capsaicinoids are mostly found in the white "ribs"/placenta that runs down the inside of the pepper, in the seeds and in the skin. Most of the overall heat is due to two capsacinoids, capsaicin and dihydrocapsaicin.

Capsaicin(C), an alkaloid, accounts for about 50% to 70% of the total capsacinoids. It gives the bite but has no odor. The other bite contributing components are 20% to 25% dihydrocapasaicin (DHC), which together with capasaicin provides the most fiery notes from mid-palate to throat); 7% nordihydrocapsaicin (NDHC), which is fruity and sweet and has the least burning sensation; about 1% homocapsaicin (HC) and 1% homodihydrocapsaicin (HDHC) both of which give a numbing and prolonged burn.

The degree of heat or pungency of chile peppers varies based on the vari-

TABLE 18. Physical description of Chile Pepper Varieties

Variety	Description
Cayenne	Red, orange, green 3–10″ long, 1″ wide, narrow and elongated, sometimes curved, ending in a sharp point; hot to very hot.
New Mexican	Green to red, smooth, elongated, 6–12″ long; mild to hot.
Jalapeno	Medium size (2–3″ long), bright green, red, or purple with tapered to blunt ends, and conical and cylindrical in shape; slightly hot to very hot. (Red, ripe jalapeno is sweeter than the immature green).
Chipotle	Wrinkled dark brown skin with a smoky, sweet chocolaty flavor.
Serrano	Dark green to red orange, yellow or brown, 1–3″ long or conical in shape; very hot.
Poblano	Green, dark green, 3–6″ conical or heart shaped; slightly hot.
Ancho	Flat, wrinkled and heart shaped, red to reddish brown; mild to slightly hot.
Mulato	Reddish brown to blackish brown with chocolaty notes.
Chiltepins	Tiny and ovoid; very hot, but the heat dissipates quickly.
Ajis	Diverse in size and shape, typically 3–5″ long, elongated, orange, red, yellow or brown; medium to hot.
Habanero	1–2-1/2″ long, 1–2″ wide, bell shaped, red, green, yellow, white or orange, fruity taste and great aroma; very hot.
Tabasco	1–1-1/2″ long, 3/8″ wide and yellow when unripe, red orange when ripe; hot.

eties, origins, growing conditions and drying conditions. Heat is measured by using organoleptic tests and/or high-performance liquid chromatography (HPLC). The capsaicinoids are measured in parts per million and are then converted to Scoville Units (Table 19).

Heat values can vary with the same type of chile pepper, such as jalapenos or cayennes, based on its origins, breeding, climatic conditions and growing conditions.

Chile peppers are high in vitamins A and C. Green chilies have double the vitamin C of a regular sized orange, while red chilies have more vitamin A than carrots. They are low in calories and sodium and contain potassium, phosphorus, magnesium and folic acid.

How Prepared and Consumed: the ancient Mayans and Aztecs used chilies as food and medicine and in religious rituals and ceremonies. Nowadays, they provide not only heat but also flavor, color and visual appeal to foods around the world. Caribbeans, Mexicans, South Americans, Mediterraneans, Asian Indians and Southeast Asians use chile peppers to add zest or flavor to their cuisines. These ethnic cuisines use chile peppers as vegetables when fresh and as a spice when dried or smoked.

Chilies are also made into hot sauces or signature spice blends that are characteristic of a region. The early hot sauces were made by the Caribs and Arawak

TABLE 19. Scoville Units of Chile Pepper Varieties.

Chile Pepper Variety	Scoville Units
Mild bell pepper, sweet banana, pimiento, paprika, Hungarian sweet pepper	0–100
Mexi-bell, cherry, romesco, nora, piquillo	100–200
New Mexican, prik num, guajillo, Hungarian hot pepper	200–500
Poblano, ancho, pasilla	1000–1,500
Mirasol, jalapeno	2500–5000
Jalapeno, chipotle, sandia, pasado, guero	2500–10,000
Serrano, jalapeno, Kashmiri, sriracha, aji amarillo	10,000–25,000
Cayennes, prik khee fah, chile de arbol, tabasco cabe merah, rocotillo, ajis	30,000–60,000
Chiltepin, aji, rocoto, santaka, piquin, malagueta	40,000–50,000
Prik khee nu, chile padi, cabe rawit, bird peppers	50,000–100,000
Habaneros, Scotch bonnets, piment, Bahamian	100,000–500,000
Pure capsaicin	16 million

Indians who mixed chile peppers with cassava juice. Today, chile peppers are pickled or pureed and are used as table condiments or sauces all over the Caribbean. The fiery Bajan sauce contains chile peppers with other ingredients such as fruits, mustard, garlic, thyme, green onions and clove. Similarly, Louisiana hot sauce, sambals of Indonesia, salsas and moles of Mexico, goit chu jang of Korea, balachuang of Myanmar, harissa of Tunisia, romesco of Spain, Matouk sauce from the Antilles, berebere of Ethiopia, shichimi togarashi of Japan, rouille of southern France and the chaat masalas of India contain chile peppers as an essential flavoring.

Many dried chilies such as tien tsin, chipotle, Guntur red or chile de arbol are ground and added to spice mixtures and made into hot chile oils. Japanese use these spice mixtures, such as schichimi togarashi, to add heat by sprinkling over foods. In Asian cooking, dried whole red chilies are fried in hot oil until they turn dark brown. This hot oil is added to Szechwan and Hunan stir-fries, and in India, a mixture of chile pepper and oil is added to spice up many curries and dals.

Chile peppers pair well with garlic, fermented beans, ginger, coconut, shallots, fermented fish or shrimp, galangal, turmeric, sesame oil and fruit sauces. In Southeast Asia, Korea and Szechwan region, they are added to fermented soybeans and seafood to make fiery hot pastes for many dishes. Dried or fresh chilies are a must in Thai curry pastes, Indonesian rendangs and Malaysian sambals. In India, the dried red chile peppers and fresh green chile peppers are popularly used. Sri Lankans and South Indians use whole cut chilies abundantly in snacks, chutneys and curries. The black curries of Sri Lanka contain bird peppers and cayennes with toasted spices.

In Eastern Europe, chile peppers are not commonly used, except in Hungary. In the Mediterranean regions, the North Africans usually use the hotter chilies, while the Spanish use the mildly hot peppers.

In Latin America, chilies add heat to bland potatoes, yuccas, salsas, ceviches, moles and condiments. Mexicans, Ecuadorians, Brazilians and Peruvians use many types of fresh and dried chilies (jalapenos, habaneros, ajis, rocotos, malaguetas) to create moles, ceviches and salsas. Ocapa (potato, peanut, cheese with ajis), stuffed rocotos, pebre (with olive oil, cilantro and red chilies), aji molida (with fresh chilies, herbs and onions), salbutes (puffed tortillas) topped with a hot sauce of habaneros, lime juice and onions, and chimmichurri with ajis, cilantro, garlic and lime juice are common applications.

Some examples of how chile peppers are used in ethnic cuisines are listed in Table 20.

TABLE 20. Chile Peppers and Applications.

Chile Pepper Variety	Description of Applications
Rocotos	Popular in Andes and Mexico—salsas, stuffed with cheese or meat, rellenos, empenadas
Chiltepin	Used in soups, ceviches, salsas, bean dishes and pickles of South and Central America, Mexico and Texas
Serrano	Popular in Mexico, Tex-Mex and Southwest cooking, good for pickling—escabeche, fresh salsas, table sauces, relishes
Tabasco	Added to Tex-Mex, Cajun, Creole cooking and Mexican sauces.
Cayenne	Popular in Indian curries, chatnis, vindaloo, in Chinese, Korean and Japanese stir-fries and condiments, in Southeast Asian and African curries and sambals, in Cajun and Creole cooking, used as sculptured toppings for Thai cuisines
Jalapeno and chipotle	Used in Mexican and Central American salsas, snacks, stuffings, toppings and pickles; In the U.S., popular in Tex-Mex and Southwest cooking—in enchiladas, nacho toppings, pickles, sauces and chile con carne
Habaneros	Popular in hot sauces of the Caribbean, Yucatan salsas, and pickles—they are the principle ingredient in Pickapeppa (Jamaica), Yucateca (Mexican), Pica rico (Belize), curry colombo, Jamaica Hell Fire, Bajan (Barbados) sauces and jerk seasoning
Ajis	Are found in many South American sauces—aji molida, salsas, ceviche, pebre, ocapa and chimmichurri
New Mexican	Used in Southwestern salsas, stuffed and in sauces; used in Chinese and Indian cooking; also used in the Southwest as decorative ristras
Piquillos	Sweet, slightly hot peppers found in Spanish cuisine—they are fire roasted or smoked and used to stuff fish, meat or vegetables
Exotics	Are Thai peppers, hontaka, santaka, red chilies—they are dried and used for Szechwan stir-fries, sambals, as chile bohs, for curries or ground into chile powder in Asia

Spice Blends: chili powder, jerk, sambal trassi blend, mole negro blend, pebre, berbere, periperi sauce, romesco sauce, jhoug, kaeng ped blend, nuoc cham blend, nam prik blend, ma pla blend, rendang blend, kochujang and kim chee blend.

Therapeutic Uses and Folklore: the Incas from South America regarded chile peppers as holy plants and used them in offerings to appease the gods. They were valued as currency and were used as decorative ornaments. They were also used to ward off spells. Chile peppers have long been recognized by many cultures around the world for their medicinal qualities. Pre-Columbian Indians mixed pepper with other ingredients for sore throats, coughs, arthritis, acid indigestion, earaches, to promote better eyesight and to expedite childbirth. The Caribs rubbed it into the wounds of would-be young warriors to strengthen their courage. The Andean cultures used it to treat severe headaches and strokes.

When chile peppers are eaten, capsaicin stimulates the release of endorphins, which give a pleasurable feeling. Chile peppers are believed to increase circulation, relieve rheumatic pain, treat mouth sores and infected wounds, reduce blood clots and aid digestion by stimulating saliva and gastric juice flow.

CHIVES

Chives are a very popular spice in French and Chinese cooking. The name "chives" is derived from the Latin word, *cepa,* meaning onion, which later became *cive* in French. Chives are a member of the onion family and are used as herb or vegetable, depending on the region.

Scientific Name(s): regular chives: *Allium schoenoprasum;* Chinese chives: *Allium tuberosum.* Family: Alliacea (onion family).

Origin and Varieties: chives are native to Scandinavia, Germany, Britain, China and Southeast Asia.

Common Names: chives and onion chives. They are also called basal (Arabic), bieslook (Dutch), civette/ciboulette (French), schnittlauch (German), irit (Hebrew), erba cipollina (Italian), asatsuki (Japanese), cebolinha (Portuguese), luk-rezanets (Russian), cebollanna (Spanish) and graslok (Swedish).

Chinese Chives: garlic chives, spring onions, kow choi (Cantonese), kuchai (Malaysian, Indonesian), chiu tsai (Mandarin) and kui chai (Thai).

Form: chives are hollow, long, narrow stems attached to a bulb on one end and tube-shaped green leaves at the other. Chives are used fresh or dried. The light purplish flowers of chives have a light, delicate chive-like flavor. The entire length of the tubular leaf is used in foods. Chinese chives have flat and wider stems than regular chives.

Properties: the stems of chives are bright green. Regular or onion chives have a mild, oniony flavor because of their sulfur content. They have a more delicate flavor than scallions. Chinese chives are slightly garlicky and stronger in flavor than onion chives. The bulbs of chives have a delicate onion-like flavor.

Chemical Components: the essential oil of chives consists mainly of dipropyl disulfide, methyl pentyl disulfide and penthanethiol.

Chives contain vitamin A, vitamin C, potassium and folic acid.

How Prepared and Consumed: chives are a popular ingredient in European cooking because of their delicate flavor. They are an essential flavoring of the fine herbs blend that is commonly used in French cuisine. The bright green leaves of chives are good as a garnish in cold salads and soups or when sprinkled over cooked sauces, soups and finished meals. The finely chopped leaves and bulb are used to garnish vichyssoise, cheese and cream sauces, gravies, dips and baked potatoes.

Chinese chives are chopped or sliced and are used as a common garnish for all dishes, stir-fried vegetables, fried noodles, fried rices, grilled meats, fish dishes and dumpling sauce. Chives pair well with lemon, tarragon, parsley, sesame oil, vinegar, garlic, soy sauce, ginger, potatoes and chile peppers. Their bulbs are also pickled. In the U.S., chives are used in cottage cheeses, egg dishes, cocktail sauces, dried soups and sour cream.

Spice Blends: remoulade blend, fines herbes, dumpling sauce blend and vichyssoise blend.

Therapeutic Uses and Folklore: chives have been used to help lower blood pressure and aid digestion. Chives also stimulate the appetite and possess some antiseptic properties.

CINNAMON

Known as "sweet wood" in Malaysian, Arab traders brought cinnamon from the Far East to the West, where it was greatly treasured. Cinnamon is one of the oldest known spices. The Chinese used it as a medicine as early as 2500 B.C. It is mentioned in the Bible and was used by Egyptians in their embalming mixtures to preserve their pharaohs.

Cinnamon's name is derived from the Greek word, *kinamon*. The Greeks and Romans used it as body perfume and medicine and burned it as incense. It was one of the ingredients in ancient European love potions. The Dutch, Portuguese and British explorers' goal in searching for a quick route to the Far East in the fifteenth Century was to control the trade in spices, including cinnamon.

There are several varieties of cinnamon, but two types are commonly used: Cassia (also called Chinese cinnamon) and Ceylon cinnamon (also called canela). The former is used widely in East Asia, Southeast Asia and the U.S. (sometimes mixed with Ceylon cinnamon), while the latter is most frequently used in Mexico, England and South Asia. Indonesian cinnamon, called kayu manis padang (sweetwood), which comes from Padang, Sumatra, is sometimes sold as Ceylon cinnamon in the U.S. because its aroma closely resembles it.

Scientific Name(s): Chinese cinnamon/cassia: *Cinnamomum cassia;* Ceylon cinnamon/canela: *Cinnamomum zeylanicum;* Indonesian cinnamon: *C. burmannii,* Vietnamese cinnamon: *C. loureirii*. Family: Lauraceae.

Origin and Varieties: cassia is native to China, Indochina, Burma and North Vietnam and is also cultivated in Indonesia and Malaysia. There are mainly four types: Indonesian cinnamon [korintje (or Padang) and cassia vera], Vietnamese cinnamon (cassia lignea) and Chinese cinnamon (Tunghing, Sikiang, Canton). Korintje, classified as cassia, has a flavor that is closer to Ceylon cinnamon. It comes in many grades; the lower grades are commonly sold in the U.S.

Ceylon cinnamon/canela is native to Sri Lanka, India, and is also cultivated in Mexico, South America, Seychelles and Madagascar. There are three main types: Ceylon, Seychelles and Madagascar cinnamon. Ceylon cinnamon is used abundantly in South Asian, English, Mexican, Central American and South American cooking.

Common Names: cassia/Chinese cinnamon: false/bastard cinnamon. It is also called darasini (Arabic), kassia (Dutch), cannelle de Chine/saigon (French), kassie (German), Vietnamesisches/Saigon zimt (German), kinamon (Hebrew), cassia (Italian), kashia keihi (Japanese), kayu manis (Malaysian/Indonesian), you kwei (Mandarin), kassia (Norwegian), korichnoje derevo (Russian), canela de la Chine (Spanish), op choey (Thai) and que (Vietnamese).

Indonesian cinnamon/kayo manis padang: Java cassia, cassia vera, Timor cassia, Korintje and Padang cinnamon. It is also known as Padang zimt (German), canelle de Padang (French) and jaavankaneli (Finnish).

Ceylon cinnamon/canela: true cinnamon. It is also called kerfa/êrfeh (Arabic), dalchini (Bengali), kaneel (Dutch), darchini (Farsi), cannelle (French), zimt (German), darchini (Hindi), cannella (Italian), seiron nikkei (Japanese), kayu manis (Malaysian, Indonesian), lavang patti (Malayalam), kanel (Norweigian), koritsa (Russian), kurundu (Singhalese), canela (Spanish, Portuguese), kanel (Swedish) and patta/illavangum (Tamil).

Form: every part of the cinnamon tree, the bark, leaves, buds, roots and flowers, has some use. The bark oil is used extensively in commercial foods. The dried scented bark, sold as whole sticks called "quills," is ground. Ceylon cinnamon has slender and thin quills with smooth surfaces, while Chinese cinnamon has thick quills with coarse surfaces. The bark of Ceylon cinnamon has a dark reddish brown interior color and a light brown or tan outer color, whereas cassia has a much darker outer surface. Cinnamon bark is sold whole, in chunks or ground.

The highest quality cinnamon comes from the trunk, and lower quality cinnamon comes from the side branches. Finely ground cinnamon is smooth, not gritty, and this is indicative of good quality cinnamon. Cinnamon is graded according to its length, breadth and thickness. The quills are the best grade, with lesser grades being quillings, featherings and chips. Cassia bark is more mucilaginous than canela bark.

The unripe fruits of cassia, called buds, look like cloves and are sweetish and less aromatic than the bark. The yellow flowers of cassia are very fragrant.

Properties: cassia has a coarser and thicker bark than Ceylon cinnamon. Cassia has a sweet spicy aroma, a harsh taste and is more pungent than Ceylon cinnamon. Ceylon cinnamon has a warm, sweet, slightly clove- and citrus-like delicate flavor. Ceylon cinnamon has a more delicate flavor than cassia. Chinese and Vietnamese cinnamon are more intense in flavor than the other cassias. Korintje cinnamon has a delicate flavor and is least bitter. Ceylon cinnamon has a lighter colored and much thinner bark than the cassia.

Indonesian or Korintje cinnamon's aroma is similar to Ceylon cinnamon: slightly intense but not coarse like Chinese or Vietnamese cinnamon. Its quills are thicker than those of Ceylon cinnamon so are less fragile, with an outer reddish brown color and a darker interior. It has lower slime than Chinese or Vietnamese cinnamon.

Chemical Components: depending on its origins, cassia has 0.9% to 4% essential oil, and canela has 0.5% to 2%. Cassia oil mainly contains cinnamic aldehyde (65–95%), cinnamyl acetate, cinnamic acid, benzaldehyde and coumarin, but it does not contain eugenol. Canela oil mainly consists of cinnamic aldehyde (65–75%) with eugenol (5–10%), linalool (2.3%), cinnamyl acetate (5%), safrol, 1,8-cineole and benzyl benzoate. The oil is yellowish brown to brown in color and darkens as it ages.

Canela leaf contains about 1% essential oil, mainly eugenol (70–95%) with linalool, cinnamic aldehyde and caryophyllene.

The oleoresin from the bark is dark reddish brown in color, and 2 lbs. are equivalent to 100 lbs. of freshly ground spice.

Cinnamon has calcium, potassium, manganese, iron, magnesium and vitamin C.

How Prepared and Consumed: European and Mediterranean regions primarily use the ground form of cinnamon. Asian and Latin American countries use whole and ground forms. Cinnamon pairs well with tomatoes, onions, yogurt, nuts, bananas, oranges, peaches, apples, grapes, chocolate, corn and cauliflower. Overcooking makes it bitter.

Cassia is the preferred cinnamon in the U.S. because of its spicy-sweet flavor. It is used in its ground form in fruit-based confections, cakes, pies, cookies, buns and breads. The cinnamon stick flavors mulled cider. In East and Southeast Asia, cassia is used in savory applications—five spice blend, sauces, curries, soups and desserts. The Chinese use the ground form in a five spice blend for braised dishes, soups and sauces. The Vietnamese add it with star anise to their long-simmering beef soups, called pho bos, which are served hot with noodles and herb garnishes.

Sri Lankan or Ceylon cinnamon is popular in European, Latin American, North African, Middle Eastern and English cooking. In England, Ceylon cinnamon is mainly used in sweet applications, such as fruitcakes, stewed fruits and pastries. In India and Sri Lanka, it is an essential spice in the fiery curries, pickles, teas, garam masalas and fragrant biryanis. For curries, cinnamon stick is fried in hot oil, whereby it unfolds to release its aroma. It is then added to the curry or rice being cooked.

Ceylon cinnamon is also popular in Arab and North African cooking: it is added with paprika in baharat, with rose petals in ras-el-hanout and with black pepper in berbere. It is also indispensable in Latin American confections and chocolate beverages and the moles of Mexico.

Cinnamon buds are dried and used for flavoring confections and pickles in China. The leaves of Ceylon cinnamon are sometimes used as a substitute for Indian bay leaf to flavor curries and are also used as a garnish in India and Southeast Asia. The cassia flowers perfume the sweets, soups, pastries, wines and teas in China.

Spice Blends: garam masala, mole negro, Panang curry blend, baharat, ras-el-hanout, berbere, Indian curry blends, pho bo blend and Chinese five spice.

Therapeutic Uses and Folklore: as early as 2700 B.C., Chinese herbalists used cinnamon to treat diarrhea and to help circulation and depression. Indonesians use a pair of cinnamon buds (growing side by side) in a drink for the bride and groom at weddings as a symbol of harmony and togetherness. Cinnamon is also used to relieve head colds and nausea. Its leaves are used as a treatment for diarrhea.

Since ancient times, cinnamon was commonly used as a flavoring for meat because of its antibacterial properties that helped to prevent spoilage.

CLOVES

Cloves are native to the Moluccus (or Spice Islands), where traditionally, when a child was born, a clove tree would be planted. If the tree grew straight and strong, then so would the child. Children were also given clove necklaces to protect them from illnesses.

Cloves are an ancient spice. The Chinese used cloves as early as 200 B.C., when court officials chewed cloves to sweeten their breaths in the Emperor's presence. Arab traders introduced clove to the Romans and Greeks, who used it in their love potions. Later, Europeans fought each other to monopolize the clove trade in the Spice Islands because of its value. The name clove is derived from the French word *clou* and Spanish *clavo,* both meaning "nail," because of its resemblance to the shape of a nail.

Scientific Name(s): Syzyium aromaticum (formerly *Eugenia caryophyllata* or *E. aromaticus*). Family: Myrtaceae.

Origin and Varieties: cloves are native to the Moluccas islands of Indonesia and are now produced in Brazil, Madagascar, Zanzibar, Sumatra, Malaysia, Sri Lanka, Grenada and Jamaica.

Common Names: cloves are also called koronful (Arabic), kruidnagel (Dutch), mikhak (Farsi), clou de girofle (French), nelke (German), tziporen (Hebrew), lavang (Hindi), garofano (Italian), choji (Japanese), grampu (Malayalam), bunga cinkeh (Malaysian, Indonesian), ting hsiang (Mandarin), nellik (Norweigen), cravo (Portuguese), gvozdika (Russian), clavo (Spanish), karabu nati (Singhalese), kryddnejlikor (Swedish), clovas de comer (Tagalog), kirambu (Tamil), kan ploo (Thai), karinfil (Turkish) and hanh con (Vietnamese).

Form: cloves are dried unopened flower buds that are sold whole or ground. The rose-colored buds are picked just before opening, and the cloves are then dried in the sun. The color of the clove changes to a dark reddish brown stem with a lighter reddish brown bud or crown. The yellowish wrinkled cloves are immature and yield a low eugenol content. Defective cloves such as "mother cloves," headless cloves or fermented cloves and stems are impurities. The ground form is from the clove crown.

Properties: cloves have a spicy, woody, burning, sweet and musty aroma. Their taste is sharp, bitter and pungent with a numbing feeling.

Chemical Components: the essential oils from the clove buds are superior to those found in clove stems or leaves. The clove buds have from 5% to 20% essential oil, which is yellowish green in color. It darkens upon aging. It contains eugenol (81%) that gives its characteristic pleasant and sharp burning flavor. The other components are eugenyl acetate (7%), β-caryophyllene (9%), α- and β-humulene (1.67%), traces of benzaldehyde and chavicol. Clove stems have 6% essential oil (about 83% eugenol), and leaves have about 2% essential oil (with 80% eugenol). Clove also has 13% tannins (quercitannic acid) and 10% fixed oils. Its oleoresin is brownish green to greenish brown in color; 6 lbs. of oleoresin are equivalent to 100 lbs. of freshly ground spice.

Clove contains vitamin C, calcium, vitamin A, sodium, manganese, potassium, magnesium and calcium.

How Prepared and Consumed: clove goes well with sweet, fruity, caramelized, chocolaty and meaty notes. In the U.S., ground cloves or oil are used in baked ham, red cabbage, carrots, pickled beets, sausages, pot roasts, luncheon meats, cured meats, salad dressings, desserts, fruitcakes and puddings. It contributes to the flavor of Worcestershire sauce and ketchup. It blends well with cinnamon, fish or meat extracts, tamarind, soy sauce, coffee, chocolate, chile peppers and fruits.

Sri Lankans and North Indians use cloves whole or ground in garam masala, biryanis, meat dishes and pickles. Clove is also an ingredient in the betel nut mixture (used as a breath freshener) that is chewed after meals.

Europeans like clove in sweet dishes. In England, it is added to apple tarts, pickles and mincemeat; in France, it is used in quatre epices, soup stocks and stewed fruits; in Germany, clove is mixed with sauerbraten and breads.

Clove is an important ingredient in Chinese five spice blend that is used for barbecued meats or pork. In North African cooking, cloves are used in many spice blends, such as baharat, berbere and ras-el-hanout, that are used to flavor meat and rice dishes. In Indonesia, cloves provide a sweet aroma to local cigarettes called kretek. They are also added to their rendangs, gulais and spicy condiments. In Mexico, clove enhances the local moles and other local sauces.

Spice Blends: Chinese five spice, garam masala, curry blends, quatre epices, baharat, berbere, ras-el-hanout, rendang blend, gulai ayam blend, Worcestershire sauce blend and ketchup blend.

Therapeutic Uses and Folklore: ancient physicians recommended clove as an aid to digestion and to promote circulation. Portuguese women in the East Indies distilled a liquor from cloves for heartaches. In India, it was combined with nutmeg and black pepper to create an aphrodisiac for men. Clove is a pain reliever, so it is used as a mild anesthetic for toothaches. Clove causes the blood vessels near the skin to dilate and bring blood to surface, thus giving a warm feeling to the skin. Clove helps to get rid of intestinal gas and fights nausea. Its leaves are used to soothe skin burns.

Clove has antibacterial and antifungal properties because of its eugenol content.

CORIANDER

Since ancient times, coriander has been enjoyed by many cultures for its culinary and medicinal values. Coriander is mentioned in Sanskrit literature as far back as 5000 B.C. and in Greek Eber Papyrus as early as 1550 B.C. About 400 B.C., Hippocrates, the Greek physician, recommended coriander for its medicinal value. Coriander was found in Egyptian tombs dating from 1090 B.C. As early as fourth century B.C., the Chinese ate it to attain immortality. The Arabs used it as an aphrodisiac, while the Romans used it as a seasoning.

Coriander includes the fruit (seeds), leaves, stems and roots, each having its own distinct flavor. Its name is derived from the Greek word *koris* meaning bedbug, due to the perception of a "buggy" odor from the unripe seeds. Coriander is a versatile ingredient, with its seed, leaf and root commonly flavoring Latin American, Asian and Middle Eastern foods. In the Americas, the leaves and stems are called cilantro, while the seeds are called coriander. Latins, Caribbeans and Asians use a more pungent variety called leafy coriander, culantro, long coriander or shado beni.

Scientific Name(s): Coriandrum sativum. Family: Umbelliferae (parsley family). Long coriander: *Eryngium foetidum.* Family: Umbelliferae (or Apiaceae-parsley family).

Origin and Varieties: native to Asia, now grown in India, the Middle East, Romania, Morocco, Mexico, Argentina, the U.S. (Kentucky), the Caribbean and Southeast Asia.

Common Names: coriander seed is also called kuzbara hab (Arabic), nan nan zee (Burmese), coriandre (French), koriander (German, Swedish, Dutch, Norwegian), zir'ey kusbara (Hebrew), dhania (Hindi, Urdu), coriandolo (Italian), koyendoro (Japanese), ketumbar (Malay, Indonesian), hui sui (Mandarin), coentro (Portuguese), koriandr (Russian), coriondro (Spanish), kothamalli (Tamil, Malayalam), met phak chee (Thai), kishnish (Turkish) and ngo (Vietnamese).

Coriander leaf—coriander leaf, Mexican parsley and Chinese parsley. It is also called kuzbara warak (Arabic), nan nan bin (Burmese), persil arabe (French), Indische Petersilie (German), kusbara (Hebrew), hara dhania (Hindi), koendoro (Japanese), daun ketumbar (Malay, Indonesian), yuen sui (Mandarin), kinza (Russian), kotthammallie (Sinhalese), cilantro (Spanish), wansuey (Tagalog), phak chee (Thai), ngo gai (Vietnamese).

Long coriander leaf—thorny coriander, Mexican coriander, Japanese saw leaf herb, fitweed (Jamaican), Puerto Rican coriander, culantro, recao leaf, shado beni (in the English-speaking Caribbean), chandon beni (French), langer koriander (German), ketumbar Java (Malay), pak chi farang (Thai) and ngo gai (Vietnamese).

Form: it is a globular-shaped seed (or fruit) with the mature form having a brownish yellow color. It is sold whole or ground. The leaf is used fresh (whole or chopped) and dried (whole or crushed). The stem and roots are used fresh or dried and are used whole, chopped, shredded or minced.

Properties: the many parts of the coriander plant exhibit different sensory characteristics. Coriander seed is spherical to slightly oblong and ribbed. Generally, the seeds have a sweet, spicy nutty flavor with a hint of bitter orange. Coriander seed has tart, cedar and floral-like undertones. Its color varies from a brownish yellow to a deeper brown. The ground Indian variety has sweeter notes and is lighter in color than the Moroccan or Romanian versions. Moroccan coriander is bigger than Indian coriander, while the Romanian type is smaller in size.

Coriander seed contains calcium, phosphorus, sodium and magnesium.

There are many types of coriander leaf (or cilantro), but two types are most commonly known and used. The flat-leaf, parsley-like variety has a refreshing, soapy, piney, anise-like and slightly lemony flavor, with mint and pepper overtones. A second variety, known as thorny coriander, Japanese saw leaf and Vietnamese ngo, is also commonly called "culantro" or "recao" by Latin Americans. It is deep green, 6 to 10 inches long with serrated edges. It has a stronger and more pungent flavor than the parsley-like cilantro.

Coriander root has stronger and more pungent notes than the leaf or seed.

Chemical Components: depending on regional varieties, coriander seed has 0.2% to 2% essential oil, mainly *d*-linalool (60–70%), α-pinenes (6.5%), βterpinene (10%), camphor (5%), limonene (1.7%), ρ-cymene (3.7%), geranyl acetate (2.6%) and geraniol (1.7%). The aldehydes, *trans*-2-tridecenal and decanal, are found in the riper seeds. The fixed oil is about 13%. Its oleoresin is brownish yellow with 3 lbs. of oleoresin equivalent to 100 lbs. freshly ground coriander seed.

Coriander leaf or cilantro has 0.1–.2% essential oil, mostly aldehydes (benzyl benzoate, cinnamaldehyde, *trans*-2-tridecenel and decanal), caryophyllene and eugenol. It has less linalool than the seed. The *trans*-2-tridecenel and decanal are responsible for the fresh taste of the leafy spice. The pungent aroma of long coriander is due to the long chain aldehydes.

The fresh leaf contains vitamin A (5200 IU/100 gm), calcium, vitamin C (250 mg/100 gm) and potassium. The dried leaf contains higher amounts of vitamin C, potassium, phosphorus, calcium, magnesium and iron.

How Prepared and Consumed: the Romans were the first Europeans to introduce coriander as a cooking spice. Coriander pairs well with lentils, beans, onions, potatoes, coconut milk, vegetables and fish. In the U.S., the ground form and oil are used in hot dogs, chilis, pastries, sausages, frankfurters, luncheon meats, confections, stews, cookies and desserts.

When young, the entire plant is used to make chutneys and sauces in India. The mature, ripe seeds, whole or ground, are used in North American, Mediterranean, North African, Mexican, Indian and Southeast Asian foods. Mexicans use the whole seeds inside a hard sugar candy called colaciones, which is used to make bread pudding. Coriander seed develops a more intense aroma when lightly dry roasted or toasted in oil. Indian and Southeast Asians roast the whole seed before using it in curries, pickles or snacks.

Coriander seed is also coarsely crushed, like peppercorns, and added to foods. Turks flavor their strong coffee with ground coriander. Coriander is a popular ingredient in Tunisian spice blends, such as baharat and tabil, used in couscous, stews and salads. Ethiopians use coriander as an ingredient in berbere to season sauces, stews and dips. Yemenites mix coriander with dried fruits, green chilies and other spices to make jhug for dips and condiments. Moroccans use a rub of ground coriander combined with garlic, butter and paprika on lamb before it is roasted.

Cilantro adds flavor and color to the cuisines of Southeast Asia, Mexico, India, the Caribbean, the U.S. (Tex-Mex) and North Africa. Cilantro is generally chopped or pureed to add flavor and color to green salsas, chutneys, samosa fillings, taco toppings or bhel puris (North Indian snacks). It can be used whole or chopped to brighten dishes such as salads, dips, curries, salsas or soups. It is commonly added to Middle Eastern and Northern African foods such as Egyptian bean puree (bissara), Moroccan tagines and soups (harira) and snacks such as bisteeya and falafel. Lamb dumplings are filled with ground lamb, pine nuts and seasoned with coriander leaves, smen (clarified butter) and garlic. Whole cilantro is generally used fresh and is added toward the end of cooking to preserve its pungency and bright green color.

In Indian cuisines, cilantro is sautéed with mint and added to chutneys or is chopped and added as garnish to snacks or dips. It is an essential coloring and flavoring ingredient in a popular seasoning of North India called chat masala. In Malaysia, it is pureed with spearmint, green chilies and other spices to make famous local kormas that accompany rice or local breads.

Cilantro pairs well with chile peppers (fresh or smoked), bell peppers, tomato, onion, nuts, lime or lemon, rice, corn, cumin, ginger, cinnamon, garlic, tortilla, couscous, chicken and yogurt. Cilantro is also chopped and served with iced tea or coffee.

Culantro is commonly used by Mexicans, Puerto Ricans and Central and South Americans as a garnish in cooked salsas, ceviches, fillings, taco toppings and soups. In Trinidad and Jamaica, it is a popular seasoning used with fish and meat marinades. Southeast Asians use it as a garnish for soups, noodles and curries. Vietnamese use culantro as a wrap to give fragrance to meats, chicken and seafood. The leaf has a great affinity to other herbs such as mint, parsley, basil and lemongrass.

Coriander root and stem are chopped and used in Thai, Puerto Rican and Asian Indian soup stocks, curries or recaitos. They are indispensable ingredients in Thai green curries and in Vietnamese soups and salads.

Spice Blends: coriander seed—curry powders, chili powder, berbere, jhug/zhoug/sug, baharat, garam masala and pickle blends.

Cilantro—chat masala/green masala, zhoug, Thai green curry blend, tabil, salsa verde blend, Malaysian kurma blend, ceviche blends and black bean blend.

Culantro/recao—recados, salsa blend, ceviche blend, recaito, sofrito, chimmichurri and fish marinade.

Therapeutic Uses and Folklore: coriander seed is an ancient spice eaten by the Chinese, Indians and Arabs for longevity and as an aphrodisiac. Early Egyptians used coriander as a medicine. European monks used coriander to perfume liqueurs and medicines. The Greeks flavored their wines and tonics with coriander.

Coriander is traditionally used to treat migraines and indigestion, to help purify the blood and to relieve nausea, pain in joints and rheumatism. It is also considered a mild sedative. The leaves are used to soothe skin burns.

CUMIN AND BLACK CUMIN

Cumin has been a popular spice in India since ancient times. Cumin got its name from the Sanskrit word *sughandan* meaning "good smelling." In Europe, cumin was a very valuable spice, and Romans and English used it to pay their taxes. They also used it to season food and to aid in digestion. The Romans introduced it to Germany, where it was named Roman caraway and is now called kummel, identifying it as a foreign variety of caraway. Its popularity in Europe slowly disappeared after the Middle Ages, surpassed by caraway.

The ancient Egyptians used it to mummify royalty. It has now become a staple ingredient in Latin American cooking.

Black cumin, a smaller variety of cumin with a different flavor, is a popular spice in North Indian, Pakistani and Iranian foods. It is more expensive than cumin. It was a popular ingredient in Muglai cooking and was given the name shahi jeera, meaning "imperial cumin," by the Moguls. The Moguls, who were Muslims, ruled North India during the sixteenth and seventeenth centuries. Muglai cooking combined Iranian and Middle Eastern ingredients with local preparation methods and influenced much of North Indian cuisine.

Scientific Name(s): cumin: *Cuminium cyminum;* black cumin: *Cuminum nigrum.* Family: Umbelliferae (or Apiaceae-parsley family).

Origin and Varieties: it is indigenous to the eastern Mediterranean, Turkey, Egypt, Syria, China and India and is now cultivated in Iran, Mexico, the Middle East and Argentina. Black cumin is grown in Kashmir, Iran and Pakistan.

Common Names: cumin—green cumin, white cumin and cummin (English). It also called kemun (Amharic), kammun (Arabic), ziya (Burmese), komijn (Dutch), sire (Farsi), cumin/cumin blanc (French), romischer kummel (German), kamun (Hebrew), jeera (Hindi), comino/comino bianco (Italian), kumin (Japanese), jintan putih (Malay, Indonesian), jeeragam (Malayalam), hsiao hui hsiang (Mandarin), spisskummen (Norwegian), cominho (Portuguese), kmin (Russian), comino/comino blanco (Spanish), spiskummin (Swedish), jiragum (Tamil), met yeeraa (Thai), kimyon (Turkish) and jirah (Urdu).

Black cumin—Kashmiri cumin, black caraway and Roman nutmeg flower. It is also called kamun aswad (Arabic), cumin noir (French), schwarzer kreuzkummel (German), kala/shahi jeera (Hindi), comino nero (Italian), karun jeeragum (Malayalam, Tamil), svartkarve (Norwegian), kaluduru (Singhalese), comino negro (Spanish) and svartkummin (Swedish).

Form: it is the dried, brown, ripe fruit or seed that is available whole or ground. Black cumin is dark brown, very thin and much smaller, about 3 mm long.

Properties: cumin seed is elongated and oval shaped with a ridged surface. It is light brown, grayish or greenish yellow in color. Cumin has a nutty, earthy, spicy and bitter taste with a penetrating, warm and slightly lemony aroma. Ground, it has a dark brown to a brownish yellow color.

Black cumin seeds are smaller, thinner and darker, almost black. They taste sweeter than cumin with lemony, caraway-like notes. When cumin is toasted in oil or dry roasted, it becomes nutty, with a richer and mellower flavor.

Chemical Components: cumin has 2.5% to 4.5% essential oil (depending on age and regional variations), which is pale to colorless. Chiefly, it contains cuminic aldehyde (33%), β-pinene (13%), terpinene (29.5%), ρ-cymene (8.5%), ρ-mentha-1,3-dien-7-al (5.6%), cuminyl alcohol (2.8%) and β-farnesene (1.1%). Cumin has 10% fixed oil.

Black cumin has 0.5% to 1.6% essential oil, mainly carvone (45–60%), limonene and ρ-cymene.

Oleoresin cumin is brownish to yellowish green in color; 5 lbs. is equivalent to 100 lbs. of freshly ground spice.

Cumin contains calcium, vitamin A, potassium, sodium, iron, magnesium and phosphorus.

How Prepared and Consumed: in ancient Rome, cumin was ground and spread on breads and was used as a substitute for black peppercorns. Cumin pairs well with coriander, onions, garlic, ginger, turmeric, chile peppers, potatoes and oregano. In Europe today, whole seeds are used mainly to flavor Swiss and Dutch cheeses or German sauerkraut, while in France and Germany, ground cumin is used in bread, cakes, pastries and a liquor called kummel.

When dry roasted, cumin develops a more intense aroma and is added to Indian and Sri Lankan curries. In South India, whole cumin is roasted in oil or ghee, ground and mixed with curry leaves, lentils and other roasted spices to make a spice blend called sambar podi. Cumin has become an essential part of Southeast Asian curries, especially where there is an Indian influence, such as Burma, Malaysia and Singapore.

Cumin is also a popular ingredient in Mexican, Southeast Asian, Cuban and Tex-Mex cooking, especially in chili powder, which is used to flavor chili con carne and enchilada and burrito fillings.

In Mexican and Caribbean cooking, cumin is a popular spice in achiote blends, adobos, black beans and many tortilla fillings.

Black cumin is popular in North African and Middle Eastern cuisines, such as Yemeni zhoug and Saudi Arabian baharat. It is used in tagines (meat stews) of North Africa and in lamb roasts, couscous, sausages, vegetables and ground meat dishes (kibbeh) of the Middle East. Iranians use black cumin to enhance many meat and lamb dishes.

In India, black cumin is used whole and roasted and is added to yogurts, chutneys, curries, biryanis, kormas, kebabs, garam masalas, lassis (yogurt drinks) and the breads of North India and Pakistan. It is an essential flavoring of meats and rices in Kashmiri cuisine, which is based on rich creamy sauces.

Spice Blends: garam masala, panch phoron, ras el hanout, chili powder, sambar podi, jhoug, baharat, curry powder and achiote blend.

Therapeutic Uses and Folklore: in ancient times, cumin was a popular flavoring that was also associated with superstition. In Europe during the Middle Ages, the Roman women drank tea spiked with cumin to give a paler complexion, and it was also used as a symbol of love. Greeks associated it with the symbol of greed, and German girls flavored bread with cumin to keep their lovers from straying.

In India, since ancient times, cumin was used for its medicinal properties to aid digestion and to treat dysentery. Egyptians and Indians took it to relieve stress and lower blood pressure. Cumin also stimulates circulation, dispels gas in the abdomen and relieves cramping.

DILL AND DILLWEED

In the southern Mediterranean region, as early as 3000 B.C., dill was popular for its medicinal properties. Dill promoted digestion and soothed the stomachs of Assyrians and later Egyptians, Romans and Greeks, and eventually northern Europeans. Dill's name comes from the Old Norse word, *dilla,* meaning "to soothe" because Norse parents used it to ease the stomach pains of crying babies and to lull them to sleep. Dill was also believed to have magical properties and was valued as an ingredient in love potions and aphrodisiacs.

Dill is still popular in northern Europe as a flavoring for breads, sauces, soups, potatoes and pickles.

Scientific Name(s): Anethum graveolens: European dill (true dill); *Anethum sowa:* Indian dill. Family: Umbelliferae.

Origin and Varieties: dill is native to the southern Mediterranean and southern Russia. It is also cultivated in the U.S., England, Poland, Scandinavia, Turkey, northern India and Japan.

Common Names: European dill, American dill and Indian dill. It also called insilal (Amharic), shabath (Arabic), samit (Armenian), dille (Dutch), shivit and sheveed (Farsi), aneth odorant (French), dill/till (German), shamir (Hebrew), suwa/suwa patta (Hindi), adas manis/adas cina (Indonesian/Malaysian), aneto (Italy), diru (Japanese), phak si (Laotian), shih-lo (Mandarin), dill (Norwegian, Swedish), endro (Portuguese), ukrop (Russian), enduru (Sinhalese), eneldo (Spanish), satakuppi sompa (Tamil) and pak chee lao (Thai).

Form: dill seeds are used whole or crushed and are sold as dried whole seeds or ground. The feathery leaves, called dillweed, are used fresh or dried and are sold whole, finely chopped or ground.

Properties: dill seeds (or fruits) and the leaves possess different flavors. The seed is tiny, oval and flat with a sweet, mild caraway-like odor and an anise-like taste. Its back notes are slightly sharp and pungent. The European dill seed is brown with a light rim and a curved, oval, flat shape. Indian dill seed is tan or light brown in color with a yellowish edge and a pale ridge. It is longer and narrower than European dill. Its flavor is less intense than European dill, but it has a harsher, caraway-like taste.

The bluish green leaves of the European variety taste anise-like and are fresher and milder than the seed. The dried leaves do not have much flavor.

Dill seeds contain calcium, magnesium, phosphorus, potassium and vitamin A. Dill leaves contain calcium, iron, potassium, sodium and folic acid.

Chemical Components: European dill seeds have 2.5% to 4% essential oil, which is pale yellow in color and consists mainly of carvone (30–60%), limonene (33%), and α-phellandrene (20.61%), including ρ-cymene, 3,9-epoxy-*p*-menth-1-ene, α-pinene and β-phellandrene and dihydrocarvone.

Indian dill seeds have 1% to 4% essential oil, which is light brown in color. It mainly consists of dillapiole (52%) but has less carvone (21%) than the European type. It also has *trans*-dihydrocarvone (16.6%), limonene (6%), dihydrocarvone (17%) and α-pinene (1%).

Dillweed has much less essential oil than the seed, about 0.35%. The essential oil is pale yellow and contains a high proportion of terpenes but contains less carvone (28–40%) than the seed. It also has limonene (30–40%), phellandrene (10–20%), myristicin, dillapiole and other monoterpenes. Its typical fresh aroma is due to α-phellandrene.

Dill oleoresin is produced mainly from the seed and also from some European dillweed. It is greenish to pale amber in color, and 5 lbs. of oleoresin are equivalent to 100 lbs. of freshly ground spice.

How Prepared and Consumed: Europeans (Germans, Scandinavians, eastern and central Europeans) and North Americans use the seeds, leaves or the oils in pickled cucumbers, dill pickles, sauerkraut, potato salad, breads, processed meats, sausages, seafood, salad dressings, soups, sausages and stews. The seeds are ground and added as a seasoning to many dishes. Scandinavians love the flavor of dillseed in boiled fish and lamb dishes and dillweed in accompanying creamy sauces.

North Indians use dill seed in braised beans, lentils and other vegetable dishes. Dillweed is used to flavor green, leafy vegetables, especially spinach. It is an essential flavoring in dhansak, a popular meat, lentil and vegetable dish of the Parsi community in Bombay. In the U.S., Indian dill seed is used in sausages, cheeses, condiments, breads, pickles and salad dressings. Dillweed is used in fish sauces and salad dressings.

Dillweed is a popular spice in the Middle East for rices, meat and vegetable dishes. Turks use dillweed with spinach, baby lamb, casseroles, bean dips and yogurt. Pureed beans are mixed with onion, sour cream, lemon juice, black pepper and dillweed and then eaten as a salad. Greeks stuff grape leaves with rice, garlic, parsley, pine nuts and dillweed for their dolmadakis. Dillweed is enjoyed by Germans who add it to potato soup with cucumber, onion and white pepper, and by Russians who use it in a sour cream sauce that is added to stuffed cabbage.

Dill seeds pair well with cabbage, onions, bread, vinegar, potatoes, cumin, chile powder, paprika and turmeric. Dillweed goes well with rice, salads, fish, eggs, mayonnaise, mustard-based sauces, yogurt, sour cream and mild cheeses. It is usually added toward the end of cooking to retain its flavor.

Spice Blends: dal curry blends, pickling spice, sauerkraut blend, salad dressing blend, pilaf blend and dolmadakis blend.

Therapeutic Uses and Folklore: Egyptians and Greeks brewed dill seed into a magic potion used against witchcraft. They also used it for treating hiccups and many ailments. Asian Indians use dill seed to relieve stomach pains and hiccups because of its soothing effect on the digestive system. It is an important ingredient in a tonic that is given to babies in England, India and Southeast Asia to relieve colic pains. Chewing on the seeds clears halitosis, stimulates appetite and induces sleep.

EPAZOTE

Epazote characterizes the taste of Mayan cuisine in the Yucatan and Guatemala. The name "epazote" comes from the Nahuatl words, *epote,* meaning disagreeable or foul, and *tzotl,* meaning "sweat," reflecting its strong aroma. The Swedes call it citronmalla, because of its lemony undertones. Another variety is used in the southern U.S as a cure for intestinal worms, thus, the origin of epazote's nickname, "wormseed."

Scientific Name(s): Chenopodium ambrosioides. Family: Chenopodiaceae (goosefoot family).

Origin and Varieties: epazote is native to central and southern Mexico and Central America. It is cultivated in the U.S., Mexico and Central America.

Common Names: wormseed, sweet pigweed, West Indian goosefoot, Mexican tea, stinkweed and Jerusalem oak (English). It is also called mexicanischer traubentee (German), katuaya modakam (Malayalam), sitronmelde (Norwegian), mastruz (Portuguese), epazote/yerba de Santa Maria (Spanish) and citronmalla (Swedish).

Form: epazote has small, serrated leaves. It is used in fresh or dried forms. Both the leaf and the stem are used.

Properties: epazote has an herbaceous, strong, bitter taste with faint lemony notes. It has a pungent, turpentine odor with camphor and mint-like overtones. The dried form has less flavor than the fresh form.

Chemical Components: epazote contains less than 1% essential oil and has mostly monoterpenes and its derivatives, such as ascaridol (up to 70%), limonene and ρ-cymene with camphor, α- pinene, myrcene, terpinene, thymol and *trans-iso*-carveol.

Epazote contains calcium, iron, phosphorus, vitamin A, vitamin C and niacin.

How Prepared and Consumed: Mexicans use epazote fresh in salads, soups and meats and especially to enhance bean- and chile-based foods such as refried beans (frijoles refritos), frijoles negros, moles or rice and beans. It is usually added toward the end of cooking to prevent bitterness in the finished product. Spaniards flavor teas with epazote.

Epazote pairs well with cilantro, lime, chipotle peppers, cheeses, pork fat, black beans, pinto beans, cumin, garlic, onion, corn and squash blossoms. It also goes well in tortilla soups, fillings and toppings, moles, quesadillas, huitlacoche, soups and stews.

Spice Blends: black bean soup blend, refried bean blend, huitlacoche pate blend, tortilla soup blend and mole blend.

Therapeutic Uses and Folklore: epazote helps prevent flatulence, and Mexicans use it in many bean dishes. Mexican mothers steep the herb in milk and sugar and serve this tonic to their children to rid them of intestinal parasites. It is also used to treat dysentery.

FAGARA/SZECHWAN PEPPER

Fagara originated in China, where in ancient times it flavored foods and wines offered to the gods. The Chinese brewed it in teas and also gave it as a gift to friends. Today, fagara is commonly referred to as Szechwan pepper. It is a popular spice in South China, Japan and North India, especially in their local meat dishes.

Scientific Name(s): globally, there are a number of varieties, but the three most popular types are as follows: Chinese pepper: *Zanthoxylum piperitum;* Japanese pepper: *Z. sansho;* and Indian pepper: *Z. rhetsa* Rutaceae. Family: Rutacene (citrus family).

Origin and Varieties: fagara is indigenous to South China, Vietnam, Japan and North and West India.

Common Names: Sichuan pepper, Chinese pepper, anise pepper, prickly ash and flower pepper. It is also called fah jiu chiao (Mandarin), Szechuan-pfeffer (German), tilfda (Hindi), tippal (Konkani), katmurikku (Malayalam), timur (Nepali) and sezchuan pappar (Swedish)

Sansho, sometimes called Japanese pepper, and tilfda are closely related to fagara. It is also called Japanischer pfeffer (German).

Form: fagara is a reddish brown dried berry that is used whole, crushed or coarsely ground. The seeds are round and hollow with a rough prickly exterior. They come as split-open berries, sometimes containing black bitter seeds that are discarded. It is the berry that is used. The leaves and young shoots are also used in cooking.

Properties: fagara is aromatic with woody, peppery and bitter notes with undertones of citrus. Sansho is tangy and peppery and is characterized by a numbing pungency.

Chemical Components: the berry contains up to 4% essential oil, mainly terpenes such as geraniol, linalool, 1,8-cineol, citronellal, limonene and dipentene.

Its pungency is due to the amides hydroxy α-sanshool and β-sanshool. Both of these amides degrade quickly after the pepper is ground. The Indian variety has mostly sabinene, limonene, α- and β-pinenes, ρ-cymene and the terpinenes.

Sansho leaves mainly have monoterpenes, such as citronellal, citronellol and Z-3-hexenal, which give them their grass-like odor.

How Prepared and Consumed: Chinese dry roast fagara, add it to salt and grind this mixture coarsely for use as a dry condiment. The fagara is roasted to provide a more intense flavor to Szechwan dishes. It is an essential ingredient in Chinese five spice blend, which is commonly used in barbecues and roasts. In Southeast Asia, five spice blend is used in Chinese-influenced meat, poultry and duck dishes. Fagara pairs well with citrus, soy sauce, plum sauce and black bean paste.

The Japanese grind sansho with nori and tangerine peel, black sesame seeds and poppy seeds to make the popular seven spice blend. This is sprinkled over grilled meats, soups and noodles. A spicier and hotter version of the seven spice blend, which is made with chile peppers, is called shichimi togarashi. This spicy blend is a commonly used seasoning and condiment. In Japan, sansho is sometimes preserved in soy sauce and is typically used with fatty foods. The Japanese garnish their soups with sansho leaves and add its shoots to misos.

In Goa, Kerala and Gujerat, fish and vegetable dishes are seasoned with the Indian variety of fagara. It is usually not cooked but is added to the cooked dish just before serving.

Spice Blends: Chinese five spice, Japanese seven spice, shichimi togarashi, Szechwan sauce blend and Chinese BBQ blend.

Therapeutic Uses and Folklore: Chinese used fagara as a traditional cure for dysentery.

FENNEL SEED

Fennel seeds resemble anise, cumin and caraway in flavor, so certain cultures, particularly Asian, use them interchangeably. In India, fennel is thought to be a variety of anise (which is called saunf in Hindi). In fact, the Hindi name for fennel, moti saunf, means big, thick anise seed. Indonesians think of fennel as a variety of cumin and call it jintan manis, which means "sweet cumin."

To the ancient Greeks, fennel represented success, and so it was called "marathon" in reference to the battle where the Greeks defeated the Persians in 490 B.C. Fennel was also a symbol of success to the Romans. They used all parts of this plant as ingredients in their foods and as medicine to fight diseases and to sharpen their vision. During the thirteenth century in England, fennel was considered a royal spice and was served to the king with fruits, bread and pickled fish seasoned with fennel seeds.

Scientific Name(s): there are many varieties of fennel, but the two main types are sweet fennel and bitter fennel. The sweet fennel, called *Foeniculum vulgare* var. dulce, is also known as French or Roman fennel. The bitter fennel is called *Foeniculum vulgare* Mill (or *F. officinalle* All).

The Finocchio or Florence (Italian-type) sweet fennel is grown for its bulbous stalk, which is eaten as a vegetable.

Origin and Varieties: sweet fennel is cultivated in France, Italy, Morocco, India and the U.S. Bitter fennel is grown in central Europe, Russia, Germany, Hungary, Argentina and the U.S. Indian fennel is reported to be a distant variety.

Common Names: the most common variety in the U.S. is bitter fennel. It is also called insilal (Amharic), shamar (Arabic), samong saba (Burmese), venkel (Dutch), fenouil (French), fenchel (German), shumar (Hebrew), moti saunf (Hindi), finocchio (Italian), uikyo (Japanese), jintan manis (Malaysian, Indonesian), perum jeerakam (Malayalam, Tamil), hui-hsiang (Mandarin), funcho (Portuguese), fenkhel (Russian), maduru (Sinhalese), hinojo (Spanish), fankal (Swedish), yira (Thai) and sonf (Urdu).

Form: fennel is available as whole seeds or ground. The green leaf, stalk and bulb of the Florence fennel are used as garnishes or vegetables in the Mediterranean region.

Properties: the seeds are oval and ridged. They are bright or pale green to yellowish brown in color. They can be slightly curved or straight. Fennel has a fresh, camphoraceous, anise-like aroma with a slightly bittersweet taste. The leaves have an anise-like taste with a slight sweetness. The celery-like stalk of Florence fennel is sweet.

Chemical Components: fennel seed contains 1–6% volatile oil and 10–20% fixed oil. The bitter fennel contains 50% *trans*-anethole, 10–15% fenchone, 10–30% limonene, 3–11% α-phellandrene, 12–16% α-pinene, with α-thujene, β-pinene, estragol (methyl chavicol), myrcene and 1,8-cineole. The sweeter variety has no fenchone, but has higher levels of limonene. Its oleoresin is brownish green in color, and 6-1/2 lbs. are equivalent to 100 lbs. of freshly ground fennel.

Fennel seeds contain calcium, potassium, magnesium, phosphorus, sodium, vitamin A and niacin.

How Prepared and Consumed: fennel is a popular spice in European and Mediterranean regions. The English use the seeds in soups, the Germans use them in breads, fish and sauerkraut, the French use them in fish soups and spice blends, Italians use them in sausages, pizzas, meatballs, salami, pepperoni and sambuca, the Spanish use them in their cakes and other baked goods, and the Arabs use them in breads and salads.

Fennel seeds complement rich fish sauces, roast pork, mutton and lamb curries, sweet and sour dishes, roast duck and cabbages. They are recommended for oily fish such as mackerel and salmon. Asian Indians roast the seeds in oil to intensify their flavor and then use them whole or ground in curries, spice blends, soups, lentils, vegetables and breads. Fennel seeds are an essential ingredient in a Bengali spice blend called panchporon. Ground fennel is an essential flavoring in egg and fish curries of Kashmir. It is also a popular spice with the Mopla Muslims of Kerala and the Chettiars of South India.

In Southeast Asian curries, fennel adds a sweetness and balances the spiciness of other ingredients. The Chinese include ground fennel in their five spice blend for flavoring roast pork, roast duck and other braised meat dishes. It balances the spicy smoked flavors.

Chopped, fresh fennel leaves garnish or flavor fish dishes, sauces, salads, stews and curries.

In Europe, the root bulb and stalk of Florence fennel is used in soups and sauces or is baked. They are chopped and eatern raw or sliced and mixed in salad, used as a stuffing or blanched and eaten as a vegetable. It goes well with cheese, butter and black pepper.

Spice Blends: curry blends, Chinese five spice, Bengali five spice (panchporon), mirepoix blend and herbes de Provence.

Therapeutic Uses and Folklore: ancient Europeans considered fennel a sacred herb. The seeds were used as a rejuvenator, to stop hiccups, sharpen eyesight and cure earaches. Egyptians, Chinese and Asian Indians used it as a remedy for snakebites and scorpion stings. Asian Indians blend fennel into a spice mixture called paan masala, used with betel leaves, which is chewed after meals to aid digestion and sweeten the breath. Fennel soothes stomach upsets and is an ingredient in "gripe water" that is given to infants to sooth colic.

FENUGREEK

Fenugreek was cultivated in Egypt as early as 1000 B.C. It was prized throughout the Middle East and in India as a flavoring, medicine and fumigant. The Egyptians and Indians soaked the seeds in water until they swelled and then used them to reduce fevers and aid digestion. By 1050 A.D., fenugreek had spread as far as China.

The name fenugreek is derived from the Latin word *foenum graecum,* which translates to "Greek leaves," although they were never used as a spice in Greek cooking. Today, North Indians and Middle Easterners enjoy fenugreek (both seeds and leaves) in their cooking, but many Westerners do not like its bitter taste.

Scientific Name(s): Trigonella foenum-graecum. Family: Leguminosae (bean family).

Origin and Varieties: native to the eastern Mediterranean, fenugreek is now cultivated in India, Pakistan, France, Morocco, Greece, Lebanon, Germany, Argentina and the U.S.

Common Names: fenugreek was called Greek hay and goat's horn or cow's horn in ancient times because of its horn-shaped seed pods. The seeds and leaves are called by the same name in many countries where the leaves are eaten. Its other names include abish (Amharic), hulba (Arabic), chaiman (Armenian), penantazi (Burmese), fenegriek (Dutch), shambelile (Farsi), fenugrek (French), bockshornklee (German), methi (seed) and saag methi (leaf)(Hindi, Urdu), garger (Hebrew), fieno greco (Italian), koroha (Japanese), halba (Malaysian), venthiam (Malayalam), feno grego (Portuguese), hu lu (Mandarin), pazhitnik grecheskiy (Russian), alholva/fenogreco (Spanish), bockhornsklover (Swedish) and vendayam (Tamil).

Form: the seeds are enclosed in long sickle-shaped pods. They are light to yellow brown, smooth yet hard and have a deep diagonal groove that divides each seed. The seeds come whole or ground. The green leaves, called sag methi in North India, are sold as fresh whole leaves or as dried whole or crushed leaves.

Properties: the seeds are bitter tasting, but when they are lightly dry roasted, their flavor mellows and some of the bitterness is removed. The aroma of roasted fenugreek is like that of burnt sugar, and its taste is comparable to maple syrup. Once roasted, the seeds are used whole, crushed or ground. The leaves taste very bitter and have a sweet hay-like aroma.

Chemical Components: the essential oil of the seed is less than 0.02% but is rather odorous due to a compound called 3-hydroxy-4,5-dimethyl-oxolane-2-one. It also includes *n*-alkanes, sesquiterpenes and alkanoles. The fixed oil is 7% to 10%. The non-volatile components contain trigonelline, choline, sterol and diosgenin derivatives and furostanol glycosides, which are responsible for its bitter taste.

The oleoresin is dark brown to greenish yellow and is high in proteins, gums, mucilage and saponin. The fixed oil has linoleic, oleic and linolenic acids.

The seeds contain fiber (a high amount), folic acid, vitamin A (1040 IU/100 gm), calcium, potassium, phosphorus, iron, magnesium and sodium. After roasting, the niacin content increases. The leaf contains calcium, iron, carotene and vitamin C (43 mg/100 g) that decrease with frying, boiling or steaming, with the last cooking method showing the least loss. Its mucilage content is 40%, which is a concentrated source of dietary fiber.

How Prepared and Consumed: fenugreek is an important spice in India, Egypt, Saudi Arabia, Iran, Armenia and Turkey. The seeds pair well with fish and lentils and, thus, are commonly used in fish curries and dals in South India and Southeast Asia. Vegetarians in India dry roast the seeds or fry in hot oil and use them whole or ground to flavor curries, sambars, fermented breads and chutneys. Fenugreek seed is an essential ingredient in Indian spice blends, such as sambar podi in the south and panchporon in Bengal. In South India, the seeds are added to rice flour and lentil flour to make fermented flat breads such as dosai and idli. It is also roasted and ground and mixed with dried, ground chile peppers and other spices to be used as a dry-based seasoning dip for local breads.

In Iran, the seeds and the leaves are used. The seeds are toasted and added to salads to provide crunchiness. In the Middle East, fenugreek seeds are ground into a paste and rubbed on salted meat, which is then dried. This salted, spiced beef is called pastourma in Iraq and aboukht in Turkey. In Yemen, fenugreek is mixed with coriander leaf, tomato, chile paste, garlic and other spices to make a hot dip called hilbeh, which is used as a spread for breads. Yemenites also use fenugreek in a seasoning called zhug, which is added as a topping to stews. Armenians use it with garlic and chile pepper in a spice called chemen to spice up a beef dish called bastirma. The Greeks boil fenugreek seeds and eat them with honey.

Ethiopians and Egyptians use fenugreek seeds in sweets and breads. In many regions of Africa, they are soaked until they swell up and then are used as legumes. In the U.S., fenugreek seeds are used in soups, baked goods, icings and meat seasonings. They are the main ingredients in artificial maple syrup. Because the seeds have high mucilage (40%), they can be used as a natural stabilizer in processed foods.

Fenugreek leaves and young shoots can be used as cooked vegetables. In North India and Iran, the leaves are cooked with garlic, potatoes and other root vegetables or used in breads. It goes well with fish, meats and chicken curries of North India.

Spice Blends: fish curry blend, sambar podi, berbere, chemen, zhug, panch phoron, hilbeh and aboukht blend.

Therapeutic Uses and Folklore: Egyptians used fenugreek paste on the body to reduce fever and to cure chapped lips and mouth ulcers. They also used fenugreek along with other spices in the embalming process. Romans ate fenugreek seeds because they were considered an aphrodisiac. The Greeks and Romans, not realizing its potential as a spice, used fenugreek as cattle fodder to increase cow's milk flow and to restore nitrogen to the soil.

In Indonesia, fenugreek is used to promote hair growth. In India, Southeast Asia and Ethiopia, the seeds are soaked in water (the seeds swell because they are coated with mucilage) and are then drained and eaten to aid digestion, as a laxative, to treat bronchitis and to cure sore throats. It is used to prevent sharp rises in blood sugar and to lower cholesterol.

In Ethiopia, nursing mothers take fenugreek to promote production of milk. In India, women take fenugreek to help with menstrual cramps and difficult childbirths and to promote lactation after the baby is born.

The leaves are believed to stimulate bile secretion and to aid in the digestion of fatty meat.

GALANGAL/GALANGALE/GALINGALE

Galangal is an aromatic, peppery, ginger-like spice. First used in Chinese and Southeast Asian cuisines, galangal has traveled around the world and is found in Egyptian, Indian and European applications. There are three types of galangal: the greater galangal, the lesser galangal and the kaempferia galangal. Greater galangal, called liang chiang in Mandarin (which means "mild ginger") and lengkuas in Malay, is globally more popular and is used in many cuisines of Southeast Asia. The lesser and kaempferia galangals are found mainly in Indonesian, Malaysian, Thai and other Southeast Asian cooking. Kaempferia galangal, called temu kunci in Indonesian, refers to the many elongated fingers of the rhizome. It is used in Chinese traditional medicine and as a seasoning in Southeast Asia.

Scientific Name(s): Alpinia galanga (greater galangal), *Alpinia officinarum* (lesser galangal), *Siphonochilus aethiopicus* (kaempferia galangal). Family: Zingiberaceae (ginger family).

Origin and Varieties: the greater galangal is indigenous to Southeast Asia and southern China and is cultivated in Indochina, Thailand, Malaysia and Indonesia. The lesser galangal is indigenous to Indonesia and is cultivated in India and Africa. Kaempferia galangal has its origins in southern China and is cultivated in Thailand, Indonesia, Singapore and other parts of Southeast Asia.

Common Names: greater galangal, sometimes called Thai ginger, is also known as kholajan (Arabic), kulinjan (Bengali), lam keong (Cantonese), djus rishe (Farsi), souchet long (French), galangawurzel (German), kulinjan (Hindi), laos (Indonesian), galanga (Italian), naukyo (Japanese), romdeng (Khmer), lengkuas (Malay), aratta (Malayalam), gao liang chiang (Mandarin), gengibre do laos (Portuguese), galang (Spanish), galangarot (Swedish), arattai (Tamil), khaa (Thai), kulanjam (Urdu) and rieng kho (Vietnamese).

Lesser galangal, sometimes called Chinese ginger, is also known as kentjoer (Dutch), kleiner galgant (German) and kencur (Malaysian, Indonesian).

Kaempferia galangal, is known as aromatic ginger, fingerroot or Chinese ginger in English. It is also called suo shi (Cantonese), temoe koentji (Dutch), Chinesischer ingwer (German), temu kunci (Indonesian), kunci (Malaysian), Chinese key (Singaporean), krachai (Thai) and ngai num kho (Vietnamese). In Indonesia, it is sometimes confused with lesser galangal and wrongly called kencur. Zeodary (called white turmeric), which has a very bitter taste, is sometimes confused with kaempferia galangal.

Form: the rhizome comes whole (fresh, frozen or canned) and dried (ground, sliced or pickled). The slices of dried galangal need to be soaked in boiling water for 20 to 30 minutes before use. Its flowers are also eaten fresh. The lesser galangal is smaller in size. Kaempferia galangal has a central smaller globular shape with 10 or more slender, long tubes sprouting in the same direction from the central core, like fingers on a hand, and, hence, it is often called fingerroot.

Properties: fresh greater galangal has a knobby look with an inner yellowish brown to pale brownish skin that has reddish brown rings. It is aromatic, spicy, peppery, ginger-like and has a slightly sour note. The interior is hard and woody in texture. Dried galangal is spicier with a cinnamon-like taste. Young rhizomes are pink in color and are more tender and flavorful.

The lesser galangal has a reddish brown skin with a soft white interior. It is crunchy and more pungent, with a gingery, cardamom-like taste and hints of eucalyptus notes. The dried version does not possess the same flavor intensity as the fresh form.

Kaempferia galangal has a pale reddish brown skin, and the inside is watery and soft. Its taste is similar to lesser galangal, but it has a more camphorous and medicinal-like taste.

Chemical Components: greater and lesser galangal have 0.5% to 1.5% essential oil, with greater galangal having mainly 1,8-cineol, α-pinene, eugenol, methyl cinnamate, camphor and *d*-pinene. The dried version has a different composition than the fresh, with lesser aroma components. Its pungent taste is due to galangol or alpinol, which are phenyl alkanones and diaryl heptanoids.

Kaempferia galangal has ethyl cinnamate, ethyl-*p*-methoxycinnmate, *p*-methoxycinnamic acid and 3-carene-5-one.

How Prepared and Consumed: greater galangal is used abundantly in Malaysia, Thailand, Vietnam, Singapore, Cambodia and Indonesia. By itself, galangal's texture is woody and its flavor is undesirable, but when it is added and cooked with other ingredients, it enhances the overall flavor profiles of applications.

Greater galangal pairs well with coconut, chile peppers, lime leaves, turmeric, fish sauce, tamarind and shallots. In Malaysia and Singapore, the Malays and Nonyas (female descendants of Chinese traders and local Malay women) use it abundantly with lemongrass, garlic, scallions and tamarind for laksas, soups and curries. In Thailand, it is used in pungent curry pastes, meat marinades and stir-fries. Indonesians use it in their fiery hot rendangs and popular one-bowl meals, such as nasi goreng and nasi padang. Thais and Indonesians use more galangal than ginger and enjoy its flowers when they are fried or pickled.

Lesser galangal is generally less popular than greater galangal. It is used mostly in the fiery, pungent spice pastes of Indonesia that season meats, vegetables and fish. It is also combined with chile peppers, ginger, turmeric and other spices to create bumbu, a seasoning used to spike up sauces and soups in Indonesia. In Bali, lesser galangal is used in a spice paste called jangkap with lemongrass, chile peppers, ginger, nuts and other spices. This spice paste is rubbed on duck that is then wrapped in banana leaf and baked or roasted.

Kaempferia galangal is used in liqueurs, bitters and beers of Russia and Scandinavia. It is popular in Thailand, where it is grated for use in meat curry pastes, soups and fish curries with chile peppers, kaffir lime leaf and coconut. It is also used in Indonesia and in pungent Malay-style curries, sambals and condiments.

Spice Blends: Thai red curry paste blend, bumbu blend, rendang blend, laksa blends, Thai stir-fry blend, Thai five spice blend, Nonya curry blend and jangkap blend.

Therapeutic Uses and Folklore: Asians use galangal for indigestion, respiratory problems and to stop nausea.

GARLIC

One of the world's most popular spices, garlic is used extensively from China to the Americas, in French aioli, Turkish cacik, Vietnamese pho bo, Indian korma and Greek skordalia. Garlic's name is derived from the Anglo-Saxon word *garleac,* meaning "spear plant." Since ancient times, garlic has been used as a cure as well as a food. Egyptians used garlic since 3700 B.C. to provide strength and prevent disease. Jews ate it on their long journeys, Romans honored garlic for providing strength and courage and Greeks used it to treat colds and coughs.

Scientific Name(s): Allium sativum (softneck); *Allium ophioscorodon* (hardneck). Family: Alliaceae (onion family).

Origin and Varieties: garlic is indigenous to Asia, and it was brought to the Mediterranean regions, Egypt, France, Spain and Italy. It is now cultivated in the U.S., Asia, Europe, the U.K., Taiwan, South America, Mexico and Hungary. There are about 200 varieties of garlic, with different colors, sizes, shapes and flavors, but only two cultivated varieties exist—hardneck and softneck varieties. The hardneck variety (Siberian, Spanish roja, Prussian white, Italian rocombole, German white, German red, Marino Xian, Persian star and Romanian red) is superior in flavor but is harder to grow and has smaller yields and a shorter shelf life. Softneck varieties have smaller cloves that are harder to peel. Approximately 90% of all garlic sold in the U.S. is of the softneck variety from California. Other softneck varieties are French white, Creole Red, Burgundy, Inchelium and Chinese.

Elephant garlic is not really garlic, but is a member of the leek family. It is mild and does not have the pungency of regular garlic.

Common Names: garlic is also called netch'shinkurt (Amharic), tum (Arabic), sekdor (Armenian), rasun (Bengali), chyethonphew (Burmese), knoflook (Dutch), sir (Farsi), ail (French), knoblauch (German), shum (Hebrew), lasun (Hindi), aglio (Italian), ninniku (Japanese), bawang putih (Malaysian, Indonesian), velluthulli (Malayalam), suan tou (Mandarin), alho (Portuguese), chesnok (Russian), sudulunu (Sinhalese), ajo (Spanish), vitlok (Swedish), velai pundoo (Tamil), kratien (Thai), sarmisak (Turkish), lehsun (Urdu) and tai (Vietnamese).

Form: each bulb contains plump and succulent egg-shaped bulblets, called cloves, within an outer skin that is white, buff, rose or purple, depending on the variety. Garlic comes fresh or dried. Fresh, it is a firm whole clove that is also available crushed, sliced, minced, chopped or roasted. Dried, garlic comes powdered, granulated, flaked, diced, ground, minced, chopped and sliced.

Properties: the flavor of freshly cut garlic ranges from mild and sweet to strongly pungent, depending on the variety. Some can be pungent at first but become milder during cooking. Hardneck varieties are pungent and strong flavored, with heat that usually lingers after cooking. They have a wider range of tastes than softneck varieties. German White has a strong pungency with heat that is experienced at the back of the throat after eaten, Prussian White has a pleasant aftertaste and Spanish Roja has a lingering taste with a bite on the tip of the tongue. Softneck varieties can be harsh to mildly sweet to pungent and sometimes contain sharp bites. Mild softneck garlic can become more pungent during storage.

Whole garlic is odorless when intact but gives a strong aroma when cut or bruised. When fresh garlic is cut or bruised, the enzyme allinase acts on alliin in garlic to produce allicin that breaks down to diallyl disulfide which gives the penetrating sulfur-type aroma. The sharp bite typically subsides after cooking or roasting. Cooking softens the flavor, while roasting gives garlic a well-balanced, delicate, nutty flavor. Dried garlic (mainly from the white skinned variety) has a very strong, persistent aroma and taste. Roasted garlic has a slightly sweet and delicate flavor.

Raw garlic contains calcium, phosphorus, vitamin C and potassium. Dried garlic has higher potassium, phosphorus and magnesium levels but has a lower calcium level and negligible or no vitamin C.

Chemical Components: garlic has 0.1% to 0.25% essential oil, formed enzymatically when cloves are crushed or cut or rehydrated. It consists of 60% diallyl disulfide, 20% diallyl trisulfide, 6% allyl propyl disulfide and diallyl sulfide. Allicin (diallyl disulfide) contributes to its strong aroma.

Oil of garlic (undiluted form) has 200 times the strength of dried garlic and 900 times the strength of fresh garlic. The oils come in many dilutions for easier blending and handling in applications. Oleoresin garlic is brownish yellow and contains 5% of the volatile oil of garlic; 8 lbs. of oleoresin are equivalent to 100 lbs. of dehydrated garlic, and 2 lbs. are equivalent to 100 lbs. of fresh garlic. Garlic salt consists of garlic powder, salt and an anticaking agent (tricalcium phosphate or starch).

How Prepared and Consumed: garlic is commonly used in French, Italian, Spanish, Middle Eastern, Asian and Latin American cooking. It is fried in oil or gently simmered before onions or other spices are added to create a wonderful base for stir-fries, curries, soups or sauces. Garlic pairs well with onions, ginger, basil, turmeric, greens, beans, spinach, chicken, pork and seafood. It rounds up and modifies other flavors, such as tomato, chiles, onions and ginger.

Garlic is an essential component of many spice blends that enhance the flavor of soups, stews and curries in Asian cooking. Chinese and Thais pickle garlic in vinegar for use in noodle dishes, roast pork and chicken. Thais fry whole garlic in oil and add this to create enhanced flavors and textures in many dishes. Mediterraneans roast garlic for many dishes, such as cassoulet, beef bourguignonne, sofritos and gambas alajillas. For Italians, it is an integral flavoring along with olive oil, basil and tomatoes in their pasta sauces, pestos, vegetables and soups. Latin Americans and Caribbeans add dried garlic to their meat marinades, seasonings, soups and stews. Mexicans use garlic in their famous mole sauces and condiments.

Spice Blends: tabil, rouille, aioli, refogado blend, adobos, sofritos, moles and chimmichurri blend.

Therapeutic Uses and Folklore: called the "stinking rose" by the English, garlic has been used since ancient times to drive away evil spirits, kill intestinal parasites and stimulate blood flow for a healthy face and body. Known in Russia as "Russian penicillin," it is one of the world's oldest medicines. In the sixteenth century B.C., the Egyptians listed 22 remedies containing garlic, using garlic to treat everything from heart disease and tumors to insect bites. The Greeks gave it to the condemned to purify them of their acts. While wealthy Romans did not eat garlic, they gave it to their soldiers to make them strong.

In Asia, Hindu priests and the strict Buddhists eliminated garlic from their diets because of its reputation as a sexual stimulant.

Garlic's curative powers have been recorded since the eighteenth century. It has been reported to stimulate the digestive system, thereby helping poor digestion, lower blood pressure and help blood circulation by reducing the amount of fat in the blood and thinning it, thereby enabling platelets to move more freely.

Recent research has focused on garlic's role in preventing heart disease, enhancing the immune system, preventing cancer and enhancing memory.

Since ancient times, garlic has been used as an antibiotic and as a fungicide.

GINGER

Asia's most treasured spice, ginger, derives its name from the Sanskrit word, *shringavera,* meaning "shaped like a deer's antlers." The Chinese have used ginger since at least the sixth century B.C., and it is highly esteemed throughout Asia for its therapeutic effects as well as for its flavor. Ginger is most noted for its soothing effect on the stomach, as it is a popular traditional cure for common stomach ailments such as nausea and motion sickness. Ginger's spicy-sweet flavors complement and enhance many Asian foods.

Arab traders brought ginger to the Mediterranean region from Asia before the first century A.D. The Greeks made gingerbread, which later gave rise to gingersnaps, cakes and biscuits. In the sixteenth century, Spanish explorers grew ginger in Jamaica and exported it to Europe.

Scientific Name(s): Zingiber officinale. Family: Zingiberaceae.

Origin and Varieties: there are many varieties of ginger—Cochin ginger, Chinese ginger, Jamaican ginger and African ginger. Indigenous to southern India, China and Southeast Asia, ginger is also grown in Hawaii, Fiji, Sierra Leone, Nigeria, Jamaica, Japan, Mexico, Sri Lanka and Australia. The names for fresh and dried ginger differ in many regions of the world.

Common Names: root ginger, green ginger, gingerroot, stem ginger and black ginger. It is also called zinzibil (Amharic), zanjabeel (Arabic), ada (Bengali), gin (Burmese), gember (Dutch), jamveel (Farsi), gingembre (French), ingwer (German), zang'vil (Hebrew), adrak (fresh), sauth (dried) (Hindi), zenzero (Italian), shoga (Japanese), saenggang (Korean), halia (Malaysian, Indonesian), inchi (Malayalam), jiang, sheng jiang (fresh) and gan jiang (dried) (Mandarin), gengibre (Portuguese), imbir (Russian), inguru (Sinhalese), jengibre (Spanish), ingefara (Swedish), luya (Tagalog), ingi (Tamil), khing (Thai), adraka (Urdu) and gung (Vietnamese).

Form: ginger is a rhizome (an underground stem that looks like a thick root) that is available as a spice in many forms: fresh, dried, black or white. Fresh ginger is preserved and crystallized. The fresh form is knobby and branched, firm and tan colored and comes whole (unpeeled), sliced, julienned, chopped, crushed or grated. Young or immature fresh ginger is juicy, less pungent and more delicate in flavor with pink shoots, a thin skin and a crispy texture. The mature fresh ginger has a tough, shriveled skin with a desirable flesh underneath. Some types of ginger are peeled before sale.

Dried ginger is used bruised, sliced or powdered. Black ginger is created by scalding or steaming the rhizome and then drying it. White ginger is made by peeling the outer layers before washing and drying it. Preserved ginger is made from fresh young rhizomes that are peeled and sliced and then cooked in heavy syrup. Crystallized ginger is preserved ginger, which is fresh ginger that is peeled, cooked in sugar syrup, dried and rolled in sugar. Pickled ginger is fresh, peeled gingerroot that is thinly sliced and pickled in a vinegar solution, which gives it a pink color.

The pink ginger bud (or torch ginger bud), called bunga kantan in Malaysian lends a floral and aromatic sensation to foods. This torch ginger bud is from the Etlingera elatior family.

Properties: ginger's flavor and color vary with its origin and harvesting, storage and processing conditions. Fresh ginger has a juicy, spicy, refreshing and slightly sweet, lemon-like aroma, along with a strong bite. It is more aromatic than dried ginger. Drying conditions change ginger's flavor and pungency. Fresh ginger is firm and plump, not shriveled, and its color varies from tan to pale brown. Dried ginger is fibrous and has less pungency than fresh ginger. Aged or older ginger is fibrous, tough and harsher in taste, whereas young ginger is tender and delicate in taste.

Jamaican ginger is light tan in color and has a delicate aroma. It is more pungent than African ginger, which is darker in color and has a weak aroma and a harsh flavor. Cochin and Calicut (South Indian) gingers are strongly aromatic and pungent with a lemon aroma. They are considered the best gingers. Japanese and Chinese ginger are weaker in aroma than the Cochin ginger. Chinese ginger is whiter in color, more fibrous, slightly bitter and is usually used in the preserved form.

Fresh ginger contains magnesium and potassium, whereas ground ginger contains calcium, potassium, phosphorus, magnesium, sodium, niacin, vitamin A and manganese.

Chemical Components: the essential oil from ginger is pale yellow in color and ranges from 1% to 4%, depending upon the variety. The oil gives ginger its characteristic aroma but not its bite. The chief aromatic constituents are zingiberene (70%), curcumene, α-pinene, sabinene, limonene, borneol, linalool, farnesene and citral. Ginger's bite or pungency is due to non-volatiles such as gingerol, zingerone, shogaol and paradol. Fresh ginger (depending on the variety) is the most pungent because of high levels of gingerol. Dried ginger has less pungency because it has less gingerol, and changes occur in its shogaol, zingerone and paradol contents.

Oleoresin ginger is dark brownish green in color and is extracted from dried ginger. It contains 20–30% volatile oil, 10% fixed oil and 50–70% pungent resinous constituents (mainly gingerol) that are responsible for its pungency.

How Prepared and Consumed: ginger can add flavor, round off flavor and enhance particular flavors. While ginger is used throughout the world, its primary popularity as a savory spice is in Asian countries such as India, Korea, China and Southeast Asia. In Asia, it is eaten fresh, dried, pickled, preserved in syrup and crystallized. Fresh ginger is peeled before use. Many Asians, especially the Chinese and Indians, like to freshly grate it into their meat, poultry, vegetable and fish dishes.

In Southeast Asia, ginger is used raw and grated or finely chopped for flavoring, tenderizing and preserving meat. East Asians and Southeast Asians like young ginger for its intense flavor. The Chinese use the fresh, pickled and preserved gingers to create spicy, sweet flavors with rice porridges (congee), soups and stir-fried meats and vegetables. Dried ginger is used for sauces and soups that require long simmering times.

In India and Sri Lanka, dried and fresh gingers are used frequently. Fresh ginger is fried or boiled for sauces and curries. Dried ginger is part of curry blends, stir-fries, snacks, teas and desserts. It has an affinity with garlic, spring onions, fish, soy sauce, chile peppers, oranges and vinegar. Ground ginger is popular in hot and cold beverages such as tea and ginger beer as well as baked goods and confections. Indian vegetarians use ginger abundantly to flavor many dishes.

The young green ginger, either pickled or crystallized, is popular in the Caribbean, Asia, Australia and Europe. Pickled ginger, sweet or hot, is commonly eaten as a snack in Southeast Asia and China. The Japanese serve thinly sliced pink or red pickled ginger as a garnish or as a side dish called gari to refresh the palate in between meals. It also accompanies their sushi dishes.

Europeans and North Americans traditionally preferred the dried, crystallized or preserved form of ginger as a table seasoning in cakes, puddings, cookies, gingerbread, pies, fish soups, marmalades, ice creams, sausages, frankfurters, puddings and teas. Today, fresh ginger is becoming more popular. Ginger has become a mainstream flavoring in North America to identify Asian-influenced applications.

Scandinavians still enjoy their gingerbread, ginger cakes and confections made with ginger. Dried ginger is commonly found in Mediterranean spice blends such as quatre epices, berbere, la kamut and ras-el hanout. Arabs use greater amounts of the dried ginger in many of their seasonings. Caribbeans use ginger in marinades, curries, stews, ginger beer and ginger ale.

The fragrant ginger flower called bunga kantan in Malaysia is used in Malay and Nonya (term given to female ancestry from Chinese traders and local Malay women) curries and laksas.

Spice Blends: curry blends, berbere, five spice, quatre epices, ras-el-hanout, stir-fry blends, congee blend, Nonya laksa blends, Nonya chicken porridge blend and Chinese soup blends.

Therapeutic Uses and Folklore: ginger is one of the oldest and more popular medicinal spices. It has a warming effect on stomachs, soothes digestion and is traditionally used to neutralize toxins, decrease acidity and increase blood circulation in the gastrointestinal tract. The Greeks prescribed ginger for stomach ailments and as an antidote to poisons.

The Chinese and Indians use it in a variety of remedies and consider it to be effective against motion sickness and nausea, to discharge phlegm and even to relieve pains during childbirth. The Chinese chewed on it to prevent motion sickness during their sea travels. Indians used ginger as a kidney and bowel cleanser and for blood circulation. In Asia, it is also brewed in teas to ease menstrual cramps, bloating and morning sickness.

Today, ginger ale is still a common home remedy for an upset stomach.

GRAINS OF PARADISE

Also called Melegueta or Malagueta pepper, guinea grain or guinea pepper, grains of paradise were prized as a spice and as a substitute for black pepper in Europe during the Middle Ages. Grains of paradise were combined with ginger and cinnamon to flavor wine called *hippocras*. Currently, they are not widely used around the world, except in northern Africa and in the western regions of Africa, such as Ghana, Benin and Nigeria. The Yorubas in West Africa call them atare.

Scientific Name(s): Aframomum melegueta or *A. grana paradisi.* Family: Zingiberaceae.

Origin and Varieties: grains of paradise are indigenous to the West African coast, along the Gulf of Guinea, from Congo to Sierra Leone. This West African coastal area is called the "Grain Coast" and owes its name to this spice. Grains of paradise are cultivated in the Ivory Coast, Sierra Leone, Guinea and Ghana, but mostly in Ghana. Grains of paradise are related to cardamom. They are grown on a small scale in India and the Caribbean, where they were brought in the 1800s on slave ships from West Africa.

Common Names: they are also called kewrerima (Amharic), hubub al janel (Arabic), graines de paradis/poivre de Guinee/malaguette (French), paradieskorner (German), grani de Meleguetta (Italian), rajskiye zyorna (Russian) and malagueta/grano de parai (Spanish).

Form: grains of paradise are small, dried, reddish brown, pyramid-shaped seeds with a shiny exterior. They are sold with or without an enclosed seedpod or fruit (60–100 seeds in a pod). The seeds are sold whole or ground.

Properties: the seeds have a whitish interior and a reddish brown exterior. They have a pungent, peppery taste with bitter notes and a faint odor that is similar to cardamom. They have less pungency than black pepper.

Chemical Components: grains of paradise contain 0.5% to 1% essential oil, which is yellow in color and contains mostly eugenol. Their mild pungency is due to paradol, shogoal and gingerol.

How Prepared and Consumed: grains of paradise go well with black pepper, ginger, cinnamon, clove, coriander and cumin. The seeds are usually ground and added toward the end of cooking in order to retain their aroma. Northern Africans use them to create aromatic stews and also add them with other spices to flavor their coffees. They are an important ingredient in the Moroccan spice blend, ras-el-hanout, and are added to tagines, lamb dishes, couscous and rice. In Tunisia, grains of paradise are used in a blend, called galat dagga (or Tunisian five spice), to flavor lamb and vegetable dishes, especially eggplant dishes. In West Africa, they are used to spice many vegetables such as pumpkin, potatoes, eggplant or okra.

In Europe, grains of paradise were used traditionally to spice wine and beer but are now used in vinegars and liqueurs.

Spice Blends: ras-el-hanout, galat daqqa, couscous blend and Nigerian vegetable stew blend.

Therapeutic Uses and Folklore: in West Africa, grains of paradise were used as aphrodisiacs and to relieve pain and cure fevers. They were also chewed on cold days to warm up the body.

HORSERADISH

A relative of turnip, cabbage and mustard, horseradish is a popular seasoning in central and northern European regions. The German name for horseradish, *meerettich,* means "more radish" and reflects horseradish's much stronger flavor than the ordinary radish.

Scientific Name(s): Armoracia rusticana. Family: Brassicaceae.

Origin and Varieties: horseradish is indigenous to eastern and northern Europe and the Mediterranean and is also cultivated in central Europe.

Wasabi is Japanese horseradish, which is a little stronger than horseradish.

Common Names: it also called fejel har (Arabic), mierikwortel (Dutch), raifort (French), meerettich (German), hazeret (Hebrew), barbaforte/rafano (Italian), seiyo wasabi (Japanese), raiz-forte (Portuguese), khrjen (Russian), ra'bano picante (Spanish) and pepparrot (Swedish).

Form: horseradish is a root with a brown outer skin and a fleshy white interior. Fresh forms are sold sliced, grated or shredded. Dried horseradish comes flaked, granulated or powdered. It is also pickled and canned. The leaves are broad and green.

Properties: the root by itself does not give much aroma, but releases a pungent, burning, mustard-like aroma upon grating, cutting or shredding, which quickly disappears. Horseradish has a sharp, acrid, burning, bitter taste.

Chemical Components: horseradish has about 0.2% to 1.0% essential oil, mainly sinigrin, sinigrin-derived allyl isothiocyanate, diallyl sulfide, phenyl propyl thiocyanate and phenethyl. Myrosinase enzyme acts on sinigrin to give allyl isothiocyanate, which gives horseradish its burning taste.

Horseradish has a high vitamin C (302 mg/100 gm) content.

How Prepared and Consumed: in Europe, the U.K. and the U.S., horseradish is used in cocktail sauces, prepared mustards, relishes, baked beans, salad dressings, meats and seafood. The leaves are used in salads and stews. The British use horseradish as a condiment that enhances the flavor of roast beef. In Austria, it is used with ham and grated apples as a relish. When mixed with ketchup, horseradish becomes a cocktail sauce or dip for seafood, such as shrimp and oysters. It pairs well with tomato, eggs, vinegar, mustard, cheese, soy sauce, scallions and garlic.

Since the allyl isothiocyanate is not heat resistant, horseradish is usually best served with cold dishes or when it is added after cooking. Vinegar and other forms of acids are added with horseradish to prevent the pungency from disappearing quickly.

Spice Blends: relish blend, cocktail sauce blend, salad dressing blend and mustard horseradish blend.

Therapeutic Uses and Folklore: horseradish was traditionally used by Europeans to treat gout, kidney stones, asthma and bladder infections. It was used in Europe to prevent scurvy before vitamin C was discovered. Grated horseradish mixed into a paste is a home remedy for chest congestion and stiff muscles because it brings blood to the surface of the skin and warms the skin.

Horseradish has antioxidant and antimicrobial properties.

JUNIPER

Juniper characterizes gin flavor and comes from the colder regions of Europe and North India. In ancient times, the berries, bark, root and other parts of the plant were used for medicinal purposes.

Scientific Name(s): Juniperus communis. Family: Cupressaceae.

Origin and Varieties: juniper is indigenous to the U.K. and is also cultivated in Italy, the U.S. and India.

Common Names: araar (Arabic), jeneverbes (Dutch), genièvre (French), wacholder (German), ar'ar (Hebrew), enebro (Spanish), enbar (Swedish), araar (Hindi), ginepro (Italian), einer (Norwegian), mozhzhevelnik (Russian) and abhal (Urdu).

Form: juniper is a bluish black, small, pea-shaped, ripe berry. The berry has three seeds inside. It is dried and used whole, crushed lightly or coarsely ground.

Properties: juniper's flavor is woody and astringent with sweet, lemon and pine-like overtones. It has a gin-like aroma with a slightly bitter aftertaste. Its flavor is released when it is lightly crushed.

Chemical Components: juniper generally has 0.5% to 2% essential oil, with sometimes as much as 2.5%. It is mainly made up of monoterpenes such as 80% α- and β-pinene, thujene, sabinene, α-terpineol, borneol, geraniol, camphene and traces of sesquiterpenes such as caryophyllene.

It contains calcium, vitamin A, potassium, phosphorus and magnesium.

How Prepared and Consumed: juniper is commonly used in England and Europe to flavor pork and remove the strong, gamey flavor from roasted or barbecued wild meat, such as venison, squab and wild boar. It is also used in liver pates, stuffings and German sauerkraut, with caraway and bay leaves. Juniper is added to cheeses, goose, pot roasts, pickled meats and seafood. It pairs well with black pepper, bay leaf, vinegar, cabbages and alcohol for marinades.

Spice Blends: bouquet garni, barbeque blends, sauerkraut blend and pickling spice blend.

Therapeutic Uses and Folklore: in ancient Europe, juniper was used to treat tapeworms, keep away evil spirits and clean the air. Long known as a diuretic, it was used to treat urinogenital tract infections and gonorrhea. Juniper was also used for gastrointestinal infections, gout and rheumatic pains. It is a home remedy for hangovers.

KAFFIR LIME (LEAF, FRUIT)

Kaffir lime is a very popular spice used in Southeast Asian foods, especially in Thai, Malaysian, Cambodian and Balinese cooking. The leaves and the rind of the fruit are similar in taste and have the strong, zesty, fruity notes that Southeast Asian cultures enjoy.

Scientific Name(s): Citrus hystrix. Family: Rutaceae.

Origin and Varieties: indigenous to Southeast Asia.

Common Names: leaf—Indonesian lime leaf and wild lime. It is also known as daun lemo (Balinese), daun limau perut (Malay), daun jeruk purut (Indonesian), bai makrut (Thai) and chanh sac (Vietnamese).

Fruit—lemo (Balinese), limettier herisse (French), kaffernlimette (German), kraunch soeuth (Khmer), limau purut (Malaysian, Indonesian), makrut (Thai) and truc (Vietnamese).

Form: Kaffir lime fruit is a dark green, knobby, pear-shaped, wrinkled fruit. The fresh rind of the fruit is sliced or grated for use in cooking. It can be used as whole pieces, chopped or julienned. Unlike the lemon, Kaffir lime fruit has virtually no juice. Kaffir lime leaves have a shiny green surface and two leaves attached together. They are sold fresh, frozen or dried.

Properties: both the fresh and dried leaves have a strong, pungent, floral and lemon-like aroma. The fruit also has a strong lemon-like flavor with bitter overtones. The dried fruit pieces are reconstituted and added to foods.

Chemical Components: S-citronellal is responsible for its distinct aroma. It also contains citral and limonene.

How Prepared and Consumed: Kaffir lime leaf characterizes Thai cooking and is used abundantly in soups, such as tom yom goong (hot and sour soup) and tom kha gai (coconut soup). It is also used with stir-fries, grilled fish or chicken, spicy salads, sauces and pork curries. The leaves are finely shredded and added to salads, while the rind is grated and added to sauces and curries. Kaffir lime pairs well with coconut, fish broth, squid, galangal, garlic, ginger, chile peppers, basil and mango. In Indonesia, Malaysia and western Cambodia, the grated fruit is used in fish and chicken dishes, while the leaf is finely shredded or added whole to minced fish and cooked sauces.

Spice Blends: tom yom goong blend, tom kha gai blend, yellow curry blend and spicy Thai salad blend.

KARI LEAF

Kari leaf (called kariveppilai in Tamil) is grown all over India and has been used for centuries in South India and Sri Lanka as a flavoring for curries, chutneys, vegetables and beverages. South Indian traders introduced it into Malaysia, Burma and Singapore. When the British were in India, they called it curry leaf after seeing it in seasoned sauces called kari by Tamils.

Scientific Name(s): Murraya koenigii. Family: Rutaceae (citrus family).

Origin and Varieties: indigenous to southern India and Sri Lanka, it is grown in Southeast Asia and in California and Florida in the U.S.

Common Names: curry leaf and curry patte. It is also called pindosin (Burmese), feuilles de curry (French), curryblatter (German), kari patta (Hindi), aley kari (Hebrew), daun kari (Indonesian/Malaysian), kareapela (Malayalam), suravi (Sanskrit), karapincha (Singhalese), kari veppilai (Tamil) and bai karee (Thai).

Form: kari leaf is very fragrant when used fresh, but it loses its flavor when dried. The fresh leaf is used whole, crushed or chopped.

Properties: the leaf has fresh, spicy, piney, curry-like notes, with a slight tangerine peel-like taste.

Chemical Components: the fresh leaf has about 2.6% essential oil, but there is a gradual decrease in volatile content with advancing maturity. The essential oils mainly consist of sabinene (34%), α-pinene (27%) and dipentene (16%) with β-caryophyllene, β-gurjunene, β-elemene and β-phellandrene.

Curry leaf has vitamin A (beta-carotene is 12600 IU/100 gm), calcium (810 mg/100 gm) and phosphorus (600 mg/100 gm). It also has high levels of oxalates (1.35%).

How Prepared and Consumed: it is an essential spice in South Indian, Sri Lankan and Malaysian curries, dals, samosas, dosai fillings, chutneys, snacks, sambars, soups, breads and vegetables. Kari leaf is popularly used in South Indian vegetarian and fish dishes and Sri Lankan meat and chicken curries. Kari leaf pairs well with mustard seeds, turmeric, ghee, cumin, coriander, fenugreek, dals, ginger, garlic, tomatoes and yogurt. It provides a certain zest to yogurt-based salad dressings and vegetable dishes, such as fried cabbage, lentils, beans, okra or eggplant. It is usually removed before the food is eaten.

Kari leaf also provides a distinct spicy flavor to cold dishes and buttermilk. It gives a more intense flavor and crunchiness when it is toasted in oil or ghee and added as a garnish to many vegetarian foods.

Kari leaf can be kept frozen or refrigerated in a plastic bag for about two weeks. Freezing better retains its flavor, but its color changes to black. To retain its fresh flavor, it is best not to remove the leaves from its branches until ready to use.

Spice Blends: curry blends, sambar podi, rasam podi, chutney blends and fish curry blends.

Therapeutic Uses and Folklore: the leaves, root and bark are used as medicinal aids in India. The leaves are used to help blood circulation and menstrual problems. The fresh leaves are taken to cure dysentery, and an infusion made of roasted leaves stops vomiting. It is also recommended for relieving kidney pains.

LEMON BALM

Lemon balm, called bosem in Hebrew (which means fragrance), originated in the Middle East. While it is a popular spice in Mediterranean cooking, lemon balm is valued more for its medicinal properties than as a seasoning.

The Latin word for lemon balm, *melisa* (which means bee) came from the Greek practice of planting this spice to feed bees.

Scientific Name(s): Melissa officinalis. Family: Lamiaceae (mint family).

Origin and Varieties: lemon balm is indigenous to Greece and the Middle East and is cultivated in Middle Asia, North Africa, Hungary, Egypt, Italy and the U.S.

Bergamot (Monarda didyma) or bee balm, which is used to flavor teas, is a relative of lemon balm.

Common Names: balm mint, sweet balm, common balm and true balm. It is also called habak (Arabic), badrang buye (Farsi), citronelle (French), zitronenmelisse (German), bosem (Hebrew) melissa (Italian), ning mon xian yu (Mandarin), limonnik (Russian) and balsamita maior/balsamo limon (Spanish).

Form: the fresh and dried lemon balm leaves are used whole or chopped.

Properties: they are light green with jagged edges. Lemon balm has a sweet lemon aroma and a slightly minty taste.

Chemical Components: it has about 0.1% essential oil, mainly citronellal, β-caryophyllene, geranial, including neral, citronellol, linalool and limonene.

How Prepared and Consumed: lemon balm is used as a garnish or is added to salads to give a lemony flavor. It goes well in teas, vinegars, stewed fruits, jellies, puddings and custards. It is also added to fish, poultry, eggs, salads and soups. Lemon balm pairs well with fresh fruits, so it is a common flavoring in fruit drinks, iced teas and fruit-based desserts. It can be made into a pesto or a sauce with olive oil, garlic, fruits and almonds.

Lemon balm is sometimes used to substitute for lemongrass and to intensify a lemony aroma in a dish that contains lemon juice.

Spice Blends: pesto blend, lamb stew blend and barbecue blends.

Therapeutic Uses and Folklore: since ancient times, lemon balm was used to heal wounds, sores and bee and wasp stings. Called the "elixir of life" in Europe, it was mixed into a drink to ensure longevity and to treat asthma, stomach ailments, indigestion, menstrual cramps and fevers.

Lemon balm has also been used traditionally to promote emotional well-being by relieving tension, helping depressive illnesses, calming nerves and preventing headaches.

LEMON VERBENA

Lemon verbena is a popular spice used in European teas, fruit drinks and desserts. It is indigenous to South America and was brought to Europe by the Spanish. It was originally used as a perfume.

Scientific Name(s): formerly *Lippia citriodora,* now called *Aloysia triphylla.* Family: Verbenaceae.

Origin and Varieties: it is indigenous to Central and South America and is cultivated in France, China and Algeria.

Common Names: lippia, cedronella, erba, verbena odorosacitronelle/verveine odorante (French), verbena (Hebrew), cedrina (Italian), verbene/zitronenstrauch (German), yerba de la princesa/vervena/verbena limon (Spanish).

Form: lemon verbena is long, pointed and is available as fresh or dried green leaves. They are used whole or chopped. The dried form retains its flavor.

Properties: lemon verbena has a fresh, strong, lemon-lime-like flavor, with a fruity and penetrating aroma.

Chemical Components: lemon verbena contains 0.1% to 0.2% essential oil, which is yellowish green. It is chiefly composed of citral, cineol, dipentene, limonene, linalool, borneol, geraniol and nerol.

How Prepared and Consumed: it is commonly used by Europeans to flavor fruit-based drinks, fruit salad dressings, soups, marinades, puddings, jams and desserts. It does not lose its flavor during cooking.

Spice Blends: fish marinade blend, soup blend and pickle blend.

Therapeutic Uses and Folklore: lemon verbena has been used traditionally by Europeans as a diuretic and a gout remedy, to treat inflammation of the liver or spleen and even to aid depression. It is also brewed in tea as a home remedy to relieve colds and fevers. Lemon verbena is a natural insect repellent.

LEMONGRASS

Lemongrass derives its name from its characteristic scent and appearance. It is a popular herb in Southeast Asian cooking including Thai, Malaysian, Singaporean, Vietnamese, Cambodian, Indonesian and Sri Lankan.

Scientific Name(s): Cymbopogon citratus (Southeast Asian type); *C. flexuosus* (Sri Lankan, Indian types). Family: Poaceae (grass family). It is a relative of citronella.

Origin and Varieties: lemongrass is indigenous to Southeast Asia and South Asia and is cultivated in India, Australia, the Caribbean, Central America and the U.S. There are many species, but the two main varieties are the South Asian and Southeast Asian types. The latter is also cultivated in the West Indies.

Common Names: Indian verbena, ginger grass and Java citronella. It is also called zabalin (Burmese), sitruunaruoho (Finnish), zitronengras (German), sera (Hindi), sereh (Indonesian), citronella (Italian), bai mak nao (Khmer), serai (Malaysian), xiang mao cao (Mandarin), capim-santo (Portuguese), zacate de limon (Spanish), tanglad (Tagalog), servi-pillu (Tamil), ta khrai (Thai) and xa (Vietnamese).

Form: lemongrass has a long stalk (or stem), about 4–6 inches long, with pale green, slender, long leaves at the top and a bulbous base that is woody and fibrous. The hard, fibrous outer sheaths of the lemongrass stalk are peeled and discarded, and the delicate inner stalk is used. It comes fresh, frozen or dried. The fresh form is whole, sliced, finely chopped or coarsely pureed, while the dried form is whole, shredded or ground. The fresh stem is bruised or macerated to release its flavor. The dried forms need to be soaked in hot water before use, otherwise they are woody and fibrous. The ground form is added directly into cooking.

Properties: its flavor is found in the lower tender part of the stem, about 4 inches from root end to 3 inches from the upper leaves. The root-end stalk gives the most flavor. The whole fresh stalk becomes aromatic when it is crushed or cut. It gives a refreshing lemon-lime-like taste with a tinge of mint and ginger. It has a delicate floral-like (rose) and fresh, grassy aroma. The dried form has very little aroma or taste.

Chemical Components: the essential oils in lemongrass vary with the variety. *C. citratus* has 0.3% to 0.4% essential oil with 70% citral; and *C. flexuosus* has 0.2% to 0.5% essential oil with 80% to 85% citral. The other components are myrcene (14%), geranyl acetate (3%), methyl heptenone (2%) and linalool (1%).

Lemongrass contains calcium, vitamin A, iron, potassium, magnesium, phosphorus, sodium and manganese.

How Prepared and Consumed: lemongrass enhances many ingredients and does not dominate the flavor profile of a dish. It pairs well with garlic, galangal, shallots, cilantro, turmeric, kaffir lime leaves, candlenuts, ginger, chicken, pork, fish and chile peppers. Lemongrass is used whole in soups or is chopped and pounded for use in soups, stews, curries, laksas, rendangs and condiments of Thailand, Indonesia, Vietnam and Malaysia.

Lemongrass is an important flavoring ingredient in Southeast Asian spice blends, especially in Nonya-style dishes of Malaysia and Singapore as well as in Thai and Indonesian cooking. Malay and Thai curry pastes and Nonya laksas frequently contain lemongrass. It provides a wonderful aroma to cooked rice containing coconut milk. In Southeast Asia, lemongrass is added to many local desserts that have rice flour, gula melaka (dark palm sugar) and coconut milk.

The stalk is sliced into small rings and added to sauces and curries, or the stalk is cut into pieces, crushed with the back of a knife or spoon and added to flavor soups or stews. The whole lemongrass stalk can be tied into a knot and added to soups or stews to give them an aromatic, fresh, citral flavor. It is discarded before serving. When cooked, lemongrass imparts a fresh floral, citrus taste to foods. In Malaysia, whole lemongrass is made into a skewer or a brush to add marinade to meat and chicken satays (barbecues). In Indonesia, it is combined with turmeric, chile peppers and other spices to make bumbu, a seasoning used to flavor the local sauces and soups.

Spice Blends: sambal blends, nasi lemak blend, rendang blend, satay sauce blend, bumbu blend, assam fish blend and Malay chicken curry blend.

Therapeutic Uses and Folklore: in Southeast Asia, lemongrass is traditionally used to improve blood circulation and ease muscle pain. It is also used in teas to cure loss of appetite and reduce fever. In Thailand, lemongrass is used as a smelling salt and to relieve headaches.

LOVAGE

Lovage belongs to the parsley family, and its seeds, leaves and roots are commonly used in Europe for flavoring foods and beverages and for their medicinal properties. The Romans introduced lovage to Europe and used it to reduce fevers and treat stomach ailments. Germans called it maggikraut because its aroma reminded them of maggi cubes (meaty yeast extracts).

Scientific Name(s): Levisticum officinale. Family: Umbelliferae (Apiaceae-parsley family).

Origin and Varieties: true lovage is native to western Asia, southern Europe and the U.S. It is also cultivated in Germany, Italy, France, Czechoslovakia and Hungary. There are two other types of lovage that grow wild. One variety, called sea lovage, Scottish lovage or shunis, grows in northern Britain and along the north Atlantic coasts of the U.S. The other type, called black lovage or alexanders, grows in Britain and around the Mediterranean.

Common Names: love parsley, garden lovage, Italian lovage, true lovage, maggi herb and old English lovage. It is also called habak (Arabic), lavas (Dutch), liveche (French), maggikraut/liebstockl (German), levistico (Italian), tuho cao (Mandarin), haulopstikke (Norwegian), ljubistok (Russian), ligustico (Spanish) and libsticka (Swedish).

Form: lovage leaves, seeds, stems and roots are used in cooking, but the leaves are the most commonly used flavoring. Lovage has green, serrated leaves and hollow stems that are sold fresh, dried, frozen or crystallized. The leaves, which resemble celery leaves, can be used whole or chopped. The younger leaves are smaller in size.

The seeds are tiny, ridged, crescent shaped, brown and aromatic. The roots are slightly thick and fleshy with a greyish brown color.

Properties: the fresh leaves have a yeasty and musky taste with a lemon and celery-like aroma. The dried leaves have a stronger flavor than the fresh leaf.

Chemical Components: the fresh leaf has 0.1% to 0.2% essential oil, while the dried leaf has 0.5% to 1% essential oil, which is yellow amber to greenish color. It consists mainly of phthalides (ligustilide, butylphthalide, sedanolide) with lesser amounts of α-terpineol, eugenol and carvacrol.

How Prepared and Consumed: ancient Greeks and Romans commonly used the seeds, leaves and roots in their cooking. Today, lovage is a favorite flavoring in Britain and southeastern Europe. It is eaten cooked or raw. The leaves are used in soups, stocks, flavored vinegars, stews and salads. In Italy, lovage is used with oregano and garlic for tomato sauces. The seeds are sprinkled over salads and mashed potatoes and are crushed for breads, pastries, biscuits and cheeses. The stems and stalks are chopped for use in sauces and stews, while the crystallized leaves and stems are used for decorating cakes. The roots are peeled to remove the bitter skin and are then used as a vegetable or are pickled.

Spice Blends: tomato sauce blend, soup blend, stew blend and stock blend.

Therapeutic Uses and Folklore: Europeans traditionally use lovage as a digestive stimulant, for stomach upsets, water retention and skin problems. It was also taken to treat poor circulation and menstrual irregularities.

MACE

Called "flower of nutmeg" in France, mace was first introduced into Europe during the eleventh century by Arab traders to flavor beer. Mace's popularity as a spice reached its height during the seventeenth century when the Dutch monopolized the nutmeg trade in the Spice Islands. Mace and nutmeg are from the same fruit, mace is the lacy covering of the seed, which is the nutmeg (with a ratio by weight of 1:25). They have different color and flavor profiles.

Scientific Name(s): Myristica fragrans. Family: Myristicaceae (nutmeg family).

Origin and Varieties: mace is native to the Banda Islands in eastern Indonesia (Moluccas) and is cultivated in the Banda Islands, Grenada, the Caribbean, South India, Sri Lanka, Malaysia, Sumatra and Brazil.

Common Names: basbasa (Arabic), foelie (Dutch), fleur de muscade (French), muskatblute (German), mess (Hebrew), javitri (Hindi), sekar pala (Indonesian), macis (Italian), kembang pala (Malaysian), jathipathri (Malayalam/Tamil), tau kuo (Mandarin), sushonaya shelukha (Russian), wasa vas (Singhalese), macia/maci (Spanish), muskotblomma (Swedish), dok chand (Thai) and dau khau (Vietnamese).

Form: mace is the deep red, lacy, net-like, leathery covering of the nutmeg seed called aril or mace blades. These blades are removed and dried in the sun, become brittle and assume a red color. Mace is used whole, broken or ground. It retains its flavor better than nutmeg when it is ground. There are different grades of mace based on regional variations.

Properties: mace blades are smooth and shiny and can be up to about 4 cm. long. Mace can be pale orange, yellowish brown, orange or reddish brown in color. The color and flavor of mace varies depending on its origins—reddish orange from Indonesia and brownish yellow from the West Indies. It is spicy and bitter with clove-like and piney overtones. Its aroma is terpeny. Mace is more aromatic than nutmeg but has more bitter notes. Ground mace is lighter in color than ground nutmeg.

Chemical Components: depending on its origins, mace has 7% to 14% essential oil and about 30% fixed oil. It contains the same aroma compounds as nutmeg but in smaller amounts, mainly monoterpenes (87.5%), monoterpene alcohols (5.5%) and other aromatics (7%). Like nutmeg essential oil, the main constituents of mace essential oil are sabinene, α-pinene, β-pinene, myrcene, limonene, 1,8-cineole, terpinen-4-ol, myristicin, γ-terpinene and safrole. Mace oil is more expensive than nutmeg oil.

The mace oleoresin is amber to dark red in color; 7 lbs. of mace oleoresin are equivalent to 100 lbs. of freshly ground spice. Mace butter, which has fixed oils and volatiles, has 60% unsaturated fats and 40% saturated fats.

Ground mace contains vitamin A, phosphorus, potassium, magnesium, sodium and calcium.

How Prepared and Consumed: in Europe and the U.S., mace is usually used in light-colored products, such as cream soups, cream sauces, chowders, crackers, pie fillings and cakes. Mace is also used with fish and in vegetable purees, meat stews and meat pies. Commercially, mace flavors frankfurters, doughnuts, pickles, preserves, ice cream, confections, icings, sausages, knockwurst, ham, soup mixes and poultry.

Mace pairs well with fruit, sugar, chocolate and milk-based products—cakes, cookies and doughnuts. It provides an intense aroma to Middle Eastern and Asian foods. In Middle Eastern, Iranian and northern Indian recipes, mace is often combined with nutmeg. Arabs add mace to mutton and lamb dishes and many spice blends such as ras-el-hanout, baharat and galat dagga. It is ground and sprinkled over North Indian pulaos, lamb and other meat dishes to give aroma. Ground mace is commonly used in Southeast Asian, Chinese and Indian foods such as garam masalas, curries, sauces, puddings, cakes, pies and cookies.

Spice Blends: curry blends, garam masala, ras-el-hanout, baharat, galat dagga and rendang blend.

Therapeutic Uses and Folklore: Asian Indians have traditionally treated stomach pains, dysentery, vomiting and the symptoms of malaria with mace. It is also chewed to prevent foul breath.

MARJORAM

In Europe, marjoram was a traditional symbol of youth and romantic love. During the Middle Ages, it was used to cast love spells and was worn at weddings as a sign of happiness. The Greeks called marjoram "joy of the mountains," and wore marjoram wreaths at weddings. A cousin of the oregano family, marjoram originated in northern Africa but is now a commonly used spice in many parts of Europe.

Scientific Name(s): sweet marjoram: *Origanum hortensis* (or *O. majorana hortensis*). There are a few varieties of sweet marjoram. There are also many wild varieties, the main one being pot marjoram: *Origanum onites* (or *O. vulgare*). Family: Labiatae or Lamiaceae (mint family).

Origin and Varieties: marjoram is indigenous to northern Africa and southwest Asia. It is now cultivated around the Mediterranean, in England, Central and Eastern Europe, South America, the U.S. and India.

Common Names: sweet marjoram, knotted marjoram and annual marjoram. It is also known as marzanjush (Arabic), marjolein (Dutch), avishan (Farsi), marjolaine (French), majoran (German), myoran (Hebrew), marwa (Hindi), maggiorana (Italian), mayorana (Japanese), maruvammu (Malayalam), ma yueh lan hua (Mandarin), manjerona (Portuguese), mayoran (Russian), mejorana (Spanish), mejram (Swedish), maruvu (Tamil) and marva kusha (Urdu).

Pot marjoram: rigani, wild marjoram, common marjoram, oregano, French marjoram, golden marjoram, curly marjoram, gold splash marjoram, amaraco and marzanjush (Arabic).

Form: marjoram is used fresh, as whole leaves or chopped, and dried, as whole leaves or broken, and ground.

Properties: sweet marjoram is small and oval shaped. It is light green with a greyish tint. Marjoram is fresh, spicy, bitter and slightly pungent with camphor-like notes. It has the fragrant herbaceous and delicate, sweet aroma of thyme and sweet basil. Pot marjoram is bitter and less sweet.

Chemical Components: sweet marjoram has 0.3% to 1% essential oil, and is yellowish to dark greenish brown in color. It mainly consists of *cis*-sabinene hydrate (8–40%), γ-terpinene (10%), α-terpinene (7.6%), linalyl acetate (2.2%), terpinen 4-ol (18–48%), myrcene (1.0%), linalool (9–39%), ρ-cymene (3.2%), caryophyllene (2.6%), and α-terpineol (7.6%). Its flavor varies widely depending on its origins. The Indian and Turkish sweet marjorams have more *d*-linalool, caryophyllene, carvacrol and eugenol.

Its oleoresin is dark green, and 2-1/2 lbs. are equivalent to 100 lbs. of freshly ground marjoram.

Marjoram contains calcium, iron, magnesium, phosphorus, potassium, sodium, vitamin A, vitamin C and niacin.

How Prepared and Consumed: marjoram is typically used in European cooking and is added to fish sauces, clam chowder, butter-based sauces, salads, tomato-based sauces, vinegar, mushroom sauces and eggplant. In Germany, marjoram is called the "sausage herb" and is used with thyme and other spices in different types of sausages. It is usually added at the end of cooking to retain its delicate flavor or as a garnish. It goes well with cabbages, potatoes and beans.

Greeks use pot marjoram in grilled lamb and meats and to complement onions, garlic and wine. Italians use it in tomato sauces, pizzas, fish dishes and vegetables.

North Africans and Middle Easterners use marjoram in lamb, mutton, barbecues, vegetables and seafood. In the U.S., it is used commercially in poultry seasonings, liverwurst, bologna, cheeses, sausages, soups and salad dressings.

Spice Blends: bouquet garni, fines herbes, sausage blend and pickle blends.

Therapeutic Uses and Folklore: Greeks used marjoram extensively to treat dropsy, convulsions and poisons. Traditionally, it was used in tea to cure headaches, calm nervous disorders and to clear sinuses. Marjoram has also been used to comfort stomachaches and muscular pains and improve circulation.

MINTS: SPEARMINT AND PEPPERMINT

Spearmint was the symbol of hospitality in traditional Europe, where it was crushed and used in baths and other leisure places. It was called Herba Santa Maria in Italy and Our Lady's mint in France. The Greeks used it as an aphrodisiac, while the Romans used its aroma as an appetite stimulant. Today, spearmint is commonly used in chutneys, curries and sauces in India and Southeast Asia. It is also a popular flavor for teas around the world.

Peppermint was discovered in the seventeenth century in England as a wild plant and was mainly used as a medicinal infusion. The familiar "after dinner mints" originated in England to soothe the stomach after a meal. Nowadays, it is widely used in confectionery, liqueurs and chewing gums.

Scientific Name(s): peppermint: *Mentha piperita;* spearmint: *Mentha spicata;* cornmint or Japanese mint: *Mentha arvensis*. Family: Labiatae.

Origin and Varieties: peppermint is native to southern Europe and is now cultivated in northern Africa, Russia, the U.S., India and England. Spearmint is cultivated in China, Southeast Asia, India and Japan. There are many varieties and cultivars, and some plants that are called mints are not real mints but tend to have the flavor of mints like Vietnamese mint or lemon mint. Spearmint (Moroccan, perilla shiso) and peppermint (white and black) are the two most commonly used mints. Other mints (and so-called mints) are lemon mint, apple mint, basil mint, Vietnamese mint (or polygonum/water pepper), pineapple mint, ginger mint, orange mint or lavender mint.

Common Names: for Mint: naana (Arabic), menthe (French), podina (Hindi), menta (Italian, Portuguese), myata (Russian) and hierba buenas/mentas (Spanish). Shiso is also called Japanese basil. Perilla leaf (la tia to in Vietnamese) is sometimes referred to as shiso in the U.S. because of its similar flavor.

Peppermint: naana (Amharic), ruh alnaana (Arabic), pudina (Bengali), pepermunt (Dutch), na'ana (Hebrew), pudina desi (Hindi), daun pudina (Indonesian, Malaysian), menthe poivre (French), pfefferminze (German), menta pepe (Italian), seiyo hakka (Japanese), bo ho (Mandarin), mjata pjerechnaya (Russian), meenchi (Sinhalese), menta/piperita (Spanish), pepparmynta (Swedish) and pothina (Tamil).

Spearmint: naana (Amharic, Arabic), menthe verte (French), grune minz (German), menta meshubelet (Hebrew), pahari podina (Hindi), daun pudina (Indonesian, Malaysian), mutthina (Malayalam), mentastro verde (Italian), menta verde (Spanish), pothina (Tamil), bai saranae (Thai) and rau thom/hang que (Vietnamese).

Form: mint is available fresh or dried. The dried form is sold whole, as flakes, chopped and fine or coarse. Fresh, it is eaten raw, pureed or cooked. The leaves can also be crystallized.

Properties: peppermint has bright green leaves, with fresh, slightly sweet, tangy and strong menthol notes. Menthol gives peppermint its cooling sensation. Spearmint has a slightly pungent, warm, fresh and herbaceous taste, with lemony and sweetish notes. Spearmint does not have menthol and so does not have the cooling sensation of peppermint.

Other so-called mints: Vietnamese mint has large purplish aromatic leaves; shiso has a decorative mint- and basil-like flavor, and perilla, which is similar to the shiso, has a lemony mint-like flavor. When bruised, these mints lose their flavor easily.

Chemical Components: peppermint has 0.5% to 5% essential oil that is pale yellow. It mainly consists of menthol (26–46%), menthone (16–36%), menthyl acetate (3.8%–7%), menthofurane (2–8%), isomenthone (2–8%), limonene (2.5%), pulegone (1.4–4%) and β-pinene (1.5–2%). Menthol and menthyl acetate are responsible for the refreshing and cooling pungent odor and are mostly found in the older leaves.

Spearmint has 0.5% essential oil, mainly 50–70% carvone and dihydrocarvone, including dihydrocuminyl acetate, dihydrocuminyl valerate, phellandrene, limonene, menthone, menthol and 1,8-cineol.

Oil of cornmint is referred to as mint oil in the U.S. and is blended with peppermint oil because it is less expensive. Cornmint oil has 28–34% menthol, 16–31% menthone, 6–13% isomenthone, 5–10% limonene and a higher content of α- and β-pinenes.

Dried peppermint contains calcium, potassium, vitamin A, niacin, sodium, magnesium, phosphorus, vitamin C and iron. Mints contain vitamin A (2700 IU/100 gm), iron, phosphorus and other minerals.

How Prepared and Consumed: Greeks and Romans used peppermint in condiments, cordials and fruits. Europeans use it in sweet products such as desserts, candy, jams, jellies, chocolates, cordials and liqueurs. In Europe, the crystallized leaves are also used as decorations in cakes and pastries. Today in the U.S., peppermint is mainly used in bakery products, teas and confectionery.

In England, spearmint is used in savory products, such as sauces for roast lamb, boiled mutton, peas, potatoes and in teas, chocolate and vinegar. Middle Easterners use chopped spearmint in many dishes—yogurt dressings, dips, salads, vegetables, grilled lamb, poultry, fish and in teas. Dried leaves are popular in many Middle Eastern cuisines including Turkish and Iranian cooking. It is used in dry blends, marinades and beverages. In Southeast Asia, Vietnamese, Malaysians and Thais use fresh spearmint as garnishes in salads (along with basil and cilantro), in fish and poultry curries, seasoned rices, sauces, biryanis and soups. The polygonum type of mint is very popular in Vietnam and is always used fresh as a garnish for nearly every southern Vietnamese dish, for soups, salads, spring rolls, stir-fries, dips and as food wrappers.

In India, spearmint is ground with coconut, green chili, onion and green mango to flavor green or chat masalas for chutneys and curries and is added with yogurt and cucumber in raita as dips and drinks. Peppermint is normally used to enhance the Mogul-style biryanis of North India.

Spearmint pairs well with cardamom, shallots, basil, cilantro, lemongrass, green chiles, oranges, lime, lamb, chocolate, green papaya, yogurt and green mango. Normally, mints are added at the end of cooking because heat rapidly dissipates their flavor.

Spice Blends: chat masala, Malaysian kurma blend, podina chutney blend, tabbouleh blend, mint sauce blend and Thai green curry blend.

Therapeutic Uses and Folklore: mints were used as a symbol of hospitality in many cultures. They are also used as a general body cleanser and strengthener. Mint has been used to relieve cold and flu symptoms, fevers, headaches, muscular aches, sore throats and toothaches. Peppermint tea helps relieve nausea and is taken to relieve seasickness. It is also used to promote digestion.

Mint has antibacterial properties.

Vietnamese mint: although called mint, it does not belong to the mint family. In Malaysia and Indonesia, it is called daun laksa (or leaf) because it is an essential flavoring ingredient in the stew-like pungent, hot noodle dishes called laksas created by the Nonyas of Malaysia and Singapore.

Scientific Name(s): Polygonum hydropiper. Family: Polygonaceae.

Common Names: it is also called polygonum or water pepper. Its other names are pakarmul (Bengali), laksa yip (Chinese), poivre d'eau (French), wasserpfefer (German), daun kesom (Malay), vodyanoii peretc (Russian), pimienta acuatica (Spanish), bitterblad (Swedish) and rau ram (Vietnamese).

Properties: it has a clean, pungent, minty taste.

Chemical Components: its essential oil components are 1,4-cineol, borneol, carvone, fenchone, ρ-cymene, poligonone, terpineol, cinnamic alcohol and cinnamic acid esters. A sesquiterpene, tadeonol, is responsible for its pungent taste, and rutin is responsible for its bitter notes.

How Prepared and Consumed: Vietnamese mint is commonly used in Vietnam, Singapore and Malaysia. The Nonyas of Singapore and Malaysia add chopped fresh mint with fermented shrimp paste, turmeric, galangal and lemongrass to lend a unique flavor to their spicy soups, curries, laksas and condiments. It is an important spice in Vietnam, where whole fresh leafy spices are a significant part of preparing or garnishing a meal or dish. Vietnamese mint leaves are sometimes used to wrap beef, seafood or vegetables that are cooked or served as salads.

MUSTARD

Mustard seed was a symbol of fertility for the ancient Indians and has been used by Greeks, Chinese, Indians and Africans since ancient times. Mustard paste making was introduced to central and northern Europe by the Romans. The Romans ground mustard seeds, which were called sinapis, with grape must to make the first table mustard, which they called mustum ardens meaning "burning or hot must." Eventually, mustum ardens became mustard, and before long, the seed took this name as well. The prepared mustard condiment was first made by the French in Dijon during the thirteenth century, and mustard flour was first made during the eighteenth century in England. Mustard oil is a popular cooking oil in North India.

Scientific Name(s): there are three types of mustard seeds: pale yellow or white mustard (*Brassica/Sinapis alba* or *B. hirta* Moench); brown and oriental mustard (*Brassica juncea*) and the black or dark brown mustard (*Brassica nigra*). Family: Brassicaceae (cabbage family).

Origin and Varieties: yellow or white mustard is indigenous to southern Europe and western Asia; brown mustard is indigenous to northern India, China and Africa; and black mustard is indigenous to Iran and India. These mustards are also cultivated in Nepal, Russia, Canada, southern Italy, northern Africa and Central and South America. Black mustard is not as popular in the U.S. or Europe because it is difficult to harvest. Brown and white mustards are the most popular types used in the U.S., but black and brown are the most popular in Asia.

Common Names: mustard (in general), khardal (Arabic), mosterd (Dutch), moutarde (French), senf (German), hardal (Hebrew), rai (Hindi, Urdu), senape (Italian), karashi (Japanese), biji sawi (Malaysian, Indonesian), chieh kai (Mandarin), mostarda (Portuguese), gorchitsa (Russian), mostaza (Spanish), senap (Swedish) and kadugo (Tamil).

Yellow/white mustard seeds: sufed rai (Hindi), moutarde blanche (French), senape biancha (Italian), biji sawi putih (Malaysian, Indonesian) and mostaza blanca/mostaza silvestre (Spanish).

Brown mustard includes two varieties: one called oriental, used mostly by the Chinese, and the other a darker, stronger brown mustard that is used by Indians. There are many cultivars of these different brown types.

Black mustard seeds are called true mustard or Indian mustard. They are also known as senafich (Amharic), zwarte mosterd (Dutch), moutarde noir (French), rai (Hindi), senape nera (Italian), biji sawi hitam (Malaysian, Indonesian), mostarda preta (Portuguese), abba (Singalese) and mostaza negra (Spanish).

Form: mustard seeds can be large yellow seeds, medium-sized brown to dark brown seeds or small black seeds, depending on the variety. The seeds are used whole, crushed, ground or as flour. Mustard also comes in wet or prepared paste forms with water, vinegar, sugar, oil, spices and herbs called prepared mustards. Ground mustard is made from yellow or brown whole seeds.

Mustard oil is the fixed oil in mustard seed and has little pungency. It is pale yellow in color and has a raw, pungent and bitter taste with an unpleasant aroma, but it becomes pleasant and sweeter during cooking. Mustard meal (that includes bran) is made mostly from yellow mustard and is sometimes blended with brown mustard. The seed is ground with the bran. Mustard flour/powdered mustard is usually a blend of yellow or brown mustard seeds. Their proportions depend on the region. Mustard seeds are milled to remove the bran. A wide variety of flours are available based on volatile oil and particle size. Prepared mustard is a smooth condiment made from mustard seeds, salt, spices and vinegar.

Mustard leaves, called mustard greens, have a radishy taste and are used as a prepared vegetable or are put into salads. The most common mustard greens are from the brown mustard variety and are commonly eaten in the southern U.S.

Properties: the smaller and darker the seed, the more intense and hot the flavor is. Black mustard is oblong and is the smallest in size. It has the sharpest flavor along with a nutty aftertaste. The brown mustard seed is spherical and medium in size and has a nutty, sweeter and mellower burning flavor. It has a taste sensation like horseradish. The white or yellow mustard is largest and has a delicate flavor that is the least burning.

The whole seed has no flavor, but can provide a pungent taste after chewing for some time. The heat experienced in yellow mustard is on the tongue, but in brown and black mustard, the heat is also felt in the nose and eyes. This latter pungency is more intense and lasts longer than the former, though bite does not build up.

The pungent aroma varies among the different mustards. The white or yellow type has a less pungent aroma than the brown mustard seeds that have a very pungent aroma. The black mustard seeds have the most pungency.

Ground mustard has no aroma. Its flavor and pungency is experienced by triggering an enzyme action that releases mustard's flavor or pungency.

Mustard pungency is due to a variety of isothiocyanate compounds that exist in mustard tissue as glycosides. The major pungent compound of black and brown mustard is allyl isothiocyanate. The release of sensation, especially in brown and black mustard, is delayed and begins at the back of the mouth, with a shooting sensation to the sinuses. This is due to the activation of an enzyme, myrosinase, that, in the presence of water, breaks down the glycoside (sinalbin) in yellow mustard or (sinigrin) in black or brown mustard to para-hydroxybenzyl isothiocyanate or allyl isothiocyanate, which gives the characteristic pungent aroma. The odor lasts until all enzymatic activity has ceased.

Acids are poor triggers of mustard's overall flavor, but they extend mustard's penetrating odor. Heat stops the flavor release, so it is important to let the mixture of mustard and water stand for 10 minutes before it is added to cooking. When water, vinegar, milk, wine or beer is added to mustard, mixed and left to stand for a few minutes, different degrees of flavor sensations are produced. With water, a very sharp and hot taste is produced; with vinegar, a milder flavor occurs; with milk, a milder, spicier and pungent flavor is created; and with beer, a very hot flavor is produced.

Whole mustard seeds are toasted in heated oil or are "popped" to give a nutty, sweetish aromatic flavor.

Mustard is a natural physical emulsifier and binder for hot dogs, salad dressings and many sauces.

Chemical Components: mustard is high in fat (35%) and protein (28%). Mustard seed is cold pressed to extract the fixed oil from the press cake. The press cake is hydrolyzed by enzymes and is steam distilled or solvent extracted to give the essential oil. The yellow seed has 2.5% sinalbin and very little volatile oil (below 0.2%). When the cells are damaged, sinalbin is hydrolyzed by enzyme myrosinase to produce the ρ-hydroxybenzyl-isothiocyanate. The brown and black seeds have 1% sinigrin and 0.5% to 1.2% volatile oil, mostly 95–99% allyl isothiocyanate. Certain types of brown mustards contain up to 2.9% essential oil.

Mustard seeds have 28–35% fixed oil, called mustard oil that is pale yellow in color and contains linoleic, linolenic and 20–30% erucuc acids. Oleoresin mustard is yellow to light brown in color, and 4-1/2 lbs. are equivalent to 100 lbs of ground yellow mustard.

Generally, mustard seeds contain calcium, potassium, magnesium, phosphorus and niacin. Also, when mustard is ground with its bran, it is high in protein.

How Prepared and Consumed: yellow mustard is mostly used for preparing mustard pastes or condiments with wine, vinegar and other spices. The vinegar in mustard paste stabilizes and mellows its pungency, while wine extends and enhances its pungency. Only a small amount of black mustard is used in preparing condiments because its pungent principle is very volatile. Therefore, yellow mustard is preferable for condiments.

Mustard seeds were sprinkled over foods by the ancient Greeks and Romans. The early French mixed it with grape must or honey and vinegar. The yellow prepared mustards have sharp tongue tastes but not pungent aromas because of their lesser pungent principles. German, Dijon, English and Chinese prepared mustards contain brown or oriental ground mustard that have more pungent aromas and bites.

Dry Coleman's mustard and the North American hot dog mustards contain the yellow type of mustard. For English mustard, brown mustard is mixed with yellow mustard; German mustard contains brown mustard. Dijon has a sharp taste and contains mainly brown mustard and spices. Chinese mustard is powdered ground brown mustard mixed with water and other ingredients. It is used as a table condiment in Chinese restaurants and for other ethnic foods.

Mustard goes well with cold meats, sausages, grilled steak, poultry, fish, herbs, wine, garlic, sauerkraut and fruits. Today, there are many customized prepared mustards that use different spices, fruits, wine and chile peppers.

Ground yellow mustard acts as a physical emulsifier and stabilizes mayonnaise and salad dressings. It is also used as a flavor enhancer and a water binder in processed meats. In the U.S., yellow mustard is commonly used as a spread for hot dogs, hamburgers and sandwiches and luncheon meats, while the seeds are used as toppings on cooked vegetables. It is an important flavoring in baked beans, combined with brown sugar, ketchup, beans, onions and bacon.

Whole yellow mustard seeds are used in pickled condiments, not only for flavor but also for their preservative function. The paste form is popular in salad dressings and hot sauces.

The English enjoy brown mustard with roast beef and ham. The Japanese use the oriental brown variety as a dip for raw fish. The Barbadians and other populations in the Caribbean use yellow or brown mustard with fruits and chile peppers for great tasting sauces, marinades and stews. In Indian cooking, especially in southern vegetarian meals, whole brown or black mustard seeds are "popped" in heated ghee or oil to bring out their nuttiness, and they are then added to sauces, chutneys, pickles, curries, sambars and dals. Black mustard is sometimes used to flavor ghee in South India. Ground mustard seeds provide flavor and consistency in Bengali fish curries.

Fixed oil from the brown and black seeds is used in many northern Indian dishes and pickles, especially in Bengali, Kashmiri and Pakistani cooking. The raw, pungent notes of this oil dissipate with cooking, and it becomes sweeter. It is becoming a popular drizzling oil in many Indian-inspired dishes.

Sprouts from the mustard seeds are used in salads in many Asian recipes.

Spice Blends: Dijon mustard blend, baked bean blend, Bajan sauce blend, pickle blend, chutney blend, sambar podi, panchphoron and Chinese mustard blend.

Therapeutic Uses and Folklore: in ancient times, the Greeks used mustard for scorpion and snakebites. Mustard plasters were used to stimulate blood circulation and to warm up cold feet, to relax stiff muscles and to treat arthritis and rheumatism. It treats skin diseases because of its high sulfur content.

Mustard also stimulates the flow of salivary and gastric juices and promotes appetite. It has been used as a laxative, as a remedy for asthma and to induce vomiting or relieve coughs.

Asian Indians use mustard oil in cooking to increase blood flow and cleanse the blood of any toxic materials.

Mustard has preservative properties.

MYRTLE

Myrtle leaves, branches and fruits have been used since Biblical times to flavor smoked or roasted meats in the Eastern Mediterranean islands of Sardinia, Corsica and Crete and western Asia.

Scientific Name(s): Myrtus communis. Family: Myrtaceae.

Origin and Varieties: myrtle is indigenous to Mediterranean regions of northern Africa, southern Europe and the Middle East.

Common Names: Corsican pepper. It is also known as hadass (Arabic), bilati mehendi (Bengali), vilayatimehndi (Hindi), mirto (Italian), myrtti (Finish), myrte (French, German), tao jian niu (Mandarin), arraya'n/mirto (Spanish), kulinaval (Tamil) and habulas (Urdu).

Form: myrtle seeds are purple-black berries that are used whole or coarsely ground. Its leaf is used whole or chopped.

Properties: myrtle berries are sweet, with juniper- and rosemary-like flavors. The leaves are astringent and bitter with a refreshing, fragrant and orange-like aroma.

Chemical Components: leaves have 0.2% to 0.8% essential oil, mainly myrtenol, myrtenol acetate, α-pinene, limonene, linalool, camphene, cineol, geraniol and nerol.

How Prepared and Consumed: in the Mediterranean, the berries were initially used to flavor wine but are now more commonly used in desserts, liqueurs and sweet dishes. The leaves are used in stews, roast meats, stuffings, salads and meat ragouts. The leaves are used to wrap wild game or roast pork before cooking.

Italians, especially Sardinians, use myrtle branches in the same way allspice branches are used in the Caribbean. They wrap meat, other game meats, birds and poultry with myrtle branches and then roast, broil or smoke them. The leaves are also stuffed in the meats and are removed before serving. The burning myrtle wood and leaves provide a fragrant note to the meat.

Spice Blends: smoked meat blends, roast blends and meat ragout blend.

Therapeutic Uses and Folklore: the berries are used as an antiseptic for bruises, and the leaves are used to relieve gingivitis and sinusitis.

NIGELLA

Nigella, called kalonji in Hindi, has become a popular spice for breads and vegetables in North India, Lebanon, Turkey and Iran. It is often confused with black cumin because of its similar size and color and, thus, they are called similar names in many cultures.

Scientific Name(s): Nigella sativa. Family: Ranunculaceae.

Origin and Varieties: nigella is indigenous to western Asia and the Middle East. It is cultivated in northern India, western Asia, northern Africa and central Europe.

Common Names: black caraway, black onion, love-in-a-mist, devil in a bush, fennel flower and onion seed. It is also called tikur azmud (Amharic), habat albaraka (Arabic), nigelle (French), schwarzkummel (German), ketzah (Hebrew), kala jeera (Hindi), nigella (Italian), svartkarve (Norwegian), charnushka (Russian), kaluduru (Sinhalese), niguilla (Spanish), svartkummin (Swedish) and corekotu siyah (Turkish).

Form: nigella is a small, irregular-shaped black seed that is used whole or ground. The pods are harvested before they are fully ripe in order to prevent the pods from bursting open and discarding the seeds.

Properties: nigella has a sharp, slightly bitter, peppery and nutty taste.

Chemical Components: it has 0.5% to 1.5% essential oil, mainly thymoquinone (54%), ρ-cymene and dithymoquinone with smaller amounts of carvacrol, carvone, limonene, citronellol and 4-terpineol. Its fat content is about 40%.

It contains calcium, vitamin C, potassium and sodium.

How Prepared and Consumed: Indians, Middle Easterners, Turks and Egyptians commonly sprinkle whole nigella seeds on breads to provide flavors and textures. In North Indian cooking, whole seeds are dry roasted or fried in oil to give a more intense aroma. They are used in curries, nans, dals, yogurts, vegetables and chutneys. In North India and Iran, nigella is used mostly to enhance vegetable dishes. It can also be used in salads, cottage cheese, lamb, poultry and pickles.

Nigella is an essential ingredient in a spice mixture of Bengal, Bangladesh and Sikkim, called panchphoron (five spice blend). It is blended with cumin, mustard seed, ajwain and black pepper and is fried in mustard oil to flavor eggplant, cabbage, squash and meats.

Spice Blends: panchphoron, dal podi, chutney blend and lamb curry blend.

Therapeutic Uses and Folklore: Indians eat nigella to reduce flatulence, treat nervous disorders and induce sweating. It is also taken by lactating women to induce milk flow.

NUTMEG

Nutmeg has its origins in the Spice Islands of Indonesia. Arab traders carried nutmeg to India and then to Europe, and eventually it was taken to the Caribbean by the Spanish. It was widely popular in Europe and in India for its flavoring and medicinal properties. During the Middle Ages, fashionable Europeans carried their own nutmegs and graters to eating establishments as a status symbol. Nutmeg's name is derived from the Latin word *nux muscatus,* meaning musky nut. The East Indian (Indonesian) nutmeg is superior in flavor to the West Indian nutmeg (Caribbean). Nutmeg and mace come from the same fruit. Nutmeg is the seed that is covered by the lacy mace (see mace).

Scientific Name(s): Myristica fragrans. Family: Myristicaceae.

Origin and Varieties: nutmeg is indigenous to the Banda Islands (Moluccas) in Indonesia. It is also cultivated in the Caribbean, South India, Sri Lanka, Sumatra and Malaysia. There are many "adulterant" species found in South India and Indonesia with flavors that sometimes lack the aroma of true nutmeg. Nutmeg's flavor varies depending on its origin. The U.S. buys Indonesian nutmeg, which has higher volatile oils than Caribbean nutmeg.

Common Names: gewz (Amharic), jos alteeb (Arabic), notemuskaat (Dutch), djus hendi (Farsi), noix de muscade (French), muskat (German), jaiphal (Hindi, Urdu), noce moscata (Italian), pala (Indonesian), nikuzuku (Japanese), buah pala (Malaysian), jo tou kuo (Mandarin), jathikai (Malayalam/Tamil), noz moscada (Portuguese), muskatniy orekh (Russian), sadikka (Singalese), nuez moscada (Spanish), muskotnot (Swedish) and luk jan (Thai).

Form: nutmeg is a light brown or grayish wrinkled seed inside a smooth, hard blackish brown nut. The nut is dried in the sun until the inner seed rattles when shaken.

Properties: nutmeg is used ground, grated or crushed. It loses its flavor easily when ground, so it should be bought whole and then ground when desired. Depending on the type, its flavor can vary from a sweetly spicy to a heavier taste. It has a clove-like, spicy, sweet, bitter taste with a terpeny, camphor-like aroma. It is sweeter in flavor than mace.

Chemical Components: nutmeg has 6.5% to 16 % essential oil, which is pale yellow in color and is called oil of myristica. Depending on the source, the essential oil has mainly sabinene (15–50%), α-pinene (10–22%) and β-pinene (7–18%), with myrcene (0.7–3%), 1,8-cineole (1.5–3.5%), myristicin (0.5–13.5%), limonene (2.7–4.1%), safrole (0.1–3.2%) and terpinen 4-ol (0–11%). Their amounts depend on whether the oil is of West Indian, Indian or Sri Lankan origin. The fixed oil is a pale yellow to golden yellow viscous oil, and 6 lbs. of oleoresin is equivalent to 100 lbs. of freshly ground nutmeg.

Nutmeg butter, which consists of fixed oil and volatile oil, is orange red to reddish brown and has the consistency of butter at room temperature. It contains trimyristin (70%), fats (4%), resins (13%) and other constituents (2%). The fats are mainly saturated (90%) with 10% unsaturated fats.

Nutmeg contains potassium, magnesium and phosphorus.

How Prepared and Consumed: nutmeg loses its flavor quickly when ground, so generally it is grated just before cooking or baking. Also, in order to retain its flavor, nutmeg is added toward the end of cooking. It complements chocolate, fruits, custards, vanilla, coconut milk, lemongrass and kari leaves. Europeans use it in mashed potatoes, rice dishes, pastas, soups, rice puddings, pies, eggnog, biscuits and milk-based drinks.

Nutmeg provides an intense, sweet, spicy aroma to pastries, cakes, sweet rolls, banana bread, pumpkin pies and apple pies, ice cream, chocolate and lemon desserts. Nutmeg is also used in cheese fondues, and it enhances savory products such as vegetable stews, bechamel sauce, tomato sauce, processed meats and pork patties.

Italians flavor spinach with nutmeg for stuffed pastas. It is also a favorite spice of the Dutch, who use it in potatoes and other vegetables. Nutmeg is an important ingredient in the French spice blend, quatre epices, that is used to flavor meats that are braised or cooked for a long time, such as ragouts or stews.

Along with mace, it is an important ingredient in spice blends of India, the Middle East and North Africa. The pungent garam masalas of North India commonly use nutmeg. The aromatic spice blends of the Middle East and Asia contain it as well. It is generally used sparingly during the cooling process. In Indonesia, nutmeg is used in sauces and curries, and nutmeg pulp is made into a local jam.

Nutmeg has become a popular spice in Caribbean cooking and is added to jerk seasoning, pastries, ice creams, fruit punches, eggnogs, breads and cakes.

Spice Blends: garam masala, bechamel sauce blend, quatre epices, baharat, ras el hanout, curry blends, jerk seasoning and Indonesian ikan pedas curry blend.

Therapeutic Uses and Folklore: traditionally, nutmeg has been used to treat digestive disorders, such as nausea and diarrhea, and kidney ailments. Southeast Asians also treat fevers, headaches and bronchial problems with nutmeg. The Chinese consider it to be an aphrodisiac.

ONION

Onion is an ancient spice, native to Asia, and it is long noted for its flavoring and pickling properties. Traditionally, Greeks valued it for its curative powers, while Egyptians ate it raw. Today, onion is an indispensable ingredient for flavoring many ethnic cuisines, whether sautéed, roasted or pickled. Indians savor it not only for its flavor but also for the texture and consistency it provides to curries. There are many types of onions that vary in color, size and flavor, such as yellow, red, purple, Italian, sweet, pearl onions and shallots.

Scientific Name(s): onion: *Allium cepa;* shallot: *Allium ascalonicum.* Family: Alliaceae (onion family).

Origin and Varieties: onion is indigenous to central Asia and is now cultivated in Egypt, Japan, North America, South America, Europe, Southeast Asia, France and Mexico. There are many types of onions and, based on their origins, they vary in size, color and flavor. The yellow onion is the most common onion. Sweet onions include the Vidalia, Spanish, Bermuda and Maui. The red types are the Italian and purple onions. Scallions are salad onions or green onions.

Common Names: common onion and garden onion. In some cultures, onion and garlic have similar names. For example, in Indonesia and Malaysia, the common onions are referred to as "red" onions (bawang merah) because of their red-colored sheath, while garlic is called "white onion" (bawang putih). It is also known as key shinkurt (Amharic), basal (Arabic), pianj (Bengali), kesunni (Burmese), ui (Dutch), pias (Farsi), oignon (French), zwiebel (German), batzol (Hebrew), piaz (Hindi, Urdu), cipolla (Italian), atasuki (Japanese), khtim slek (Khmer), yan chong (Mandarin), ulli (Malayalam), bawang merah (Malaysian, Indonesian), cebola (Portuguese), luk (Russian), lunu (Singhalese), cebolla (Spanish), lok (Swedish), vengayam (Tamil), hua hom yai (Thai) and khan ko (Vietnamese).

Shallot: eschalot, schalotte and Spanish garlic. It is also called key shinkurt (Amharic), ciboule (French), ascalonia (Spanish), schalotte (German), scalogna (Italian), sjalott-lok (Norwegian), cebolha roxa (Portuguese), schalottenlok (Swedish), kanda lasum (Hindi), bawang besar (Indonesian, Malaysian), hong tsong (Mandarin), vengayam (Tamil) and horm daeng (Thai).

Scallion: spring onion, green onion, jiao tou (Mandarin) and ton horm (Thai).

Form: the common onion is a white, yellowish white or brownish red bulb, depending on the variety and source. Onion comes fresh or dried. The fresh form is chopped, sliced or diced. The fresh onions can be roasted, grilled or pickled. Dried, it comes granulated, powdered, ground, minced, chopped or toasted. Pearl onions are tiny, red onions that are larger and round to oblong, and shallots are a cluster of small- to medium-sized bulbs with a reddish brown to orange brown outer skin. Scallions, also called spring onions or green onions, are long, slender onions that are immature yellow onion bulbs. Both their bulbs and stems are consumed.

Properties: onions are mild or pungent depending upon the variety. Onions do not give any flavor until they are cut or bruised. These actions cause an enzymatic action and give rise to a mixture of sulfides. The regular white or yellow onions have a strong, pungent and penetrating odor when raw but become sweet when cooked. The red onions are pleasant and sweet. Pearl onions do not have much aroma but taste sweet when cooked. Vidalia onions are sweet with a high sugar content and have less sulfur than regular onions. The shallot is more delicate and has a slightly pungent, sweet flavor. Scallions are mild and slightly sweet.

Chemical Components: regular onions contain 0.01% to 0.015% essential oil that has a dark brown color. They consist mainly of sulfur compounds, such as *d-n*-propyl disulfide, methyl-*n*-propyl, disulfide, vinyl sulfide and thiols. The flavor-contributing chemicals are methyl propyl disulfide, methyl propyl trisulfide and dipropyl trisulfide that are released through enzymatic action when the cells are damaged. The oleoresin is brownish in color, and 1 lb. is equivalent to 100 lbs. of dehydrated onions or 400 lbs. of fresh onions.

How Prepared and Consumed: onions add flavor, color and crunchiness to foods. They are added whole and glazed for stews; chopped or minced for soups and sauces; chopped or sliced raw as garnish for salads, omelettes and dips; and deep fried, simmered, stir-fried, roasted or pickled.

Yellow onions are good for stews, soups or sauces that require long cooking times. Sweet onions are tasty when baked, battered and fried (as onion rings), sliced and fried as toppings or eaten with broiled or roasted meats. Red onions are used raw in sandwiches and salads and can be pickled or curried. Pearl onions are glazed, pickled or added whole to soups and stews.

Onion-filled dumplings called pierogi are a popular Polish food in the U.S. that were brought to central Canada by the Ukranian immigrants. In classic French cooking, pearl onions and shallots are glazed, pickled or roasted and used in salads, roasted meats, poultry and seafood. In France, shallots are very popular and are used with red wine and are mostly braised or boiled. They are an important ingredient with tarragon in béarnaise sauce and many other French sauces. Fried onion rings are used as appetizers in the U.S. and in Germany as garnish with mashed potatoes.

In India, onions are the base of many sauces and curries, providing not only flavor but also body. They are chopped and fried in oil or ghee until golden brown, before other ingredients, such as garlic, ginger and spices, are added. When fried, onions become sweet and aromatic. Onions pair well with garlic, ginger, tomatoes, yogurt, coconut and many other spices.

Shallots are an essential ingredient that adds flavor and consistency to Southeast Asian cooking. They are sliced and mixed raw into sambals or are served alongside satay. They are pounded and added to spice blends and condiments and cooked in sauces or rendangs. They are also fried in oil and added as toppings to stir-fried noodles and nasi goreng (fried rice) to give a crunchy texture and caramelized taste.

In Malaysia, Singapore and Indonesia, shallots are pureed with spices and chile peppers and are "tumised," or fried in oil, until a fragrant aroma develops to create cili boh. Similarly, in Burma, onions are blended with vinegar, chilies, and ginger into a smooth paste and then fried in sesame oil until the oil separates, and the sauce becomes fragrant.

Scallions are chopped or sliced raw and are commonly used in soups, stir-fries or pickles in East Asian cooking. In Chinese and Southeast Asian cooking, they are chopped or sliced and added raw as toppings over cooked noodles, laksas and rices, or mixed into condiments and dumpling sauces. The Afghans have a ravioli that is stuffed with scallions and served with yogurt sauce and onions.

Spice Blends: cili boh blends, rojak blend, curry blends, pickle blend, pierogi filling blend and onion soup blend.

Therapeutic Uses and Folklore: onions have been used traditionally as an expectorant, a diuretic and to lower blood pressure. Raw onion has been used to fight colds and flues and prevent edema. Shallots are traditionally used to cure earaches and fevers.

OREGANO

Oregano's name comes from the Greek word *oros ganos,* meaning joy of the mountain. There are many varieties of oregano, including Greek, Mexican, African, Italian, Spanish and Moroccan. The word "oregano" in Spanish means marjoram and, thus, oregano is sometimes called wild marjoram. Originally used to flavor ale and beer in Europe, oregano is an essential spice in Italian and Mexican cooking. The Greek or wild oregano is usually used fresh or dried in North American cuisine. Mexican oregano is popular in Mexican-style cuisine in the U.S.

Scientific Name(s): common oregano (wild European): *Origanum vulgare;* Italian oregano: *Origanum viride* and *O. virens.* Family: Labiatae (or Lamiaceae, mint family). Mexican oregano (or Mexican sage): *Lippia graveolens.* Family: Verbenaceae.

Origin and Varieties: oregano is indigenous to the Mediterranean (Italy, Greece, Turkey, France, Israel, Morocco), India, Mexico and South America.

The U.S. uses Mediterranean oreganos (Turkish, Greek, Italian and Spanish) and Mexican oregano. Mexican oregano is not botanically related to the common oregano and has a more intense taste than the common oregano.

Common Names: common oregano—wild marjoram, winter sweet, origanum, wild European oregano, origany, pizza herb, common marjoram and Spanish marjoram. It is also called mardakosh (Arabic), wilde marjolein (Dutch), avishan kuhi (Farsi), origan (French), dosten/oregano (German), oregano (Hebrew), sathra (Hindi), oregano (Japanese), origano (Italian), ouregao (Portuguese), dushitsa (Russian), oregano (Spanish), oregano (Swedish) and mirzan josha (Urdu).

Form: oregano's dark green fresh leaves are larger than marjoram leaves. Fresh oregano is available whole, chopped or minced. The dried light green leaves are available whole, flaked or ground.

Properties: Oregano can be mild or strong depending on its origins. Typically, it has a phenolic, herbaceous, slightly floral and bitter taste, with slight lemony and pungent notes. It has an aroma resembling that of marjoram. European or Mediterranean oregano is milder and sweeter in flavor than Mexican oregano that is darker in color and stronger and more robust in flavor. Oregano has a more potent flavor when it is dried.

Chemical Components: fresh Greek oregano has about 0.9% essential oil, which is yellowish red to dark brown, and contains 60% to 70% phenols, mainly thymol and carvacrol. Greek oregano has terpinen 4-ol (46%), α-terpinene (7%), α-terpineol (7.6%), sabinene (2.8%), linalool (1%), *cis*-sabinene (7.6%), ρ-cymene *z* (3.2%), linalyl acetate and β-caryophyllene (2.6%).

Mexican oregano has 3% to 4% essential oil, which is higher than the Mediterranean types. The amounts of the two types of phenols, carvacrol and thymol, vary depending upon the origin of the oregano. The common or Greek oregano has 5% thymol and 7.5% carvacrol. Turkish oregano has 83% thymol and 0.8% carvacrol, Italian oregano has 60% thymol and 9% carvacrol and Spanish oregano has 18% thymol and 4.5% carvacrol.

Oleoresin oregano (made from Spanish oregano) is a dark, brownish-green liquid, and 4 lbs. of oleoresin are equivalent to 100 lbs. of freshly ground oregano.

Dried oregano contains vitamin A, potassium, iron, magnesium, calcium and phosphorus.

How Prepared and Consumed: the ancient Greeks and Egyptians used oregano to flavor fish, meats, vegetables and wine. Today, oregano is still a popular spice for the Greeks, who use it in salads, chicken and seafood dishes. It is an essential flavoring in Italian pizza sauce. It is also used in Italian, Spanish and Latin American dishes—meat stews, lamb and pork roasts, veal dishes, stuffings, salad dressings, soups, taco fillings, chili, corn, omelettes and bean dishes. It has great affinity for garlic, cumin, cilantro, sweet basil, lemon, chile peppers, tomatoes, eggplant, peppers, zucchini, capers, olives, vinegar and cheeses.

Mexican oregano is used in authentic Mexican cuisines, in sauces, bean dishes, empanada fillings, burritos or tacos. It is full-bodied and pairs well with ground meat, garlic, onion, cumin and chile powder. Oregano is also used with paprika, cumin and chile peppers in chili powder, which is a common U.S. seasoning used for Tex-Mex-style beans, meat stews, sauces, roasts, soups and in chile con carne. Oregano is also popular in Southwestern foods, such as green chile blends, soups, chowders, black beans and sauces.

Spice Blends: chili powder, pizza sauce blend, pasta sauce blend, bouquet garni, chili con carne blend, green chile blend, frijole blend, Sante Fe seasoning and Southwestern blend.

Therapeutic Uses and Folklore: ancient Egyptians used oregano for healing and as a disinfectant; Greeks used it to treat spider and scorpion bites and Romans used it to stimulate hair growth and give a clear complexion. It is added to tea and taken to treat exhaustion, nervous disorders and tension. Oregano is used as a traditional remedy for stomach upsets, low blood pressure, whooping cough, skin irritations, throat inflammations, toothaches, headaches and asthma.

PAPRIKA

It comes from the Hungarian word, paparka, a variation from the Bulgarian piperka, which in turn comes from the Latin *piper,* meaning pepper. In the U.S. and the rest of the world, paprika is the term for the ground or powdered form of the dried peppers of the non-pungent to slightly pungent red varieties of *Capsicum annum*. In Hungary, there are many varieties of dried peppers that are used to produce paprika, from sweet bell peppers to hotter ones, and so paprika is available sweet, semisweet and pungent. In Hungary, the term paprika includes all of the fresh varieties of peppers, including green bell peppers as well as the ground form of the mild to pungent capsicum peppers.

Paprika was introduced from the New World into Spain and Turkey. Then, it was taken into Hungary in the sixteenth century when the Ottoman Turks invaded Hungary. The Hungarians came to call it paparka, which means Turkish pepper. Through different climates, soil conditions and crossbreeding, milder capsicum annums were obtained from earlier, hotter American relatives.

Scientific Name(s): Capsicum annum. Family: Solanaceae.

Origin and Varieties: paprika peppers are indigenous to Central and South America but are now cultivated in Hungary, Morocco, Mexico, Spain, Turkey and the U.S. There are four types: American, Spanish, Mexican and Hungarian, all varying in pungency and sweetness. Climate and environment affect the color of paprika. In the U.S., the paprika used is the brilliant red, sweet type.

Common Names: pimiento, sweet red peppers, pod pepper, Hungarian paprika, tomato pepper, pimienton, sweet paprika, noble paprika and Spanish paprika. It is also called fleyfla hilwa (Arabic), piment doux (French), deghi mirch (Hindi), paprica (Italian), horn giaro (Mandarin), `apapr (Serbian), pimiento dulce/paprika (Spanish) and paprika (Swedish, German, Dutch).

Form: paprika is the brilliant red, ground powder from the thick flesh of dried red capsicum annum pods, with most of the seeds and veins removed. The ripe fruits are dried, ground and sometimes cured. Paprika has a wide range of colors and pungencies. Depending on its origin and grade, it can be sweet to pungent.

Properties: paprika has a deep red or russet to a reddish orange and brownish red color. Spanish paprika is slightly more pungent than the paprika consumed in the U.S. U.S. paprikas have mild flavors. Hungarian paprika has a deep, round flavor and can range from mild, very sweet with caramelized notes to more pungent flavors. Hungarians classify their paprika as fiery (eros), semisweet (feledes) and premium sweet (edesnemes). Paprika loses its aroma and develops bitterness when exposed to high heat. It loses its color quickly (from a russet red to tan red and finally to a brownish red) when exposed to light and air. The paler reds and the brownish reds are the most pungent paprikas.

Chemical Components: paprika has very little essential oil. Its oleoresin is dark red with 8 lbs. equivalent to 100 lbs. of freshly ground spice. Its coloring matter is due to carotenoids (0.3–0.8% in fruit) and consists of 35–60% capsanthin, 18% capsorbin, 8–23% α- and β-carotene, 8–10% zeaxanthin, 3–5% cryptoxanthin and 8–10% lutein.

Paprika contains vitamin A, vitamin C, niacin and potassium, magnesium, sodium, phosphorus and calcium.

How Prepared and Consumed: paprika is an indispensable spice in Hungarian cooking and many eastern European countries. It is used in Hungarian and Serbian soups, stews, barbecues, omelettes and sausages. In Hungarian cuisine, it is an essential flavor in porkolt (thick stew of pork, goose, beef, duck and game with onions), tokany, szekelgulyas (sauerkraut and goulash), chicken paprikas (creamy chicken stew) and beef gulyas/goulash (beef stews). It also gives the red coloring in Hungarian salami and is used as a tabletop condiment, combined with other spices and lard.

Paprika is used in cream-based sauces in European cooking, mainly to add color. Spanish paprika, called pimento, is a favorite ingredient in Andalusian cooking and is used in fish dishes, sofritos and in berza, a thick bean soup with vegetables, black sausage (morcilla), chorizos, ham hocks, clove, nutmeg and cumin.

In the U.S., paprika is valued for its coloring and mild flavor. It provides visual appeal for lighter colored foods. Paprika is used to color and garnish potatoes, cheeses, eggs, salad dressings, fish, meat, soups and vegetables. It is used commercially in sausages, condiments, salad dressings and other meat products. It flavors stuffings containing olives, fish and rice. Southwestern and Texas cuisines enjoy paprika in their chilis, soups, stews and sauces. It goes well with the added cumin, oregano, coriander and cilantro.

Paprika is a popular spice in North India, Turkey, Morocco and in the Middle East. In North India, paprika is fried with other spices to provide color for curries. In the Arab regions of the Middle East, paprika forms an important ingredient in many spice mixtures, such as ras-el-hanout, berbere, galat dagga and baharat. Baharat is a spice mixture that is fried in oil or clarified butter for use in mutton and other dishes. Chermoula, a puree of paprika, garlic, saffron, cayenne, lemon juice and cilantro, is a marinade or sauce used for meats and fish in Morocco, Yemen and other neighboring regions.

Paprika also pairs well with onions, sour cream, chile peppers, black pepper, corn, black beans, tomato, turmeric and garlic and pork. It develops a brilliant reddish orange color and a distinct sweet, pungent flavor when fried in oil. Due to its high sugar content, it should not be overheated, otherwise it will turn bitter.

Spice Blends: goulash blend, chermoula blend, baharat, curry blends, paprikash sauce blend, sofrito, ras-el-hanout, berbere, galat dagga, chili blend and Sante Fe blend.

Therapeutic Uses and Folklore: paprika is traditionally used to help prevent colds and influenza and aid circulation problems.

PARSLEY

Parsley is polular as a garnish that is used globally. It is an important component of bouquet garni and fines herbes of French cooking. It is also used by Europeans as an after-dinner breath cleanser. Throughout the centuries, parsley was not always appreciated as a culinary ingredient. At one time, it was associated with ill luck by the English and the Greeks. Later, the Greeks and Romans associated it with speed and strength. It has also been used for centuries in the Jewish Passover meal as a symbol of new beginnings.

Scientific Name(s): flat leaf parsley: *Petroselinum satiuum;* curly leaf parsley: *P. crispum;* root parsley (Hamburg): *P. crispum tuberosum.* Family: Apiaceae.

Origin and Varieties: parsley is indigenous to the eastern Mediterranean regions, Greece, Turkey and Sardinia. It is now grown throughout Europe, especially in Hungary, Germany, France, Holland, the Middle East and the U.S. There are many varieties, but the most commonly used types are flat leaf (or Italian parsley), curly leaf and Hamburg parsley.

Common Names: garden parsley, common parsley and Chinese coriander. It is also called makdunis (Arabic), azadkegh (Armenian), peterselie (Dutch), persil (French), petersilie (German), petrozil'ya (Hebrew), ajmood (Hindi), seledri (Indonesian), prezzemolo (Italian), paseri (Japanese), kothambelari (Malayalam), pasli (Malaysian), ho lan qin (Mandarin), persille (Norwegian), salsinha (Portuguese), petrushka (Russian), perejil (Spanish), persilja (Swedish), pak chee (Thai), maydanoz (Turkish) and rau mui tay (Vietnamese).

Form: the leaves come fresh or dried. They are used whole, flaked, chopped, minced or pureed. The seeds, root (Hamburg parsley) and stems of the parsley plant (Neapolitan parsley) are also used in cooking as flavorings.

Properties: flat leaf parsley has a less harsh flavor than the seed. Curly leaf parsley has no flavor, so it is used as a garnish. Flat leaf parsley has a grassy, green taste with a herbaceous and slight lemony aroma. Dry parsley has no flavor. The greyish brown parsley seed has a bitter, harsh and terpeny taste. The root has a strong aroma.

Chemical Components: commercial parsley oil is manufactured from the whole herb or seed. Parsley leaf has 0.06% to 0.1% essential oil and has a more desired aroma than the seed oil. The leaf has mainly myristicin (20.6%), apiole (18.3%), α-pinene (5.1%), β-phellandrene (12.1%), myrcene (4.3%), limonene (3.6%), *p*-mentha-1,8-triene (9.2%), α-*p*-dimenthylstyrene (7.2%), aldehydes, ketones and phenols.

The seed has 1.5% to 3.5% essential oil, with mature seeds containing up to 6%. Its apiole content is 36.2%, and its myristicin content is 13.3%. Apiole gives its characteristic odor and taste. It is toxic at higher levels, so only small quantities of fresh parsley should be eaten at any one time.

Oleoresin parsley is prepared from the seed and blended with the leaf oil. The oleoresin from the leaf is deep green; 1/3 lb. is equivalent to 100 lbs. of fresh parsley, and 3 lbs. are equivalent to 100 lbs. of dried parsley.

Fresh parsley has high levels of vitamin A (8230 IU/100 gm) and vitamin C (125–300 mg/100 gm), with folic acid, iron, potassium, sodium, magnesium and calcium. The dried form has higher levels of vitamin A (23,340

IU/100 gm), calcium, iron, sodium, phosphorus with vitamin C, niacin, magnesium and manganese.

How Prepared and Consumed: curly leafed parsley is used primarily as a great garnish to add color and visual appeal to many foods. Flat leafed parsley has a mild flavor that can harmonize mild herbs, such as chives and chervil, can tame more assertive herbs such as cilantro or tarragon and can extend pungent herbs such as basil or mint. Flat leafed parsley is used as a garnish (for salads, soups, boiled potatoes and egg dishes) and is chopped and blended in dips, quick cooked sauces and stews. Parsley is typically added toward the end of cooking, because heat destroys its flavor.

Parsley is chopped and used as a popular garnish in Central Europe. It is commonly found in many European foods such as soups, sauces, fish, meats, pies and poultry dishes. Parsley is a popular herb in Middle Eastern cuisines and is used in appetizers, beans, taboulleh, couscous and stews. The French finely chop parsley and garlic, then saute this mixture and add it to broiled lamb, chicken, beef or fried fish.

Parsley pairs well with garlic, shallots, chicken, butter, lentils, beans, vinegar, potatoes and legumes. It is finely chopped and mixed with garlic or shallots to make persillade which is added toward the end of cooking to flavor stews, vegetables and meat dishes. Parsley root is used as a vegetable to enhance soups, stews and condiments of Europe. It retains its flavor after long cooking times.

Spice Blends: bouquet garni, fines herbes, persillade blend, gremolada blend, pesto blend and chimmichurri blend.

Therapeutic Uses and Folklore: traditionally, parsley has been used to treat stomach ailments and menstrual problems, arthritis and colic. It is believed to maintain the elasticity of blood vessels, clean the blood and speed oxygen metabolism. With its high chlorophyll content, it absorbs odors and is used to remove halitosis.

PEPPERS: BLACK, WHITE, GREEN, LONG/PIPPALI, CUBEB

Black pepper has long been known as the "King of Spices." The word "pepper" comes from the Sanskrit word, *pippali,* the name of the Indian long pepper. Native to the Malabar Coast of southwest India, black pepper was introduced to Egypt and Europe through Arab traders.

Black pepper was a precious commodity in ancient times. Measured like gold, it played a vital role as a medium of exchange, whether to pay taxes, rents, dowries or ransoms. In ancient times, the Romans lavishly used it to spice their foods, and it became a status symbol of fine cuisine in Europe. Black pepper became so highly prized that it became a quest for European nations to discover the fastest routes to use to obtain the spice.

Long pepper or pippali has origins in India and Indonesia. Long before the use of black pepper, this pepper was highly regarded in Europe and, thus, became the source of the word pepper. Today, black pepper is used abundantly all over the world, while long pepper's use has disappeared, except in India.

Cubeb pepper, also called Javanese pepper, derives its name from the Indonesian word "cabe" meaning pepper. It is indigenous to Indonesia and West Africa. These are small berries of a pepper relative and are similar in appearance to black pepper except that they have a "tail." They are also sometimes confused with cassia buds. Cubeb pepper was first brought to Africa and the West by Arab traders who used it to season their meat dishes. Later, it became a popular spice in North Africa. In Benin, West Africa, it is referred to as "piment pays," meaning "pepper of the country." Ashanti peppers are false cubeb peppers.

"Substitute" peppers, such as grains of paradise, negro pepper, long pepper and cubeb pepper, were used abundantly in ancient times, because the price of black pepper was high. Nowadays, the former two are rarely used, except in West Africa, while the latter two are still used in India and North Africa.

Szechwan pepper or fagara (discussed earlier in this chapter) from China, and sansho, Japanese pepper and the pink and green peppercorns from South America are also popular peppers emerging in the U.S. Pink peppercorns from South America, one of the peppercorns in mixed peppercorns, are not true peppers. Many peppers exist that have varying flavors and colors depending on their origins and how they are processed.

Scientific Name(s): black/white/green pepper: *Piper nigrum;* long pepper: *Piper longum* (India) and *P. retrfractum* (Indonesia); cubeb pepper: *Piper cubeba* (Indonesia, North Africa). Family: Piperaceae.

Origin and Varieties: black and white peppers are native to the southwestern Malabar Coast of India and Indonesia. There are many varieties of black pepper, depending on its origins. It is now cultivated in Thailand, Sarawak (in Malaysia), Sumatra (in Indonesia), Madagascar, Vietnam, Singapore, Sri Lanka and Brazil. Long pepper is native to India and Indonesia.

Common Names: black pepper is called black gold. It is also called k'undo berbere (Amharic), fulpol aswad (Arabic), zwarte peper (Dutch), poivre noir (French), schwarzer pfeffer (German), pilpel shahhor (Hebrew), kali mirchi (Hindi, Urdu), merica hitam (Indonesian), pepe nero (Italian), kosho (Japanese), lada hitam (Malaysian), kuru mulagu (Malayalam), hay hu chiao (Mandarin), pimienta preta (Portuguese), chyorny pjerets (Russian), pimienta negra (Spanish), svartpeppar (Swedish), karu mulagu (Tamil) and prik (Thai).

White pepper: witte peper (Dutch), poivre blanc (French), weiber pfeffer (German), pilpel lavan (Hebrew), mirch (Hindi), merica putih (Indonesian), pepe bianco (Italian), lada putih (Malaysian), bai hu chiao (Mandarin), byely pjerets (Russian), pimienta blanca (Spanish), vitpeppar (Swedish), paminta (Tagalog), vella mulagu (Tamil/Malayalam) and prik kao (Thai).

Green pepper: poivre vert (French), gruner pfeffer (German), pilpel yarok (Hebrew), merica hijau (Indonesian), pepe verde (Italian), lada hijau (Malaysian), pimiente verde (Portuguese), zelyony pjerets (Russian) and gronpeppar (Swedish).

Long pepper: Balinese pepper and Bengal pepper. It is also called timiz (Amharic), fleyfla taweela (Arabic), pipool (Bengali), poivre long (French), langer pfeffer (German), pipli (Hindi), cabe bali (Indonesian), thippali (Malayalam), tipli (Singhalese), langpeppar (Swedish), tippili (Tamil) and pipul (Urdu).

Cubeb pepper: Java peppercorn and tailed pepper. It is also called hab alaru (Arabic), poivre du Java/cubebs (French), kubebenpfeffer (German), kabab chini (Hindi), cabe java (Indonesian), kubeba (Russian) and kubebapeppar (Swedish).

Form: pepper is a dried berry called peppercorn and comes whole, cracked or ground. Ground pepper is available coarse or fine and is a mixture of light and dark particles. The black, white, green and red peppercorns are berries picked at different stages of growth and processed using different methods.

Black peppercorn is a globular, wrinkled, dried, unripe berry or fruit with its outer hull or skin intact. This green or greenish yellow peppercorn (the immature berry) is picked, fermented for a few days and then dried in kilns. During drying, it shrivels and becomes wrinkled and black. The most popular black peppers are as follows Tellicherry pepper, from the coast of southwest India, is the most aromatic black pepper, with a fruity clean bouquet and less pungency than the Southeast Asian types. Lampong pepper, from Sumatra, Java and Borneo in Indonesia, is small and grayish in color and is pungent and not as aromatic as Tellicherry pepper. Sarawak pepper, from Malaysia, is hotter than the Lampong pepper. Other varieties include Chinese black pepper (light in color with a mild flavor), Brazilian, Madagascan, Vietnamese, Ceylon and Singapore.

White peppercorns are berries stripped of the outer hull and picked when near ripe, when they are yellowish red or red in color. The berries are then soaked in water or steamed to soften and loosen their skin. The outer hulls or skins are removed by rubbing, leaving a smooth, light-colored berry. They are then bleached, rinsed and sundried. White peppercorns can also be prepared from black peppercorn by mechanically removing the outer hulls, a process called decortication. This latter type of white pepper tastes more like black pepper. So, white pepper is actually the inside seed and not the whole fruit like black pepper. The more popular varieties of white peppers are Muntock (from Banda Island near Sumatra, Indonesia), Sarawak (Malaysia), Brazilian and Chinese.

Green peppercorns are the unripe tender berries that are picked and air dried or freeze dried and then packed in brine or wine vinegar to retain their color.

Red peppercorns are the ripe matured berries that are dried at high temperature. They are also pickled.

Long pepper/pippali are tiny berries that merge into a single rod-like structure that is about 1-1/2 inches long and is slightly tapered. Its taste is similar to black pepper, but it has a slight sweetness and a stronger taste. In India, it is an essential ingredient of the spice mixture used with betel nuts in paan leaf, which is chewed after a meal.

Cubeb pepper is a berry that is slightly larger than the peppercorn and has a furrowed surface. It comes with the stem attached, and thus, it is also called tailed pepper. Most cubeb berries are hollow. Cubeb pepper is sold whole, crushed or ground.

Properties: pepper provides aroma as well as a "bite" that is different from chile peppers, ginger, mustard or other pungent spices. The essential oil gives the aroma, whereas the non-volatiles, such as piperine, give the pungency. Each pepper has different flavor characteristics.

Black pepper has a hot, biting taste with a wonderful bouquet—a penetrating pungent and woody aroma. It has slight lemony and clove tones. White pepper has a sharp, winey and biting taste with less harsh notes. It has little aroma. It lacks the bouquet of black pepper. Both peppers lose flavor easily when ground. Green peppercorn is aromatic, with less of a sharp flavor than black or white pepper. Red peppercorns have a delicate sweet taste. Long pepper is hot with sweet tones. Cubebs are pungent and tea-like with slightly musky and bitter notes. They have a terpeny aroma. When ground, cubebs release cumin-like notes, and when cooked, they release an aroma that combines allspice with curry.

Chemical Components: it has a colorless to pale greenish colored essential oil. Black pepper generally contains 1–2.6% essential oil, going up to 5%, while white pepper and long pepper contain less than 1.0%. Fixed oil is 2% to 9% in black and white peppers. Essential oil consists primarily of monoterpenes (80%), such as sabinene, α-pinene, β-pinene, limonene and 1,8-cineol, all of which are responsible for the pungent aroma. During storage, ground black pepper can lose these monoterpenes. Sesquiterpenes make up the other 20% (beta-caryophyllene, humulene).

Black pepper has 10% to 15% oleoresin, which is dark green to olive green. In white pepper, the essential oil is about 8%.

The bite and pungency in black and white peppers is primarily due to the non-volatile alkaloids, piperine and chavicin. The ratio of these two alkaloids varies in different peppers, thus giving rise to the different pungencies. White pepper lacks chavicin (which is present in the epicarp) and most of the essential oil (present in the outer mesocarp), but contains piperine, which gives the bite but not the aroma. Piperine accounts for about 98% of the total alkaloid compounds in both of these peppers, others being piperittine and piperyline.

Black pepper has aroma and bite. Its piperine content varies with its origins. Indian peppers are very aromatic, whereas Malaysian and Indonesian peppers are less aromatic but have more bite; Brazilian peppers have a milder bite than other black peppers because of their low piperine content.

Long pepper has 1% essential oil, mainly phellandrene and limonene. It has a slightly higher piperine level than black pepper.

The dried fruit of the cubeb pepper has up to 10% essential oil, mostly sabinene, 1,4-cineol and carene. Pungency is mainly due to cubebin with other related compounds.

Black, white, long and green peppers contain potassium, calcium, sodium, magnesium and iron.

How Prepared and Consumed: peppers were used in ancient times to preserve meats. Today, pepper is the most important table spice throughout the world. It cuts across many ethnic cuisines. Peppers are used in marinades or spice mixtures, added during cooking or sprinkled on foods at the meal table to adjust flavor. Cracked black pepper is popular in marinades, salad dressings and spice blends. Pepper pairs well with garlic, ginger, coconut milk, lemon, vinegar, pork, beef, basil, cilantro, seafood, eggs, creamy sauces and vegetables, red wine, fermented soybeans and soy sauce.

Black pepper is popular in North American, European, Southeast Asian and Indian cooking, while white pepper is popular in European, Chinese, Japanese and Thai cuisines. In Europe, pepper is added during and after cooking to foods including steaks, soups, soup stocks, vinaigrettes, cheeses (example, boursin), vinegars, pickles or salads. With European cooking, black pepper is traditionally added to more hearty and stronger flavored foods such as steaks, pot roasts, hamburgers, soup stocks, meatballs, gravies, sausages, pepperoni and cold cuts.

The Thais and other Southeast Asians use white and black pepper, coarse or finely ground, in many stir-fry dishes, grilled meats, soups and curries. Jamaicans enjoy black pepper in goat curries and jerk dishes. In France, black pepper is added to steaks, charcuterie dishes, cold cuts and pates and is the main ingredient in quatre epices. Black pepper can be cooked for a long time without losing its flavor, as seen in béarnaise sauce and black peppercorn sauce.

The Arabs monopolized the pepper trade for years and carried it from the Far East to Europe. It is no surprise that it is a popular spice in their cooking and is found in almost all of their spice blends, including zhoug/zhug, baharat, ras-el-hanout, berbere and galat dagga.

White pepper is popular in Europe and is usually used in delicate white sauces, clear salad dressings, cream soups, fish dishes and wine, and is added more for visual reasons. In Thailand, it is added with garlic, chile pepper and other spices to fried chicken and pork dishes.

Long peppers are used in curries and spicy pickles (or achars) of India and Indonesia. They are also popular in North African and East African cuisines. The leaves and roots are used to make locally brewed beers.

Cubebs were a popular substitute for black pepper in Europe in the seventeenth century, but they are not any more. Today, they are an essential ingredient in the spice mixtures of Morocco and stews of Benin, in West Africa.

Spice Blends: mignonette pepper, berbere, ras-el hanout, baharat, galat dagga, black peppercorn sauce blend, jerk seasoning, curry blends, achar blend and boursin blend.

Therapeutic Uses and Folklore: in ancient times, pepper was used for bartering, as a bride's dowry, a form of currency and rent money. It was also used to preserve meats, because it prevents spoilage. Black pepper was traditionally used to treat headaches, constipation and diarrhea. In India and China, pepper is widely used to improve circulation and to treat colds, coughs, asthma, kidney inflammations and muscle and joint pains.

Cubeb oil is an ingredient used in throat lozenges.

POPPY SEED

Poppy seeds have been valued since ancient times. In the Ebers Papyrus, the Egyptians described them as a sedative. Later, the Egyptians produced an edible oil from poppy seeds and mixed it with honey to make a flavorful bread. As early as 1400 B.C., the Greeks valued poppy seeds as a pain reliever. Today, whole or ground poppy seeds are used in confectioneries, baked goods and sauces. Muslim missionaries introduced it into India and the Far East. It comes from the same plant as opium, but there is no narcotic content in poppy seed. The poppy seed does not fully form until the plant matures, by which time, the plant has lost all of its opium potential.

Scientific Name(s): Papaver somniferum. Family: Papaveraceae.

Origin and Varieties: it is indigenous to the Eastern Mediterranean and West Asia. It is now cultivated in India, China, Turkey, Holland, France, the U.S. and Canada.

Common Names: opium seed and maw seed. It is also called khesh kash (Arabic), posto (Bengali), bhainzi (Burmese), slaaphol (Dutch), pavot (French), mohn (German), pereg (Hebrew), khas-khas (Hindi, Urdu), papavero (Italian), keshi (Japanese), biji kas kas (Malaysian, Indonesian), ying shu (Mandarin), papoula (Portuguese), mak (Russian), ababa/adormidera/semilla de ampola (Spanish), kashakasha (Tamil, Malayalam) and vallmo (Swedish).

Form: poppy seeds are tiny, kidney-shaped seeds with creamy, brown or bluish black color, depending on their origins. When the seeds are ripe, they are used as flavorings for thickening. Poppy seeds are used whole, ground, as a paste or as a pale yellow nutty oil.

Properties: they have a nutty and slightly sweet taste with a slight smokey aroma.

Chemical Components: narcotic alkaloids are not present in the mature or ripe seeds that are used in foods. Poppy seeds contain no volatile oil but have

40–60% fixed oil that is high in unsaturated fatty acids, such as linoleic acid (30%), oleic acid (30%) and linolenic acid (3%). It is low in saturated fatty acids (about 10%).

Poppy seeds contain protein and good amounts of lecithin, minerals (magnesium, zinc, iodine, copper and manganese) and oxalic acid.

How Prepared and Consumed: the Egyptians mixed poppy seeds with honey as a dessert for their pharaohs, the Turks and Germans made a bread with poppy seeds and flour and the ancient Indians mixed it with sugarcane juice for a confection. The paste is used in desserts and sauces in Turkey and in the Mogul cooking of northern India. Poppy seeds are roasted and ground to a paste that forms the basis of many Turkish dishes.

The brown seeds are roasted to obtain a more intense flavor and are then ground with other spices to thicken and add flavor to meat and fish curries in India. The whole seeds are added to vegetables and curries to provide texture. In Japan, they are used as an ingredient in a popular seasoning called shichimi togarashi. Poppy seed oil is used as a cooking or salad oil in Europe and Asia.

In Europe and the U.S., the seeds are sprinkled on breads, buns, bagels, cookies and cakes, and they are mixed with honey or sugar to make cakes and fillings and to make strudel and croissants. In the U.S., the seeds are used as toppings to provide nuttiness to rolls, breads, cookies, pastries and cakes. Austrians grind poppy seeds and combine them with melted butter and powdered sugar and serve them with a yeast-based dumpling called germknodel. Mediterraneans and northern Europeans use poppy seeds in tuna fish and macaroni salad, sour cream and cheese dips, salad dressings, soups, cooked vegetables, boiled potatoes and coleslaw. The Jewish holiday cake known as hamentachen is made with crushed poppy seeds, beaten egg, lemon juice, honey, salt and water or milk.

Spice Blends: shichimi togarashi, hamentachen blend, fish curry blend and germknodel filling blend.

Therapeutic Uses and Folklore: Egyptians, Greeks, Arabs and Asians used the unripe seeds for their narcotic properties to relieve pain and act as a sedative.

ROSEMARY

Rosemary is a flavoring that is deeply rooted in European cultures. It was native to the Mediterranean region and has been used since 500 B.C. It was noted for strengthening the memory, and ancient Greek students wore garlands of rosemary when taking exams so as to retain their knowledge. It was also a symbol of fidelity. During the Middle Ages, its name was changed from the Latin word *rosmarinus,* which means "dew of the sea," to *Rosa Maria* or rosemary, in honor of Mary, mother of Jesus. It was used in religious ceremonies, weddings and funerals and as a symbol against evil. Today, it is still a popular spice in Europe, especially in Italy and France.

Scientific Name(s): Rosmarinus offinialis. Family: Labiatae (or Lamiaceae) (mint family).

Origin and Varieties: native to the Mediterranean, rosemary is cultivated in Algeria, Britain, France, Germany, Italy, Morocco, Portugal, Russia, Spain, Turkey and the U.S.

Common Names: it is also known as ekleel aljabal (Arabic), rozemarijn (Dutch), old man (English), iklil eklil kuhi (Farsi), rosmarin (French), rosmarein (German), rozmarin (Hebrew), rusmary (Hindi), ramerino (Italian), mannenro (Japanese), mi tieh hsiang (Mandarin), alecrim (Portuguese), rozmarin (Russian), romero (Spanish), rosmarin (Swedish) and biberiye (Turkish).

Form: rosemary leaves are used fresh or dried, whole, chopped, crushed or ground. They are small, narrow and pine-needle-like. Fresh leaves are leathery, shiny and dark green. Dried leaves are curved with dark green to brownish green colors.

Properties: when the leaves are crushed, they provide a strongly aromatic piney and tea-like fragrance with a slightly sweet taste. Rosemary leaves have a slightly minty, sage-like, peppery, balsamic and camphor-like taste with a bitter, woody aftertaste. The stems and flowers are also aromatic.

Chemical Components: rosemary has 0.5% to 2.5% volatile oil, mainly 1,8-cineol (30%)(which gives rosemary its cool eucalyptus aroma), borneol (16–20%), camphor (15–25%) bornyl acetate (2–7%) and α-pinene (25%). Different varieties differ in flavor depending on their constituents. Its oleoresin is a greenish brown semisolid, and 5 lbs. is equivalent to 100 lbs. of freshly ground, dried rosemary.

It contains vitamin C, vitamin A, potassium, sodium, calcium, magnesium, iron and phosphorus.

How Prepared and Consumed: rosemary is a very popular spice in the Mediterranean region. It is used by Europeans in roast chicken, lamb or veal, pot roasts, stews, potatoes, stuffings, marinades, vinegars, soup stocks, fruit-based desserts, breads, cold beverages and sausages. Italians commonly use it with wine, honey, garlic or chile peppers on grilled or roasted lamb, goat veal, fish, shellfish and rabbit dishes. Rosemary pairs well with olive oil, lemon, cinnamon, clove, garlic, lamb, pork, chicken, tomato- and cream-based sauces, pizzas and grilled vegetables.

In Tunisia, a strongly aromatic rosemary called kilil is used in a lamb tagine dish with paprika, cinnamon, onion and black pepper.

Rosemary does not lose its flavor with long cooking. The whole leaves are crushed before use to release their aroma. Whole sprigs used in cooking are removed before serving. The stems (stripped of leaves) are used as aromatic skewers for broiling or barbecuing chicken, shrimp or vegetables. The flowers are used in salads or in fruit fillings and purees.

Rosemary oleoresin is used as a natural antioxidant and helps preserve the color of processed meats. It is used sparingly because of its strong aroma.

Spice Blends: bouquet garni, herbs de Provence, chicken roast blend, pizza sauce blend and lamb tangine blend.

Therapeutic Uses and Folklore: rosemary has been used traditionally to treat whooping cough, fluid retention, poor circulation, jaundice, migraine, mental fatigue, panic attacks, irritability and aching joints. The Romans used it for insect bites, the Greeks used it for curing jaundice, the Arabs used it to refresh the memory and the French used it to sanitize the air in hospitals.

In ancient times, it was used to preserve meat and as a fumigant in hospitals to kill bacteria. Commercially, rosemary oleoresin is used to preserve the color of meat.

SAFFRON

Saffron derives its name from the Arab word, *zaafaran,* which means yellow. The orange yellow saffron color is sacred and bears importance to many cultures: the saffron paste in Hindu temples, the Buddhist monks' saffron-colored robes and the saffron cloaks of the Irish kings. Saffron originated in Turkey and was taken by Arab traders to Spain, Iran and Kashmir (India), where it became a popular coloring and flavoring. In the U.S., the Pennsylvania Dutch brought saffron farming from Germany. They continue to grow saffron and use it in their cooking.

Saffron is one of the oldest and world's most expensive spices because it is harvested and processed by hand. The flowers have to be picked by hand, then stigmas must be separated from the rest of the flower and finally it is dried. A pound of saffron consists of 225,000 dried stigmas or 5 pounds of fresh stigmas. Currently in the U.S., one pound of saffron was reportedly selling retail for $3000 to $5000.

Scientific Name(s): Crocus sativus. Family: Iridaceae.

Origin and Varieties: saffron is indigenous to the eastern Mediterranean—the Middle East, Turkey, Iran and Greece. Now, many varieties are cultivated in Spain, Iran, Greece, Turkey, France, Kashmir, India and the U.S. Most U.S. saffron comes from Spain. The most prized saffron comes from Kashmir, India and Iran, and the least expensive comes from Mexico. U.S. saffron is emerging as equal in value to Iranian saffron. Because saffron is very expensive, whole threads are often adulterated with safflower (also called false or bastard saffron or Mexican saffron), or the ground saffron is adulterated with turmeric.

Common Names: Spanish saffron, hay saffron and crocus. It is also called zaafaran (Arabic), jafran (Bengali), saffraan (Dutch), safran (French, German), za'afran (Hebrew), zaaffran/kesar (Hindi), kunyit kering (Indonesian, Malaysian), zaffarano (Italian), safuran (Japanese), fan huang hua (Mandarin), kesari (Nepalese), acafrao (Portuguese), shafran (Russian), azafran (Spanish), saffran

(Swedish, Norwegian), kashuba (Tagalog), kungumapu (Tamil), ya faran (Thai), zafran (Urdu) and nghe (Vietnamese).

Form: saffron is dried stigmas of the flower, which are dark red to reddish brown wiry threads. Saffron is sold as whole threads or ground. Each saffron flower has three stigmas. For flavor release, whole saffron threads need to be soaked in warm water for at least 10 minutes. Ground saffron has no flavor and is often adulterated with turmeric, marigold and other ingredients.

Properties: saffron is deep burgundy, reddish orange, bright orange or yellowish orange in color. It has a warm floral bouquet and a delicate, earthy, honey-like taste with bitter back notes. The flavors are stronger in the orange and red varieties from India, and milder in the yellow varieties from Spain. The deep red or orange types of saffron are of the best quality. The false types provide the yellow color but not the taste.

Chemical Components: saffron has 0.5% to 1.0% essential oil, mainly monoterpene aldehydes (such as safranal), terpenes (such as pinenes and cineol), and isophorone-related compounds. Its odor is due to 2,2,6-trimethyl-4,6-cyclohexidienal. The bitter taste is due to the colorless glycoside, picrocrocin. Its color is due to carotenoid pigments, especially the bright orange yellow crocin. The other carotenoid pigments are alpha- and beta-carotene, lycopene and zeaxanthin.

Saffron contains vitamins B1 and B2, vitamin A, calcium, sodium, manganese, potassium and phosphorus.

How Prepared and Consumed: saffron is used sparingly for coloring and flavoring foods. The threads are crushed and then infused or steeped in hot water or milk for about 20 to 30 minutes for its color to be fully extracted. The colored infusion is then added to foods. Saffron can also be toasted dry and added directly to rice dishes. Saffron pairs well with savory and sweet dishes, such as curries, shellfish, garlic, chicken, cream sauces, soups, polenta, rices, seafood, puddings, scones, cakes, flans, ice cream and milkshakes. Cumin, coriander, mints, rose essence, nutmeg, almonds, cashews, brown sugar, cardamom, milk, cinnamon, cloves and raisins complement saffron. Acidic ingredients such as vinegar or lemon juice inhibit saffron's color development. If used at high levels, it can create bitterness in foods.

Saffron is an important ingredient in the fish-based dishes of the Mediterranean, such as zarzuela de pescado from Spain and bouillabaisse from France. It is an essential ingredient in many flavored rice dishes: paella in Spain, biryani and pilaos in northern India, zafran pollous in Iran and risotto Milanese in Italy. Depending on the level of saffron used, the color of these foods varies from a deep orange to a turmeric-like yellow. Also, saffron is commonly used in desserts of India and Southeast Asia, such as kheer, kesari and ras malai, and in the yogurt drink lassi. In the U.S., saffron is used in sauces for flavoring lobster and shrimp, added to stewed potatoes and other vegetables or used as a marinade for roasts.

Spice Blends: paella blend, curry blend, biryani blend, bouillabaisse blend, rouille, kheer blend, kesari blend and pollou blend.

Therapeutic Uses and Folklore: in India, saffron is regarded as a sacred color, and, as a wet paste, is placed on foreheads when praying in temples. It is also used to symbolize the high castes in India. Other ancient cultures such as Greeks, Romans, Arabs, Chinese and Persians have used it to perfume their baths. The yellow color was a symbol of royalty as well as piety, and it is still used as a natural dye to color robes for religious and royal occasions. It is also used as a hair coloring for royalty and as a symbol of hospitality for the wealthy.

In India, saffron is used for urinary problems, as a digestive stimulant and for cramps, fevers, liver problems and rheumatism. It is eaten with ghee to prevent joint and uterus pains.

SAGE

A member of the mint family, sage has been used since ancient times in Europe as a seasoning and as a medicine. Italians and other Europeans associated it with wiseness, strengthening the memory and longevity. Sage was called *herba sacra,* meaning "sacred herb."

Scientific Name(s): Dalmation (English sage): *Salvia officinalis;* clary sage: *S. sclarea;* Spanish sage: *S. lavandulaefolia;* Greek sage: *S. triloba*. Family: Lamiaceae.

Origin and Varieties: sage is native to the northeastern Mediterranean region. It is cultivated in Eastern Europe, China, Turkey, Greece, Italy and the U.S. There are many varieties—narrow leaf sage, broad leaf sage, garden sage, tricolored sage, purple leaf sage, golden sage, red sage, clary sage (France, Russia and Morocco), Mexican sage, Spanish sage and pineapple sage.

Common Names: Dalmatian sage, English sage, garden sage, common sage and true sage. It is also known as maryaneya (Arabic), bhui tulsi (Bengali), salie (Dutch), sauge (French), salbei (German), marva (Hebrew), salvia (Italian, Spanish), seji (Japanese), ching chieh (Mandarin), salvie (Norwegian), salva (Portuguese), shalfey (Russian) and salvia (Swedish).

Form: common sage is slender, shiny, velvety or furry green when fresh. It turns into a silvery gray when dried. The leaves are oblong or spear shaped and are covered with short, fine hairs. Sage is used fresh or dried, whole, minced, chopped, crushed, rubbed (ground coarsely) or finely ground. Greek sage has larger and thicker leaves.

Properties: sage's flavor varies from a mild, balsamic to a strong camphorous and herbaceous-like taste, depending on the variety. It provides a cooling sensation in the mouth. Common sage is astringent, spicy and bitter with a camphor-like aroma. Greek sage has stronger, camphor-like notes. Spanish sage is less bitter with a flavor in between the flavors of Dalmatian sage and

Greek sage. Clary sage has muscat-like notes. Dried sage has a stronger flavor than fresh sage.

Chemical Components: sage has 1.5% to 3% essential oil, which is pale yellow to greenish yellow. Its composition differs in different varieties. Dalmatian sage has mainly thujone (28%), 1,8-cineol (12%), borneol (4%), camphor (23%), camphene (7.4%), α-humulene (5.31%), limonene (3.24%), β-caryophyllene (3.3%) and bornyl acetate (1.3%). Spanish sage lacks thujone but has more 1,8-cineol (27%) and camphor (20%). Greek sage has high levels of 1,8-cineol (39–67%) with smaller amounts of thujone and camphor.

Sage oleoresin, which is dark green to brownish green and very viscous, is usually extracted from Dalmatian sage, and 7 1/2 lbs. are equivalent to 100 lbs. of freshly ground spice. It is used as a natural antioxidant because of its high phenol content.

Dried sage contains vitamin A, calcium, potassium, niacin, calcium, magnesium, phosphorus and iron.

How Prepared and Consumed: sage is a popular spice in Italy, Greece and other regions of Europe. It tends to dominate the flavor of a dish, so it is used sparingly in European cooking. Sage pairs well with onions, butter, tomatoes, garlic, fatty fish, game, pork, poultry, gravies, breakfast sausages and turkey and goose stuffings. Italians use sage with veal, prosciutto and red wine sauce in the the well-known saltimboca alla Romana dish. Sage is also used in pizzas or with meat, pork and calf livers. Germans use sage in eel soups, the English use it in stuffed duck or pork, while the French use it in charcuterie, sausages and stuffings.

Middle Easterners use sage with fagioli in herb salads and in teas. In the U.S., sage is used in chowders, pork, poultry stuffings, sausages, baked fish and cheese.

Its oleoresin along with a small amount of ascorbic acid is used as a natural antioxidant to retain the color of processed meats.

Spice Blends: lamb tagine blend, charcuterie blend, saltimboca alla Romana blend and goose stuffing blend.

Therapeutic Uses and Folklore: the Romans considered sage to be a sacred herb. The Greeks used it for snakebites. Persians and Europeans associated it with long life, and the British associated it with prolonged memory. Traditionally, sage is considered good for throat and respiratory infections, calming nerves and reducing fevers. Sage helps soothe digestion, particularly after a rich, heavy meal. Sage also strengthens gums and whitens teeth.

SASSAFRAS

Sassafras is a native North American spice and is popular in the Creole and Cajun cooking of Louisiana. French settlers in Louisiana introduced many local ingredients, one of them being sassafras, into traditional French cooking to create spicier French flavors. The Choctaw Indians in Louisiana used ground or dried sassafras leaves (which are the major ingredient of file powder) to thicken stews and soups.

Scientific Name(s): formerly under *Sassafras variifolium,* now called *S. albidium.* Family: Lauraceae.

Origin and Varieties: sassafras is native to the eastern and southeastern U.S.

Common Names: ague tree, cinnamon wood, saxifrax, file powder, gumbo file and file (Creole). It is also called sasfras (Arabic), sassafras (French), fenchelholzbaum (German), sasafras (Hebrew), sassfrasso (Italian), haeng zhang (Mandarin) and sasafras (Spanish).

Form: sassafras is light green in color and is available as fresh or dried leaves of varying sizes that are used whole or ground. Ground leaves are called file powder. The dried bark of root is also used as a flavoring.

Properties: the leaf has a pleasant odor that is astringent, with a bitter and anisic-like, lemony taste. The root has a woody and very bitter taste with a camphor-like odor.

Chemical Components: the leaf has about 2 –3% essential oil, mainly α-pinene, myrcene, phellandrene, citral, geraniol, linalool and safrole. The bark has about 2% essential oil, mainly safrole (90%), which is not permitted as a food in several countries because it causes liver cancer. Safrole is removed from the leaf before it is used commercially in foods. The root has a high oil content of 6–9%, mainly α-pinene, phellandrene, eugenol, camphor and thujone.

How Prepared and Consumed: Cajuns and Creoles use file powder (which may contain other spices such as coriander, allspice and sage) to flavor and thicken meat, poultry, vegetable and seafood dishes, soups and stews. It is used to thicken gumbo, a traditional Louisiana dish made with seafood, andouille sausage, okra, crawfish or chicken seasoned with thyme, black pepper and paprika. File powder is added toward the end of cooking and blended in well. With any additional cooking, file powder will give the product a stringy texture.

Young sassafras leaves are chopped and used in salads.

Its essential oil, obtained from the safrole-free root, is used for flavoring meat products, confectionaries, soft drinks and beer.

Spice Blends: file powder and gumbo blend.

Therapeutic Uses and Folklore: the Choctaw Indians chewed sassafras roots to bring down fevers. Early European settlers in America used it in teas and cordials as a pain reliever and for high blood pressure, arthritis, gout, kidney problems and eye inflammations.

SAVORY

Savory was introduced to England by the Romans, and it was named by the Anglo-Saxons for its strong taste. Also called sweet savory, it was used with vinegar in ancient times as a seasoning, an aphrodisiac and a cough medicine.

The German word for savory, bohnenkraut, means "bean herb" because they found that it reduces their flatulent properties and complements green beans, dried beans and lentils. There are two types of savory, summer savory and winter savory, but most commercial applications use summer savory.

Scientific Name(s): summer or garden savory: *Satureja hortensis;* winter or mountain savory: *S. montana;* Spanish type: *S. thymbra.* Family: Lamiaceae (mint family).

Origin and Varieties: summer savory is cultivated in France, Germany, Spain, Great Britain, Canada, North Africa and the U.S. Winter savory is cultivated in Spain, Eastern Europe and Morocco.

Common Names: summer savory: tosinyi (Amharic), nadgh/zaatar barry (Arabic), bonenkraut (Dutch), sariette (French), bohnenkraut (German), santoreggia (Italy), sabori (Japanese), hsiang po ho (Mandarin), sar (Norwegian), segurelha (Portuguese), chabjor (Russian), sabroso/tomillo salsero (Spanish) and kyndel (Swedish).

Winter savory: sariette and savouree (French), santoreggia (Italy) and saborija (Spanish).

Form: savory leaves are small, narrow and green when fresh and brown green when dried. Summer savory leaves are slightly larger than those of winter savory. Savory is used fresh or dried and whole or crushed.

Properties: summer savory has a fragrant herbaceous aroma with a sharp and slight peppery and bitter aftertaste. It is reminiscent of thyme, mint and oregano. Winter savory has stronger, sharper and spicier notes than summer savory.

Chemical Components: summer savory has 0.1% to 0.25% yellow to dark brown essential oil, while winter savory has 0.2% to 0.25% essential oil. The essential oil of summer savory mainly consists of carvacrol (3.4–50.4%), thymol (trace to 22.5%), γ-terpinene (2.1–60.3%), ρ-cymene (3.7–5.3%), limonene (0.2–5.3%), myrcene (0.5–2.8%), camphene, α-thujene (1.8–4.2%), borneol (trace to 34%). These chemical constituents vary widely, depending on its origins and types. European types have very different flavor profiles than the North African or Canadian types.

Savory contains niacin, vitamin A, calcium, iron, potassium, phosphorus and magnesium.

How Prepared and Consumed: in the Mediterranean, savory is used mostly for vegetables, such as beans, cabbage, lentils, potatoes and mushrooms. The French use it as part of their bouquet garni blends to flavor sauces, game meats and stuffings. It pairs well with other sweet herbs, such as thyme, sage and rosemary. Savory is used in scrambled eggs and other egg dishes, salads, vegetable juices, stuffings, charcuterie and legume dishes. It enhances pork, chicken, hamburgers, fish chowders, soups and sausages. Savory is normally added toward the end of cooking so its flavor is not destroyed. Commercially, savory is used in liqueurs, bitters, vermouths, condiments, gravy, soup mixes and confections.

Winter savory goes well with stronger flavors such as game meats, liver pates and mutton.

Spice Blends: fines herbes, herbs de Provence, charcuterie marinade, hamburger marinade, cabbage sauce blend.

Therapeutic Uses and Folklore: in ancient times, savory was used to reduce pain from bee and wasp stings. It was believed to stimulate the appetite and promote digestion, especially of difficult to digest foods. It was called "herb of amor" and was said to increase the sexual and love feelings of those who eat it.

Savory has antioxidant properties.

SCREW-PINE LEAF/PANDANUS LEAF

Pandan is the Malaysian and Indonesian name for this fragrant leaf. Screw-pine leaf was the name given by English traders who traveled to Asia. In Southeast Asia, pandan leaf is used to wrap chicken, meat, fish and desserts before they are barbecued or steamed. They add distinct, sweet, floral-like notes to these products.

Scientific Name(s): Pandanus odoratissimus (North India); *P. amaryllifolius* (Southeast Asia); *P. latifolius* (Sri Lanka); *P. veitchii.* Family: Pandanaceae (screw-pine family).

Origin and Varieties: screw-pine is native to South Asia and Southeast Asia. There are a few varieties of screw-pine that differ in flavor and appearance, depending on their origin. The most aromatic types are from Southeast Asia and Sri Lanka. Kewra is the flower of the South Asian variety of screw-pine.

Common Names: pandan leaf. It is also known as kathey (Arabic), ketaky (Bengali), schraubenpalme (German), kewra (Hindi), pandano (Italian), kaitha (Malayalam), daun pandan (Malaysian, Indonesian), skrupalme (Norwegian), keora (Punjabi), rampe (Sri Lankan), thazhai (Tamil), bai toey hom (Thai) and skruvpalm (Swedish); flower—kewra or keora (Hindi).

Form: Screw-pine leaf is long, thin, narrow and green. It is sold fresh, frozen or dried. The leaves and flowers also come as extracts.

Properties: the dried leaves are less fragrant than the fresh leaves. The leaves have to be bruised or boiled in order to release their flavor. The leaves have a rose-like, almondy and milky sweet, vanilla-like flavor. The flowers are golden yellow and have a fragrant, strong, sweet aroma.

Chemical Components: screw-pine has low levels of essential oil, including 2-acetyl-1-pyrroline (which also gives the aroma in Thai and Basmati rice), styrene, linalool and β-cayophyllene. It also contains piperidine-like alkaloids (pandamarine, pandamarilactones) that give screw pine its milky, floral-like taste.

How Prepared and Consumed: the leaf is used in curries of Sri Lanka and in Malaysian, Balinese and Thai cooking. It is commonly used as a flavoring and coloring in Malaysian and Singaporean cooking, especially in Malay- and Nonya-style dishes. The screw-pine or pandan leaves are tied in a knot and placed in soup or stews that are being cooked. The leaf is also raked with the tines of a fork to release its aroma, pounded to release its aromatic juice or even boiled to obtain its flavor. Pandan leaves are used as wrappers in Southeast Asian cooking to provide a distinct flavor to the foods. They are wrapped around chicken, pork, glutinous rice, fish and desserts before grilling, roasting, barbecuing or steaming.

Pandan leaves also enhance the flavor of seasoned rices, puddings, beverages and curries. Nasi lemak, nasi kuning and nasi padang are some of the fragrant pandan-flavored rices eaten in Malaysia and Indonesia. It pairs well with coconut milk, glutinous rice, lemongrass, milk, brown sugar and turmeric. It also provides color to Malay- or Nonya-style glutinous rice-based desserts, candies, puddings, soups and coconut drinks.

Screw-pine flower, which is more delicate and fragrant than the leaf, is used in North India to perfume biryanis. It goes well with rices, coconut, lemongrass, brown sugar, star anise, cumin and nutmeg. Its extract, called kewra, is also commonly used to flavor Indian desserts such as ras gulla (cottage cheese in syrup), gulab jamun (fried cottage cheese in syrup), rasmalai (cottage cheese with condensed milk), cakes and beverages.

The commercially available pandan leaf extract is much too bright green and does not totally capture its true flavor and color profiles.

Spice Blends: nasi lemak blend, nasi kuning blend, biryani blend, kueh lapis blend, rendang blend, rasmalai blend and gulab jamun blend.

Therapeutic Uses and Folklore: in India, screw-pine leaves are sacred to the Hindu God, Shiva. In many Indian villages, the leaves are also tossed into open wells to scent the drinking water.

SESAME

Sesame originated in India and is one of the oldest spices known. It has been used in the Middle East, Asia, Southern and Eastern Mediterranean and northern Africa. The Assyrians believed their gods drank sesame wine, while the Greeks and Romans used it on their long journeys to conquer new lands. There are three types of sesame, white, brown and black, each being preferred by specific cultures.

Scientific Name(s): Sesamum indicum. Family: Pedaliaceae.

Origin and Varieties: sesame is indigenous to the Near East and northern India and is cultivated in the Middle East, China, Egypt, East Africa, Greece and Mexico. There are many varieties, and, depending on their origin, they come with different colors and flavors.

Common Names: seli't (Amharic), benne (Wolof language of Senegalese), sem sum (Arabic), sousma (Armenian), til (Bengali), sesamzaad (Dutch), gingelly (English), till (French), sesam (German, Swedish), sumsum (Hebrew), gingili (Hindi), sesamo (Italian), goma (Japanese), chitellu (Malayalam), bijan (Malaysian/Indonesian), zi ma zi (Mandarin), gergelim (Portuguese), ajonjoli (Spanish), ellu (Tamil), nga (Thai), susam (Turkish), til (Urdu) and vung (Vietnamese).

Form: sesame seeds are small, flat, oval, unhulled seeds that are creamy white, brown or black. Hulled seeds are pearly white in color. Sesame is used whole, ground or as a paste or oil. The oil can be cold pressed or hot pressed and/or refined.

Properties: sesame has a nutty sweet aroma with a rich, milk-like buttery taste. After roasting or toasting, the white seeds become golden colored and acquire a delicate almond-like flavor. Sesame oil is deep brown, aromatic and nutty. Black sesame has a stronger, earthier and nuttier taste.

Depending on regional preferences, sesame seeds are roasted or left unroasted before the oil is extracted, giving rise to different flavor profiles.

Chemical Components: sesame has no perceptible volatile oil. It has 45–65% fixed oil, mainly olein, stearin, palmitin, myristin, sesamin and sesaminol, with 30% protein. The oil contains mostly oleic acid (40%), linoleic acid (45%) and saturated acid (10%). When the seeds are toasted, 2-furylmethanthiol, guaiacol, phenylethanthiol and furaneol are formed, which give toasted sesame oil its characteristic flavor.

Sesame has antioxidant properties due to sesamol.

How Prepared and Consumed: Romans, Arabs, Turks, Greeks and Indians have eaten sesame with honey, jaggery, rice, dates or licorice. African slaves brought it to the U.S. White sesame seeds are used whole to add texture to bread and rolls (hamburger buns), biscuits, crackers, pastries, cooking oils, salad dressings and margarine. Sesame pairs well with chicken, pork, breads, noodles, soy sauce, shallots and mustard.

Middle Easterners use sesame to flavor a wide variety of foods. It is used in dips, sauces, breads, cakes and sweets (halva). Untoasted white sesame seeds are ground to make a paste called tahini (sesame butter) that is used to flavor hummus or baba ghanouj. Tahini is also used as a spread for pita breads, a dip for vegetables, an accompaniment to kebabs and fruits and as a sauce in cooking. Zahtar, a spice blend of toasted sesame seeds with thyme and sumac, is used as a dip and to add zest to vegetables, grilled meats and breads.

Asians prefer the more intense flavor of roasted sesame seeds. Indians dry roast the seeds to develop a more intense flavor and a crunchier texture before they are added to many Muslim-style savory dishes. Black sesame seeds are popular in Japanese and Chinese cooking. Chinese and Japanese dry roast sesame seeds and sprinkle them as a garnish on confections, rice, meats or steamed vegetables. They also grind them into a paste or extract the oil to flavor cold sesame noodles, sauces and confections.

The hot-pressed unroasted oil, called gingelly oil, is light and is different in flavor from the roasted sesame oil. It is used in Indian pickles and as a cooking medium in India, Burma and Southeast Asia. Sesame oil from roasted sesame seeds adds flavor to many stir-fry dishes and baked goods in East Asia and Southeast Asia. This darker sesame oil is commonly used in Korean and Szechwan cooking with chile peppers, garlic and ginger. For the best sesame oil flavor, the oil is added to a dish just before serving or is added toward the end of cooking. In Korea, dumplings called mandu contain fillings made of garlic, cloves, sesame seeds, roasted sesame oil, kim chee and soy sauce.

Spice Blends: zahtar, tahini blend, shichimi togarashi, sesame noodle blend, mandu filling blend, stir-fry sauce blends and dumpling sauce blend.

Therapeutic Uses and Folklore: in Asia, sesame is thought to have magical powers and bring good luck. Sesame was traditionally used as a laxative, to relieve constipation and hemorrhoids. Gingelly oil (called nalanai in Tamil), commercially available in India and Southeast Asia, which possesses no sesame flavor, is used for hair and body before baths. It is believed to rejuvenate the body and mind.

Sesame has antioxidant properties.

SORREL

The name sorrel is derived from the Germanic word, *sur,* and the old French word, *surele,* both meaning sour. It is an ancient herb used by Egyptians and Europeans to impart acidity to foods. Today, it is a popular flavoring for whitefish, soups and salads in French cooking. Another sorrel called Jamaican sorrel is from the Hibiscus family.

Scientific Name(s): there are a few cultivated varieties of sorrel—garden, French, round leafed and spinach dock. Garden sorrel: *Rumex acetosa* (6 inches broad arrow-shaped leaf); French sorrel: *R. scutatus* (buckler leaf); sheep sorrel: *R. acetosella* (smaller than garden sorrel but arrow shaped). Family: Polygonaceae.

Origin and Varieties: native to Europe and west Asia, sorrel is now cultivated in France, Egypt and parts of Europe and the U.S.

Common Names: garden sorrel, wild sorrel, French sorrel, sourgrass and little vinegar. It is also called oseille (French), sauerampfer (German), hamtzitz (Hebrew), acetosa (Italian) and acedera (Spanish).

Form: sorrel has large, light to dark green, oblong-shaped, spinach-like leaves. It comes as fresh or frozen and chopped or whole.

Properties: its taste ranges from a refreshing, sharply acidic or astringent spinach-like taste with bitter notes to a milder, lemony taste. The younger leaves are less acidic. The French variety has slight citrus-like notes. The dried leaf loses its citrusy-like flavor.

The fresh leaf is high in vitamin C, potassium, phosphorus calcium and magnesium.

Chemical Components: sorrel has a high level of binoxalate of potash, which gives it the acidic taste.

How Prepared and Consumed: Romans and Egyptians used sorrel in ancient times to offset rich, heavy foods. It is used typically in French and Egyptian foods, such as soups, sandwiches, salads, poached salmon, stewed or braised meats and poached eggs. It goes well with fish, onions, pepper, potatoes, meats, pork, veal, eggs, salads, cream-based sauces and goat cheese. Sorrel is pureed to flavor goose, fish or soups or for use in condiments for meats. It is also used in teas. Tough meats can be wrapped in sorrel leaves to tenderize them before cooking.

Sorrel is cooked for a minimum time to preserve its fresh flavor. To prevent sorrel from blackening and developing a metallic taste, only stainless steel knives and non-iron pots are used. Sorrel is a natural acidifier and can be a substitute for fresh lemon in salads, stews and sauces.

Spice Blends: green sauce blend, potato soup blend and meat marinade.

Therapeutic Uses and Folklore: the Romans and Greeks used sorrel to aid digestion and temper the effects of rich foods, to treat liver problems and for throat and mouth ulcers. It is traditionally used to treat scurvy and chronic skin conditions and to lower fevers.

Jamaican sorrel (red sorrel, Florida cranberry or rosselle) is not a true sorrel. It was introduced to Jamaica from Asia in the eighteenth century.

Scientific Name(s): Hibiscus sabdariffa. Family: Malvaceae.

Common Names: it is called rosella in Spanish, bissap rouge in Senegalese, karkadeh in Egypt, rozelle in French, karkadi in German, and carcade in Italian.

Form: Roselle is the fleshy young calyces surrounding the immature fruits. It also comes in dried form.

Properties: Jamaican sorrel is very acidic. It resembles cranberry in color and acidity.

How Prepared and Served: it is used to flavor drinks, jams, jellies, wine and sauces in the Caribbean, Mexico, West Africa and Egypt. Roselle is used fresh in salads, especially fruit salads, with cooked vegetables, and in sauces, stews and pies or tarts. Roselle is also dried and used as natural coloring. The Caribbeans enjoy it as a traditional Christmas drink (also called sorrel) that is mixed with spices and rum. Africans add sugar to their roselle drink to tone down the sourness. Roselle is also used in curries and chutneys in India and Southeast Asia and in foods and beverages of Cuba, Central America, Mexico and Florida.

Therapeutic Uses and Folklore: Indians, Mexicans and Africans use it as a diuretic, to thin blood and to lower blood pressure.

STAR ANISE

Star anise, a native spice of China, is a dominant ingredient in Chinese five spice blend, and it is an essential flavoring for many Chinese and Vietnamese foods. It is called jiao huei hsiang in Mandarin, meaning "eight horned fennel," since the Chinese perceived its taste to be similar to fennel. It is often adulterated with other "false" types of star anise from Japan.

Scientific Name(s): Illicium verum. Family: Illiaceae.

Origin and Varieties: star anise is native to southwestern China and northern Vietnam. It is now cultivated in China, Vietnam, India, Japan and Korea.

Common Names: Indian anise, Chinese anise and Badian anise. It is also called albadyan (Arabic), sternanijs (Dutch), anis etoile/anis de la chine (French), sternanis (German), chakriphool (Hindi), bunga lawang (Indonesian, Malaysian), anice stellato (Italian), jiao huei hsiang (Mandarin), anis estrelado (Portuguese), badjan (Russian), badian/anis estrella (Spanish), stjarnanis (Swedish), anasipoo/lavangaipoo (Tamil), poy kak bua (Thai) and cay hoy (Vietnamese).

Form: star anise is an irregular star-shaped fruit with eight carpels joined around a central core, each carpel containing a seed. It is used in the following forms: dried whole, broken pieces or ground. The fruits are picked when green, before they ripen and become sun dried.

Properties: the carpels are reddish brown and hard. Star anise has a licorice-like, sweet and pungent flavor, similar to fennel and anise. It can leave a bitter aftertaste if used at high levels. Star anise gives a spicy, sweet flavor that becomes more intense as it is cooked. The seeds have less flavor than the pods, and the broken pieces of pods are less aromatic.

Chemical Components: the fresh fruit has 5% to 8% essential oil, and the dried fruit has 2.5% to 3.5%, mainly anethole (85–90%), α-pinene, phellandrene, ρ-cymene, 1,4-cineol, limonene, and *d*-terpineol. It has about 20% fixed oil.

How Prepared and Consumed: star anise characterizes Chinese, Vietnamese and other Chinese-style cooking in Asia. It is used in marinades, barbecues, roasts, stews and soups that require long simmering. The Chinese have introduced it to every region where they have settled. Star anise goes well with roast poultry, roast pork, braised fish and beef and steamed or roast duck.

In Southeast Asia, star anise is used in Chinese-style dishes, such as simmered beef, stir-fried vegetables and steamed chicken. It is a popular spice in North Vietnamese beef noodle soups called phos. Malaysians and Singaporeans add it to their curries, soups and sauces to give them unique tastes.

Brought to India by Chettiars of Tamil Nadu who traded in China and Southeast Asia, it is a popular spice in Kashmiri, Goan and Chettinand cooking. It is also used in Caribbean masalas and spice blends.

The Europeans use star anise to flavor cordials, liqueurs, syrups, jams and confectionaries.

Star anise pairs well with mint, cinnamon, ginger, soy sauce, orange, rose essence, curry leaves, chile peppers, curry powder and black pepper. Typically, it is discarded before the dish is served.

Spice Blends: Chinese five spice, hoisin, barbecue blend, Singapore pork curry blend and Chettinand chicken curry blend.

Therapeutic Uses and Folklore: east Asians use star anise to relieve colic and stomach pains. They chew it after a meal to promote digestion and sweeten breath. Star anise is also used in teas to cure sore throats and coughs. Its oil is used as an ingredient in cough lozenges.

SUMAC

The Romans liked sumac's sour taste and called it Syrian sumac. Native to Iran, sumac is now a popular spice in Turkey, Iran, Egypt and the Middle East. It is found growing in the mountains around the Mediterranean.

Scientific Name(s): Rhus coriaria. Family: Anacardiaceae (cashew family).

Origin and Varieties: sumac is native to the Mediterranean region. There are many varieties, and some are poisonous if eaten or are toxic to the skin. The bright, purplish red berries are safe, while the white or green white types are poisonous to touch.

Common Names: summak/summa (Arabic), kankrasringi (Bengali), zuurkruid (Dutch), shumac/Sicilian sumac (English), somagh (Farsi), sumac (French), sumach (German), kankrasing (Hindi), sommacco (Italian), sumakh (Russian), zumaque (Spanish), karkhada garchingi (Tamil) and sumak/somak (Turkish).

Form: sumac are small, red, dried berries that are sold whole but are usually ground and sometimes mixed with salt. The whole berries are soaked in water for about 20 minutes and are then squeezed to extract their juice before they are dried. The wild variety has green white berries that are poisonous.

Properties: sumac is reddish purple in color with a fruity, tart and astringent taste.

Chemical Components: sumac has about 4% tannins and high acid content, especially malic acid.

How Prepared and Consumed: the ground spice is a popular spice that is sprinkled over salads, meats, fish stews and rice in Turkey, Iran and other Middle Eastern regions. It is also commonly eaten with sliced onions, as an appetizer and to season kebabs. Sumac is an important ingredient in spice blends of these regions, such as zahtar and dukkah, that perks barbecued meats in the Middle East. It is mixed with sesame, thyme and other spices in Lebanon and Syria, boiled, and the thick extract is added as a lemon substitute for flavoring drinks, salads, vegetables, meats and fish stews.

Sumac goes well with pine nuts, cilantro, parsley, garlic, chile pepper, cumin and coriander. It forms an important ingredient in spicy dips that are eaten with flat breads that are sold in the streets of Egypt and other parts of the Middle East. It is rubbed on roasted lamb, which is then grilled to give m'choui (Morocco) or khouzi (Egypt).

The Native Americans make a sour drink from ground sumac.

Spice Blends: zahtar, dukkah and khouzi marinade.

Therapeutic Uses and Folklore: Middle Easterners use it to ease an upset stomach.

TAMARIND

Tamarind is thought to have originated in Africa or India and was called tamr Hindi in Arabic, meaning the "date of India." Called asam koh, puli or tamarindo, it is now an important spice in many Asian and Latin American regions. Tamarind is a sweet acidulant used in many South Indian foods such as sambars, fish curries, chutneys and rasams and in Latin American beverages and candies. It is also one of the essential flavorings in Worcestershire sauce.

Scientific Name(s): Tamarindus indica. Family; Caesalpiniaceae (closely related to Leguminosae family).

Origin and Varieties: tamarind is indigenous to East Africa and India and is now cultivated in India, the Middle East, Africa, Southeast Asia and the Caribbean.

Common Names: Indian date. It is also called tamr Hindi (Arabic), tentu (Bengali), ma-gyi-thi (Burmese), asam koening (Dutch), tamarindi (Finnish), tamarin (French), tamarinde (German), tamrhindi (Hebrew), imli (Hindi, Urdu), ampil khui (Khmer), tamarindo (Italian), assam jawa (Malaysian, Indonesian), lo won tzu (Mandarin), tamarind (Russian), siyambula (Singhalese), tamarindo (Spanish), tamarind (Swedish), sampalok (Tagalog), puli (Tamil, Malayalam), makham (Thai) and me (Vietnamese).

Form: tamarind refers to the dark brown, dry, ripe fruits/pods or the fleshy pulp inside the pods. The pods are cinnamon colored with an inner dark brown, fleshy pulp that surrounds the black seeds. The pulp, which is fibrous and sticky, is the part used in cooking. The pods are dried, peeled, deseeded and packed into blocks. Tamarind is sold as fresh whole pods, dried pulp slices, dried pulp concentrate or paste or dried as a solid block. The concentrate has no seeds or fiber. It is also sold dried and ground. Tamarind juice is the liquid obtained when the paste is combined with warm or hot water.

Properties: the pulp has a sharp, sweet, sour, fruity taste and a sweetish, brown aroma. It has acidic, molasses-like notes.

Chemical Components: the concentrate has 20% organic acids, mostly tartaric acid (12%), malic and succinic, and sugars (up to 35%). The sugars are mainly glucose and fructose. It also has small amounts of terpenes (limonene, geraniol), methyl salicylate, safrol, cinnamic acid and pyrazine. Tamarind is high in pectin.

Tamarind has calcium, phosphorus, potassium and sodium.

How Prepared and Consumed: before use, tamarind pulp is soaked in hot water for about 10 minutes and is then squeezed and strained to obtain juice. The seeds and pulpy material are discarded. Tamarind enhances fish, meat, poultry and vegetables, and goes well with chile peppers, legumes, shallots, vinegar, brown sugar, curry powder and mushrooms. The young leaves and flowers are cooked as vegetables in India and are used in soups and salads in Africa.

Tamarind juice to an Easterner is like lemon juice to the Westerner. It is used to provide sweet and sour notes to many Indian and Southeast Asian dishes. It tends to complement fiery hot dishes, so it is used in abundance in South Indian fish curries, sambars (stews), rasams (soups), chutneys and Gujerati lentils, vegetables and jams. Tamarind dip is made for samosas and dahi wada (dumpling of North India) that is smothered with yogurt and tamarind sauce.

Tamarind is also popular in Southeast Asian cooking because it moderates the fiery notes of sambals, curries and satay marinades. Many dips and condiments in Southeast Asia are prepared with tamarind juice. Tamarind is also used in Thai salads, tom yom soups and stir-fries, Chinese hot and sour soups, Malaysian and Indonesian satay marinades, sauces and curries and Vietnamese soups, stews and curries.

Tamarind goes well with soy sauce, palm sugar, galangal, ginger, garlic, turmeric, clove, fermented shrimp paste, peanuts and black pepper. British love of Indian spices resulted in another popular condiment, Worcestershire sauce, which contains tamarind and other characteristic Indian flavors.

Tamarind is popular in Mexican candies, Jamaican rices and desserts. It is used in Puerto Rican juice drinks and as a cooling drink in India, Thailand, Latin America and the Caribbean. The Muslims take a tamarind drink at Ramadan, their holy month. It is also used in Angostura bitters and sherbets.

Because tamarind is high in pectin, it can be used as a natural stabilizer in foods and beverages.

Assam gelugor is used in Southeast Asian and South Indian cuisines. It is not tamarind even though it has a similar flavor.

Scientific Name: Garcinia atroviridis.

How Prepared and Consumed: assam gelugor is dried and thinly sliced, and during cooking, it swells up and is discarded before the food is served. It is commonly used in Indonesian, Malaysian and Singaporian cooking. Assam gelugor should also not be confused with asam kandis fruit (scientific name: *Garcinia globulosa*). Asam kandis is another popular souring fruit used in Indonesia and Malaysia, whose slices are smaller or kokum (scientific name: *G. indica*) that is commonly used in Gujerati and Kerala cooking.

Spice Blends: sambal blends, south Indian curry blends, rasam blend, vindaloo blend, rendang blend, Worcestershire sauce, tom yom blend and Nonya laksa blend.

Therapeutic Uses and Folklore: in Thailand, traditionally, the juice is used to prevent dysentery and reduce blood pressure. In India, tamarind is used to relieve asthma and fevers and is also used as a mild laxative. According to the Ayurvedics, it heals wounds and joints, sore throats and bronchial disorders.

In India, tamarind is also mixed with salt and becomes an excellent brass polisher.

TARRAGON

Also called estragon, tarragon derives its name from the Greek word *drakoon,* the Arabic word *tarkhun* and the Spanish word, *taragoncia,* all of which mean little dragon and which describe its coiled, serpentine root. It is native to Southern Russia and Western Asia and the Mongols introduced it to the West. Today, it is a popular herb in southern European cooking, especially in French cuisine.

Scientific Name(s): Artemisia dracunculus. Family: Compositae (or Asteraceae).

Origin and Varieties: tarragon is native to eastern and central Europe, southern Russia and western Asia. It is now cultivated in southern Europe, Russia and the U.S. There are two varieties: French tarragon (sometimes called German tarragon) and Russian tarragon, also known as false tarragon. In the U.S., French tarragon is commonly used.

Common Names: estragole and little dragon. It is also called tarkhun (Arabic), ai hao (Mandarin), estragoa (Dutch), dragon (Dutch, Swedish), estragon (French, German, Italian, Norwegian, Russian), taragon (Hebrew), estrago'n (Spanish), targone (Italian), esutoragon (Japanese) and estragao (Portuguese).

Form: tarragon has long, narrow, grayish green leaves. Fresh, it is used whole, chopped or minced. Dried, it is used whole, crushed or ground. Sometimes the stems are included with the leaves.

Properties: French tarragon has a cool, sweet, licorice-like aroma with slight bitter tones. Its taste is green and herbaceous, with anise- and basil-like notes. It is more delicate in flavor than the Russian variety. The Russian variety has larger leaves, lacks the anisic taste and is slightly bitter and harsh in flavor. Dried tarragon has hay-like tones and loses its flavor quickly.

Tarragon contains niacin, vitamin A, manganese, potassium, calcium, iron, magnesium, phosphorus and sodium.

Chemical Components: fresh French tarragon has 0.5% to 2.5% essential oil, which is pale yellow to amber in color and consists mainly of 60–75% estragol (also called methyl chavicol). It also has 10% anethole (which mainly contributes to tarragon's aroma), α- and β-pinenes, camphene, phellandrene, limonene and myrcene. There is less volatile oil (about 0.3–0.8%) in the dried herb. Russian tarragon does not have the estragol and, thus, is considered inferior to French Tarragon.

The oleoresin is dark green and viscous, and 2 lbs. are equivalent to 100 lbs. of ground dried leaves.

How Prepared and Consumed: in Europe, tarragon is popularly used to flavor many sauces. It is a favorite herb in France and characterizes French Dijon mustard and sauces based on sour cream, eggs and mayonnaise such as tartar, béarnaise and hollandaise. It is also used in cream soups, salads, omelettes and gravies. These sauces are added to broiled, baked or fried fish, meat and chicken. Tarragon is an important component of fines herbes and is one of the optional ingredients of herbs de Provence of France.

Armenians use tarragon on vegetables, fish and meat dishes. In the U.S., it is used in vinegar, tartar sauce, eggs, chicken and seafoods.

Tarragon enhances roast chicken or turkey and can also be used in stuffings. It is combined with vinegar and capers in salad dressings. Tarragon pairs well with salmon, chicken, veal, lobster, scallops, lentils, vinegar, lemon juice and mild herbs such as parsley and chives. Cooking intensifies and changes its flavor, so it is usually added to a dish toward the end of cooking to retain its characteristic aroma and taste.

Spice Blends: fines herbes, béarnaise sauce blend, Dijon mustard blend, tartar sauce blend and herbs de Provence.

Therapeutic Uses and Folklore: the Arabs used tarragon in ancient times to cure insomnia and to dull the taste of medicines. In traditional medicine, tarragon is used to increase appetite, to flush toxins from the body and as a digestive stimulant, especially by heavy-meat-eating cultures. Tarragon has been used as an anesthetic for aching teeth, sores or cuts.

THYME

The name thyme comes from the Greek word, *thymon,* meaning courage, and it was a symbol of elegance and grace for the Greeks. The Egyptian word for thyme, tham, means strong smelling, and it was mainly used in embalming. During the Middle Ages, thyme was a symbol of courage and love. In the U.S., it is an important leafy spice in New England clam chowder.

Scientific Name(s): garden (common thyme): *Thymus vulgaris;* white (or Spanish thyme): *T. zygis;* wild (or creeping thyme/serpolet): *T. serpyllum;* Spanish origanum: *T. capitatus;* lemon thyme; *T. citriodorus;* Moroccon thyme: *T. satureioides;* broad leaf thyme: *T. pulegioides;* herba barona, caraway thyme. Family: Labiatae (or Lamiaceae).

Origin and Varieties: thyme is native to the Mediterranean regions and is now cultivated in France, Spain, Italy, Portugal, Germany, Morocco, Algeria, Egypt, England, the Caribbean and the U.S. There are numerous varieties of thyme, each different in flavor: garden thyme, wild thyme, lemon thyme, orange thyme, anise thyme, caraway thyme and Moroccon thyme. The U.S. mainly uses the Spanish and Moroccon varieties.

Common Names: garden thyme, red thyme and French thyme. It is also called za'tar (Arabic), tijm (Dutch), awishan shirazi (Farsi), thym (French), thymian (German), timin (Hebrew), banajwain (Hindi), timo (Italian), taimu (Japanese), pai-li-hsiang (Mandarin), tomilho (Portuguese), timjan (Russian), tomillo (Spanish), timjan (Swedish) and hasha (Urdu).

Wild thyme: creeping thyme. It is also called serpolet (French), quendal (German), serpillo (Italian), kryptimian (Norwegian) and serpoleto (Spanish).

Form: The fresh leaves are green, pointed, oval, slightly rolled and covered in fine hairs. They can be broad leafed or narrow leafed or large or small based on the variety. They are used fresh or dried. The dried form is sold whole, minced or ground.

Properties: Thyme has a piney, phenolic taste with a herbaceous and slightly floral aroma. Some have bitter, slightly lemony and minty notes. It turns black in an acidic medium (tomato sauce) and loses its aroma quickly with heat. Its flavor is retained in the dried herb. Moroccon thyme is broader and shorter leafed with a higher volatile oil content than Spanish thyme.

Chemical Components: The essential oil content of thyme ranges from 1.5% to 5%. Common thyme yields about 1.5% to 2.0% essential oil, which is colorless to pale yellowish red and predominantly contains thymol. Spanish thyme (which provides 90% of the world's thyme oil) mainly consists of phenols with 12–61% thymol and 0.4–20.6% carvacrol, 1,8-cineole (0.2–14.2%), ρ-cymene (9.1–22.2%), linalool (2.2–4.8%), borneol (0.6–7.5%), α-pinene (0.9–6.6%), and camphor (0–7.3%). Most other varieties, such as Moroccan thyme, have lesser thymol.

Wild thymes yield less volatile oil, about 0.5%.

Its oleoresin is green to brownish green and is viscous, and 4 lbs. of oleoresin are equivalent to 100 lbs. of freshly ground thyme.

Thyme contains vitamin A, niacin, calcium, sodium, potassium, magnesium, phosphorus and sodium.

How Prepared and Consumed: thyme is added during cooking or toward the end of cooking or is even sprinkled over the dish at the serving table. It pairs well with fatty foods, such as mutton, goose, duck, pork, tomato sauces, stuffings, roasts, fish and wine.

Thyme is a very important herb in European cooking, especially in the southern areas of Europe, for stuffings, vegetable soups, mutton stews, beef bourguignonne, long-simmered poultry, fish, meat and game. It is used to flavor pickled olives and the liqueur benedictine, and it is combined with marjoram in many dishes. It is an important ingredient in herbs de Provence and bouquet garni. The French tie up thyme with other herbs in bundles and add this to soups, sauces and stews. The bundles are removed before serving.

In the Middle East, thyme is found in spice mixtures such as zahtar from Jordan and dukkah from Egypt. It is used with roasted sesame seeds, black pepper and other spices to flavor meat or is eaten with bread and olive oil.

In the U.S., thyme is a prominent flavoring in Manhattan and New England clam chowders, other seafood dishes and in poultry stuffing and sausages. Cajun and Creole cooking use thyme as part of a coating mix for fish or meat before they are blackened. The blackening spices become dark brown but not charred, so their flavor is retained. It is also a popular spice in the Caribbean, where the broad-leafed variety of thyme is combined with allspice and other spices to make jerk seasoning, stews and curries.

Spice Blends: bouquet garni, dukkah, zahtar, Cajun spice, herbs de Provence, curry blend and jerk seasoning.

Therapeutic Uses and Folklore: thyme was used by the ancient Greeks to purify temples and by Romans to infuse bath waters. In traditional European medicine, thyme was used to treat infections, digestive complaints and respiratory ailments, such as bronchitis, laryngitis and whooping cough. It has been used to treat circulation problems, relieve muscle pains and as a tranquilizer, as an expectorant and to help hangovers.

Thyme is also used as a mild insect repellent.

TURMERIC/TUMERIC

The name turmeric comes from the Latin *terra merita* meaning deserving earth. The use of turmeric dates back to 1500 B.C. when turmeric was mentioned in the Vedas, the sacred scriptures of the Hindus. Turmeric is an important spice in India where it is used as a cosmetic and a coloring agent. It is used in religious ceremonies and is a traditional medicine for treating digestive complaints, colds and wounds.

Also called "Indian saffron" by Westerners, turmeric is an important ingredient in the curries, stews and sauces of India and Southeast Asia.

Scientific Name(s): Curcuma longa. Family: Zingibearaceae.

Origin and Varieties: indigenous to South and Southeast Asia, turmeric is cultivated in India (Alleppey and Madras), Sri Lanka, Java, Malaysia, China, Peru and Jamaica. There are many varieties, but the Indian type is the most valued turmeric. The U.S. uses the Alleppey type from India, which has the best color value and flavor.

Other types of turmeric, including a variety called zeodary, are used abundantly in Southeast Asia. Zeodary is discussed separately.

Common Names: Indian saffron. It is also called ird (Amharic), kurkum (Arabic), anwin (Burmese), yu-chin (Chinese), geelwortel (Dutch), curcuma (French, Italian, Portuguese), cu'rcuma (Spanish), gelbwurz (German), kurkum (Hebrew), haldi (Hindi), ukon (Japanese), romiet (Khmer), shuva

(Malayalam), zholty imbir (Russian), kaha (Singhalese), dilaw (Tagalog), manjal (Tamil), kunyit (dried)/kunyit basah (fresh) (Indonesian, Malaysian), haladi (Urdu), khamin (Thai) and cu nghe (dried), bot nghe (fresh)(Vietnamese).

Form: turmeric is a rhizome with a brownish yellow skin and a bright orange yellow interior. It is cured (boiled or steamed) to intensify its aroma and color, then dried and sold whole or ground into powder. It is used fresh (chopped, grated, cut) and dried (whole or powder). Its color varies depending on its origin. The fresh form looks like ginger except it is thinner and its inner flesh is yellow. In Indonesia, the fresh leaves are used whole as wrappers or are chopped and used in local dishes.

Properties: fresh turmeric has a bright orange flesh. It has a musky and earthy aroma with gingery, slightly bitter, peppery notes. It loses its flavor when exposed to light. The dried rhizome is lemon yellow to orange yellow in color. Madras turmeric is lighter and brighter yellow, while Alleppey tumeric is a brownish yellow color on the outside with a deep yellow color to orange yellow on the inside.

Turmeric leaves are large and aromatic. Turmeric contains vitamin C, potassium, iron, sodium, phosphorus, magnesium and calcium.

Chemical Components: turmeric contains three components: an orange yellow colored essential oil (1.5–6%); a yellow coloring matter called curcumin (3–4%) and starch. The essential oil contains turmerone (30%), dihydrotumerone and zingiberene (25%), sabinene (58–60%)), 1,8-cineole (1.0%), borneol, sesquiphellandrene, curcumene, bisabolene and α-phellandrene. The essential oil is highly aromatic. The Alleppey turmeric has 3.5% to 6.9% essential oil, with 5.3–6.5% curcumin, while the Madras turmeric has 2.2% essential oil, with 2.1% curcumin.

In addition to its use as a natural colorant, there is growing interest in using turmeric's constituent, curcumin, as an antioxidant and antibacterial agent.

The oleoresin is a deep red to orange red viscous oil and contains 15–25% volatile oil and coloring matter. About 8 lbs. are equivalent to 100 lbs. of freshly ground turmeric.

How Prepared and Consumed: turmeric pairs well with cilantro, kari leaf, galangal, ginger, mustard seeds, nutmeg, coconut, pandan leaf, bay leaf, lemongrass, cumin, clove, coriander, mustard seeds, dill and black pepper. It is an essential ingredient in all curry powders and is used in most Indian dishes to help digest the complex carbohydrates.

In Asian Indian and Southeast Asian cooking, turmeric is a popular spice for providing color and flavor to curries (especially vegetarian curries), pickles, soups, lentils, vegetables (especially potatoes and cauliflower), fried fish, pullaos and desserts.

In Thai, Malaysian and Indonesian cooking, turmeric is commonly used freshly grated in yellow and red curries, yellow rice (nasi kuning), stews, laksas and vegetables. It is combined with chile peppers, lemongrass, lesser galangal, cinnamon, clove, ginger and coriander to create bumbu, a spice blend used for rendangs, gulais and other fiery sauces and soups.

In western countries, turmeric imparts a bright yellow coloring to mustard condiments and sauces. It is also used in deviled eggs, Worcestershire sauce, chicken stock, potatoes, cheeses, sausages, egg salads, pickles, relishes and spreads.

Today, in the U.S., it is used commercially as a natural colorant.

Spice Blends: curry blends, laksa blend, pulao blend, nasi kuning blend, vegetable spice blend and bumbu blend.

Therapeutic Uses and Folklore: turmeric is often used as a dye in the Middle East, in China to color Buddhist robes and by Indonesians as body paints and to ward off evil spirits. Chinese herbalists and Indian Ayurvedics use turmeric to treat liver problems, high cholesterol and digestive problems. Turmeric is also used for healing bruises and sores and to prevent hair growth. Turmeric helps inhibit blood clotting, strengthens the gall bladder and treats skin diseases.

Traditionally in Europe, turmeric is not recommended for pregnant women because it may induce uterine stimulation, but in the Indian culture, it is given to heal the uterus after childbirth and to help restore loss of blood. In Ayurvedic medicine, it is used as an anticancer agent and to lower cholesterol.

WASABI

Wasabi is a popular flavoring in Japan. It is served with raw fish or tempura and is an important flavoring ingredient in Japanese snacks.

Scientific Name(s): Wasabia japonica. Family: Brassicaceae. (cabbage family).

Origin and Varieties: wasabi is indigenous to Japan and is now also grown in the U.S.

Common Names: Japanese horseradish. It is also called raifort du japan (French), japanischer kren (German) and japansk pepparrot (Swedish).

Form: wasabi is a thick rhizome that is creamy white when young and brownish green when mature. It comes freshly grated, as a pale green dried powder or as a green paste.

Properties: dried wasabi root does not have a pungent taste but develops it when it is mixed with water. The flavor is sharp, clean, burning, pungent and nutty with a biting aroma like horseradish, but it comes through with a sweetness. The heat perceived is instantaneous. Dried wasabi is less aromatic but more pungent than the fresh form.

Chemical Components: the essential oil contains sinigrin and glucocochlearin. When the tissues are cut or bruised, or when water reacts with wasabi powder, enzymatic reactions occur that break down the sinigrin to allyl isothiocyanates which give the penetrating odor.

How Prepared and Consumed: wasabi is used coarsely grated or as a smooth paste, often alongside soy sauce, with Japanese foods such as sashimi, sushi, soba and tofu. At the meal table, sashimi or raw fish is first dipped in soy sauce and then into wasabi paste before it is eaten. Sushi, a glutinous rice seasoned with vinegar and sugar and wrapped in dried toasted seaweed (nori) or stuffed with raw fish, other seafood dishes, fresh or pickled vegetables and omelettes are served with wasabi paste. Sometimes, wasabi paste is placed between the raw fish or vegetables and glutinous rice in sushi.

Wasabi pairs well with soy sauce, brown sugar, glutinous rice, chicken, ginger and fish. It does not retain its flavor in cooked foods, so fresh wasabi is used toward the end of cooking or is added to cold foods. The fresh paste also loses its flavor quickly. Wasabi powder is mixed with warm water and let to stand for 10 minutes to develop its flavor. It can be used in marinades, barbecue sauces, creamy salad dressings, soups and steamed fish.

Spice Blends: wasabi snack blend, Japanese sauce blend and pickle blend.

Therapeutic Uses and Folklore: the Japanese believe it can kill parasites.

ZEODARY

Zeodary, a rhizome, is thought to have originated in India or Southeast Asia. In Southeast Asia and South India, it is used to flavor curries and is also used as a medicine. In Thailand, it is prized as an aromatic vegetable.

Scientific Name(s): Curcuma zedoarea. Family: Zingiberaceae.

Origin and Varieties: zeodary is related to turmeric and is native to China and Southeast Asia.

Common Names: white turmeric. It is also called zadwaar (Arabic), zedoaire (French), zitwer (German), kachur (Hindi), kencur zadwar (Indonesian), zedoaria (Italian), kunchor (Singhalese), cedoaria (Spanish) and khamin khao (Thai).

Form: zeodary is used fresh, dried and sliced or ground.

Properties: it is a pale yellow, large and fleshy rhizome with large, long leaves. The inner slices are dark brown and have a rough texture. Its flavor is similar to ginger but with a bitter and camphoraceous taste and a musky aroma.

Chemical Components: the essential oil is viscous and dark green with about 79% sesquiterpenes (mainly germacrone-4, 5-epoxide, furanodienone, curzerenone, curcumenol, isocurcumenol and zederone), 1,8-cineol, camphor and α-pinene.

How Prepared and Consumed: zeodary goes well with the poultry, lamb and fish dishes of South India and Indonesia. In Indonesia, the aromatic leaves and shoots are used in fish dishes or to wrap foods to be grilled or baked. It is used at low levels because of its bitter flavor.

In Thailand, the young rhizome is eaten as a vegetable or used in curries and spice mixtures. In China and Japan, it is used in liquors and as a medicine.

Spice Blends: pickle blend, fish marinade and chicken curry blend.

Therapeutic Uses and Folklore: zeodary is used to treat colds in India and is used as a digestive aid in Indonesia and Thailand.

CHAPTER 5

Emerging Flavor Contributors

INTRODUCTION

OTHER significant flavor, color and texture contributors in addition to spices are gaining in importance. This chapter describes the roots and rhizomes, nuts, seafoods, flowers, wrappers made from edible and inedible leaves and husks and processed "skins" from beans and grains that can be used to contribute taste and texture. Flowers impart visual appeal and flavor to applications. Roots and rhizomes are versatile ingredients that can add intense, pungent notes or simply provide texture to a product. Wrappers are natural cooking vessels that enhance the taste of the foods they contain.

Spices can also be combined with many ingredients, such as fish sauce, shrimp pastes, fruits, vegetables, legumes and nuts to provide characterizing tastes or textures to a product. In the cuisines of many cultures, these flavorings are paired with chilies, coriander, lemongrass, onions and spices to create distinct flavors. These flavorings can also be salted, fermented, dried or pickled and added with spices to create unique seasoning blends (some of which are discussed in Chapter 6, "Emerging Spice Blends and Seasonings").

Other flavorings such as vanilla, chocolate, tea and coffee contribute sweetness and bitterness to foods and beverages. They have been used to balance savory and spicy profiles in some traditional ethnic cuisines. Finally, preparation and cooking techniques help create new flavor profiles and contribute to a product's final taste.

Table 21 gives examples of these other flavorings from around the world.

These "other flavorings" also provide nutrition and valuable phytochemicals that are being tapped as therapeutic aids and natural cures. Fruits, vegetables, nuts and seeds, seafood and other flavorings have phenols, isoflavones, isothiocyanates, antioxidants, glucosinolates, omega 3 fatty acids and flavanoids.

TABLE 21. Other Flavorings Used Around the World.

Flavorings	Product Name/Region
Roots and rhizomes	Lotus root (Asia), licorice (China, Southeast Asia, Korea), parsley root (Mediterranean), daikon (Japan, Korea), taro (Southeast Asia, Latin America, Hawaii, Caribbean)
Flowers	Rose petal (India, Southeast Asia, Mediterranean), orange blossoms (Mediterranean), jasmine (Southeast Asia), lotus (China, Southeast Asia, Sri Lanka), zucchini blossoms (Mexico, Italy, France)
Wrappers	Corn husk (Latin America, Caribbean), banana leaf (India, Southeast Asia, Caribbean, Latin America), lotus leaf (China), salam leaf (Indonesia), hoba leaf (Japan)
Soybean	Fermented pastes (jiangs); salted beans (China, Southeast Asia); misos (Japan); taucheong, timcheong (Malaysia, Singapore); twoenjang (Korea); oyster sauce (China, Southeast Asia, Korea)
Soy sauce	Tamari, Shoyu (Japan); jiangjou (China); kecap soy masin, kecap soy pekat, kecap manis and kecap asin (Indonesia, Malaysia, Singapore, Philippines); kanjang (Korea)
Fish/shrimp flavorings	Nam pla (Thailand); nuoc mam (Vietnam); trassi (Indonesia); blacan, petis (Indonesia, Malaysia, Singapore); bagoong, patis, (Philippines); mam cho (Laos); pra hoc (Cambodia); dashi, shotturn (Japan); saeujot (Korea); yu lu, yu jiang (China); ngapi ye (Burma); Bombay duck (India); maldive (Sri Lanka); bacalao (Mediterranean, Caribbean, Brazil); liquamen (France)
Fruits	Ponzu (Japan); orange peel (China); plum, lichee (Southeast Asia); olives, pomegranate (India, Mediterranean); mango, kokum, coconut (India, Southeast Asia, Caribbean)
Vegetables	Pickled/salted: radish, cabbage, burdock, mustard greens (China, Japan); kimchee (Korea, Japan); mushrooms (Japan, U.S., China, Southeast Asia); tomato (Mediterranean, Asia, U.S.)
Nuts, seeds	Pumpkin seeds/pepitas (Mexico); sesame (Asia, Mexico); candlenuts (Malaysia, Singapore, Indonesia); almonds, pine nuts/pinones (Italy, U.S.); walnuts (China); almonds (India, U.S.)
Dairy products	Ghee, yogurt, buttermilk (India, Southeast Asia, Middle East, Africa)
Sweet and bitter flavorings	Vanilla, cocoa, chocolate, coffee, tea (U.S., Europe, Africa, Mexico, Asia, Middle East)
Cooking techniques	Broil, grill, steam, stir-fry, braise

ROOT/RHIZOME FLAVORINGS

Roots and rhizomes provide flavor and color to foods. Some roots provide intense, pungent notes, such as licorice, ginger, horseradish, wasabi, coriander and turmeric. Others serve merely as a bland starchy base, such as cassava (manioc/yuca), arracachia (Peruvian carrot) or yam.

Roots are a good source of complex carbohydrates and easily digestible starches. They can be used as natural thickening agents or natural starch modifiers. Many roots such as jicama, yam or lotus root are boiled, roasted, stir-fried or stewed. Others are pureed and act as non-descript bases for sauces and as thickeners for soups and curries.

Traditionally, many cultures, especially Native Americans, used roots as cures. Native Americans used American ginseng to ease childbirth, treat nosebleeds and as a tonic for mental problems. Roots and tubers are a good source of phytochemicals. Sweet potato (also called batata and boniata) contains carotenoids; yam has plant sterols; licorice contains triterpenoids and glycyrrhizin; horseradish and wasabi contain isothiocyanates; burdock contains high levels of iron and vitamin A and inulin, an easily digestible starch that provides a medium for the gut bacteria; and jicama has high levels of potassium, iron, vitamin A and calcium.

ROOTS AND RHIZOMES IN ETHNIC CUISINES

Roots and rhizomes were traditionally used to add flavor, color and texture to foods by many indigenous groups such as Mayans, Aztecs, Asian Indians and Africans who could not afford or procure meat or fish. They used potatoes, yams, sweet potatoes and malanga to provide texture and add bulk to soups, stews and sauces. Yuca was mashed and made into flour for cakes and breads.

The Amazonian Indians from South America used roots as a staple flavoring in their foods. The Arawaks and Caribs in the Caribbean extracted juice from grated cassava root, boiled it with chile peppers and flavored many of their dishes. This was called "coui" by the Caribs, and today, cassareep by Brazilians and Caribbeans. It is mixed with chile peppers, cinnamon, clove or brown sugar to flavor sauces, rice and beans and stews (such as pepperpot) in Brazil, Jamaica, Trinidad and Barbados.

In Puerto Rico and Thailand, the coriander root is used to create pungent recados and sauces and curries.

Many roots are popular in China and Southeast Asia as foods, especially in soups and beverages, such as ginseng, ginger or galangal. The Balinese grind the root of salam leaf to create intense spice pastes called jangkap. Chinese use galangal for their five-spice sauces, and Thais use coriander root for their spicy red curries.

Table 22 is a chart of the more popular roots and rhizomes, their flavoring properties and their ethnic preferences.

SPECIFIC ROOTS AND RHIZOMES AS INGREDIENTS

Several of the more important roots and rhizomes are explained in detail below.

Liquorice/Licorice/Chinese Sweetroot

Licorice comes from the root of a legume and is indigenous to China and the Middle East and is also cultivated in Spain, Sicily, Turkey and South Russia. There are two varieties, Spanish licorice and Russian licorice. It has a medicinal, bittersweet taste, and its aroma is similar to anise or fennel, but stronger.

The scientific name for licorice is *Glycyrrhiza glabra*. Glycyrrhiza is derived from the Greek word *glukos,* meaning sweet. The Romans changed the name to *liquiritia,* which subsequently evolved into licorice.

Licorice is also called gancao (Chinese), reglisse (French), jethimadh (Hindi), liqueriziz (Italian), arpsous (Arabic) and orozuz (Spanish).

Licorice is an essential ingredient of the Chinese five spice blend, which is used in sauces and barbecues. It is also used in soft drinks, ice cream, candy, smoothies, drinks and beer. In the U.S. and Europe, licorice is used in cough syrups, confectionaries and lozenges. It also masks bitterness in medicines.

TABLE 22. Global Roots and Tubers and Their Flavoring Properties.

Roots	Flavoring Property	Ethnic Preference
Licorice	Bittersweet flavor	China, Southeast Asia, Korea
Burdock	Chewy texture	Japan, Mediterranean
Celeraic/celery root	Chewy, crunchy texture	Mediterranean, U.S.
Wasabi	Pungent flavor	Japan, U.S.
Lotus root	Crunchy texture, sweet flavor	Asia, U.S.
Parsley root	Aroma	Middle East, Mediterranean
Daikon	Crispy texture	Japan, Korea, U.S.
Sweet potato/boniato	Sweet flavor, mealy texture	Southeast Asia, Latin America
Taro dasheen/eddo	crispy, creamy texture	Southeast Asia, Hawaii, Latin America, Caribbean
Yam	Mealy texture	Africa, Asia, U.S.
Yautia/malanga	Creamy texture, color	Latin America, Caribbean
Cassava/manioc/yuca	Mealy texture	Caribbean, Southeast Asia, Latin America

Licorice can be a noncaloric sweetener but must be used at very low levels in foods, so it does not impart a licorice taste.

The root has about 20–30% of water-soluble extractive and 4% of a glycoside called glycyrrhizin, which is 50 times sweeter than cane sugar. It has the sweet and slightly astringent-like flavor of anise. Licorice is sold as peeled or unpeeled dried whole root, in pieces or powdered. "Block juice" is the black brittle concentrated extract from the root that is free of insolubles. It consists of sugars, starches and gums in addition to 12–20% glycyrrhizin.

Licorice contains vitamin C, niacin, sodium, potassium, magnesium and calcium. It has high phenols and triterpenoids. The Chinese have used licorice since ancient times to treat many ailments, such as ulcers, sore throats or circulation problems.

Lotus Root

Lotus root is used by Chinese, Indians and Southeast Asians for soups, curries, stews and as a stir-fried vegetable. It has a potato-like skin and is cream to salmon colored. When it is cut crosswise, it shows a star-like pattern with symmetrical round openings. It is slightly sweet and has a texture like water chestnuts. Lotus root is available fresh or dried. It can be seasoned and eaten as an appetizer or ground and made into flour or sauces for Japanese and Chinese dishes.

Lotus root is also eaten pickled or as a garnish for salads. It is grated or sliced for stews and soups or candied and eaten as a sweet.

Ginseng root

The ancient Chinese believed that ginseng root had the spirit and power of God to strengthen the body and cure illnesses. Ginseng is from the Chinese word *jen shen,* which means "essence of the earth from the form of man."

East Asians take ginseng as a "cure." The Chinese add ginseng to chicken and vegetable dishes, soups, stews and teas to relieve stress, reduce serum cholesterol and strengthen the nervous system. Tong shui, which are sweet snack-like tonics containing ginseng, are taken to increase energy.

There are three major varieties of ginseng—Asian, American and Siberian. Asian ginseng (panax ginseng) from Japan and Korea is aromatic and has a sweet licorice-like taste. Asian ginseng can be boiled or roasted. Traditionally, the root is chewed or brewed into a tea or tonic. In the U.S. and Europe, Asian ginseng is used in beverages, pasta sauces and candies. It is eaten to boost energy, stimulate the immune system and increase stamina. The other two ginsengs are Siberian and American, the latter is used by Native Americans to increase mental strength, to induce childbirth and to treat nosebleeds. In Chinese traditional medicine, Asian ginseng is classified as "yang" and is used to increase energy and strength and to promote appetite. American ginseng is classified as "yin" and is used to reduce fatigue and increase immunity. Siberian ginseng is used to increase energy and to treat lower back pains.

FLOWERS

Many cultures around the world, including Mediterranean, Latin American and Asian, use flowers whole, dried or in essence form to give unique flavors and colors to foods and beverages. Flowers with vivid colors and fancy shapes, such as violets, pansies, carnations, primroses and orchids, create visual effects for entrees and desserts. Other flowers, such as jasmines, roses, chamomiles, elderflowers, geraniums, nasturtiums and marigolds have unique aromas.

Flowers obtained from spices, fruits and vegetables are fried, stuffed, crystallized or chopped to provide floral, sweet, bitter and other unique tastes to many dishes. Lime, plum and orange flowers (or blossoms) with strong tastes or delicate aromas are commonly used in desserts, teas and liqueurs. Spices such as rosemary, chive, dill, thyme, ginger and garlic have blossoms that add a floral scent to foods and beverages.

Flower petals are served fresh in salads, lightly fried as appetizers, crystallized for cakes, infused for teas or dried to blend with spices. Flowers tend to blend well with many spices, each flower complementing certain spices. For example, rose complements star anise, mint, turmeric, caraway, coriander and chili pepper, while orange blossom complements clove, bay leaf, ginger, cinnamon and nigella.

Since ancient times, fruit blossoms have been popular not only for flavoring but also for healing. Jasmines, marigolds, violets and roses have been traditionally used for their medicinal properties. Marigold has been used as a digestive stimulant and to heal tumors and wounds; rose hips, which are high in vitamins A and C and bioflavanoids, have been used to treat cold symptoms and relieve rheumatic pains; and violets, which are high in vitamins A and C, have been used as expectorants. Flower blossoms and oils have also been popular in massage and aromatherapy.

FLOWERS IN ETHNIC CUISINES

In Europe, since medieval times, flowers have been distilled for their aromas and candied or boiled into jams and syrups. The English and Persians have favored their extracts. Today, in Asia, Latin America, the Middle East and Europe, flowers serve important culinary purposes. They provide color, visual appeal or fragrances to a wide number of applications such as cordials, butter, ice creams, salads, soups, terrines, syrups, jams, jellies, sorbets, cakes, confectionaries and sauces.

Flowers such as rose petals, marigolds, zucchini blossoms, pansies, banana blossoms, and jasmines perk up salad dressings, vegetables, curries, stews, chutneys, puddings, sweets, liqueurs and custards of India, Thailand, Mexico, the Middle East and Europe. They are used fresh, battered or cooked.

In China, jasmines and chrysanthemums flavor teas and wine; lotus petals are added to soups, noodles and salads; and dried lily buds add texture and mild aroma to stir-fries and soups. In northern Africa and India, rose petals are added to teas, desserts and spice blends.

SPECIFIC FLOWERS AS INGREDIENTS

Nasturtiums are orange or yellow in color and have a slight pepper-like flavor. They add visual appeal to salads and sandwiches and add flavor to vinegars and marinades of the Mediterranean. They enhance the color of spreads and margarine.

Europeans and Mexicans color rice, salads and sweet dishes with marigold and violets. Marigold is crushed to give a saffron-like color and a slightly bitter herbal note to salads, soups and entrees of Europe. Violets are highly fragrant, and in Europe, they are used in sweets, chocolates and chilled soups. In Mexico and many other regions, marigold oil is added to chicken feed to provide a yellow color to chicken and eggs.

The fragrant bunga kantan (or wild ginger bud) is a characteristic flavoring in Nonya and Malay sauces and curries of Malaysia and Singapore. In Japan, it is called mioga bud, and it garnishes meals to provide visual enhancement.

Dried lily buds (dok mai jeen) are commonly used in Cantonese and Southeast Asian soups, stir-fried noodles and sauces.

Elder flowers have a muscatel aroma and are used in England to flavor wines, cordials, stewed fruits, jams, jellies and syrups.

The scientific name for Rose is *Rosa damascena,* and it belongs to the Rosaceae family. Rose is also called gul (Arabic), gulab (Hindi), rosa (Italian), bunga ros (Malaysian), ros (Swedish) and irosa (Tamil). Its petals emit a perfumed, sweet and floral aroma. Rose is a popular flavoring among many ethnic groups. It is sold as dried petals or as a distillate called rose water.

Rose has about 1% or less essential oil consisting of geraniol (75%), citronellol (20%), neroli and damascenone. Its odor is due to 2-phenyl ethanol.

Rose water (or rose essence) is greatly enjoyed in the Middle East and India. In Turkey, it is used in candies, a drink called loukoum, liqueurs and jams. Called ma' el-ward, Arabs commonly add it to meats, sweets and sauces. It is a common flavoring in desserts and preserves around the world, such as kheer, gulab jamun and rasgullah in India, jams and rice puddings in Iran and Turkey and in many Malaysian, Singaporean and Thai rice desserts.

Rose petal is used in spice mixtures, sweets, coffees, wine, vinegars and yogurt drinks of the Middle East, North Africa, Iran, North India and Southeast Asia. It is a common ingredient in Indian sherbets and sweet desserts, in curries and biryanis of North India and in some pork dishes of northern China.

Orange blossoms, called zhaar in Moroccon and ma'el zahr in Arabic, are added to fruit salads, stews, sorbets and candies. They have a slightly bitter taste so are added sparingly. The essential oil of orange blossom, called orange oil (or neroli oil), is used in blancmange, puddings and pastries of the Mediterranean. In the Middle East, it is given to revive persons from fainting spells.

Lavender's scientific name is *Lavendula angustifolia,* and it is a member of the Lamiaceae family. It is also called lavand (French), lavendel (German), lavanda (Russian) and lavendel (Swedish). It has a strong, perfumed aroma and is popular as a garnish. Lavender is indigenous to western Mediterranean regions and is commonly found in French spice blends.

Lavender has 1% to 3% essential oil, mainly linalyl acetate (30–50%), linalool (20–35%), cineol, beta-ocimene, camphor and caryophyllene.

Jasmine and chrysanthemum are popular flavorings of teas and beverages in Japan, China and Southeast Asia. In Thailand, jasmine essence is mixed with coconut milk to flavor many cakes and desserts.

Roselle or Jamaican sorrel (hibiscus family) is very acidic and flavors jams, cold beverages and wine in the Caribbean and Mexico. It is also used in Central America, India and Southeast Asia for sauces, curries, chutneys and soups.

Lotus flower is used to scent Chinese and Vietnamese soups or Sri Lankan curries. To Buddhists, it is a symbol of beauty and purity.

Squash or zucchini blossoms are found in Mexican salads and appetizers and are becoming trendy in U.S. restaurants. They go well with spices such as cardamom, clove, nutmeg, cinnamon, mint, basil and sorrel. They also tend to lose aroma after a few days, especially with prolonged heat, and so are typically added to foods toward the end of cooking or after cooking. Zucchini blossoms are also commonly used in Italian and French cooking. They are braised, stuffed, lightly fried or battered and deep-fried. In Mexico, Oaxacans chop the flowers and add them to soups and sauces or sauté them with poblano peppers, cilantro and onions and add them to quesadillas, tacos and enchiladas. In Italy, these blossoms are added whole or chopped to risottos, pastas and salads and are an essential ingredient in frittata alle erbe.

The beautiful shape, thin papery texture and subtle flavor of zucchini blossoms lends them to "stuffed" concepts. In France, they are stuffed with cheese, spices and herbs, dipped in milk and flour and sautéed or stuffed with cooked seafood and herbs and steamed or deep-fried. They are also used as wrappers, especially in the U.S., to provide texture to other foods. They can be battered lightly and deep-fried to provide crunchiness in a meal.

WRAPPERS

Many cultures from Asia, Latin America and the Mediterranean use wrappers as cooking and serving utensils. Wrappers can be edible or nonedible and provide distinct flavors, textures and colors to the foods they contain. Some wrappers make the food succulent and tender. Others give crunchiness and crispiness. These wrappers include a variety of leaves, flowers, corn husks, cactus fiber, seaweeds, hollow bamboo stems, dried rice paper, and soy milk skin. They are used with meat, poultry, fish, vegetables, rice, cheeses or spice pastes before foods are barbecued, deep-fried, grilled, baked, roasted or steamed. This is the Asian, Latin American or Native American equivalent of cooking en papillote.

Edible wrappers are used to wrap fresh ingredients and are eaten whole as salads or appetizers. Examples of such wrappers are lettuce, grape leaves, nori and shiso. Others, such as corn husks, lotus leaves or banana leaves are inedible.

Depending on the type of wrapper, varying amounts of color, flavor or texture are contributed to the finished product. Today, these wrappers are becoming a trendy restaurant item to provide fresh-cooked aromas and textures at a meal table. They are being used to create authenticity or an enhanced visual appeal to a meal.

There are several basic categories of wrappers: Corn husks: they are sold fresh or dried. Before use, they are brought to a boil in water and left submerged for about an hour to make them more pliable. They are then pat dried. Leaves: leaves used as wrappers include pandan leaf, lotus leaf, banana leaf, hoba leaf, turmeric leaf, grape leaf, lettuce leaf, spinach leaf, cabbage leaf, paan leaf, hoja santa, salam leaf and papaya leaf. Seaweeds: in Japan, nori (seaweed) is lightly toasted and used to wrap glutinous rice, which is called sushi. In China, seaweed is used to wrap fried foods such as tofu, fish or vegetables. They provide texture, color and a seaweed-type flavor. Maguey/cactus fiber provides succulence and moistness to steamed pork and other meats in Mexico. Hollow bamboo segments are commonly used by indigenous groups in Southeast Asia to cook rice and meats. Flowers: zucchini blossoms, rose petals, ginger flower and banana blossoms provide fragrant aromas and colors to meals in many parts of the world.

WRAPPERS IN ETHNIC CUISINES

Different types of wrappers are used in ethnic cuisines. Mexicans use corn husks and a variety of leaves, such as maguey, banana or hoja santa to prepare foods. The Japanese use hoba leaf and nori to wrap beef, chicken, sticky rice, seafood and vegetables. Southeast Asians and Indians commonly use banana, pandan or salam leaves to wrap fish, rice, meats or vegetables before steaming, grilling or barbecueing. The following is a list of geographic regions that use different wrappers.

Greece

Tender grape leaves are blanched and used to stuff rice, lentils or beans with nuts, spices and raisins, with the resulting food called dolmades. They give a lemony tang to the food.

France

Zucchini blossoms are used to wrap cheese or fish with herbs and spices.

Middle East

Meloukhia leaf is a deep green, spinach-like leaf, popularly eaten in Egypt, Palestine and Tunisia (in North Africa). It has a glutinous texture that gives the finished product a gelatinous, viscous texture. It is mainly used with meat, chicken or duck soups, as a main vegetable entree with boiled rice or chopped finely and cooked as a side dish, accompanied by onions, hot chilies, meat and rice. It is sold frozen or dried and can be used as a wrapper.

India

Banana leaf is a large green leaf that provides moistness and a delicate flavor to foods. It is cut up in pieces and used to wrap whole fish or fillets to be baked, chicken to be grilled or rice before it is steamed. In Asian Indian cultures, especially with the vegetarian communities, it is used as a plate for serving a whole meal. Hindus serve meals at temples on banana leaves because they signify cleanliness and purity.

Paan leaf is used to wrap a dry filling of chopped betel nuts, pieces of coconut kernel, spices and sugar. It is like an "after dinner mint," to freshen breath as well as aid digestion. The leaf is wrapped around the filling and tied with a clove.

China

Lotus leaf is a large and round edible leaf. It is commonly used in Cantonese-style cooking for stuffing rice, dumplings and fish that are then steamed or grilled. It provides a sweet tea-like flavor to foods. It is sold dried and is soaked before use.

Bamboo leaf is a greenish ribbed leaf with a magenta tinge that is used to wrap sliced fish, meat or rice before steaming, boiling or grilling. It gives a smoky aromatic flavor to foods. Before use, it is soaked in warm water until it softens.

Soymilk skin, called yuba, fuchok or fu zhu, is dried bean curd that is creamy yellow in color. It is used in Chinese cuisine to wrap seasoned fish, rice vegetables or meats before baking or grilling. It provides a firm yet chewy consistency to foods. Dry yuba needs to be soaked until fully hydrated before use. Similarly, rice paper is commonly used to wrap freshly cooked rice, vegetables and meats before serving. It provides a tender, smooth, slightly chewy texture to the finished product.

Japan

Hoba leaf is a large leaf used in Japanese cuisine to wrap beef and chicken for grilling.

Southeast Asia

Thais, Vietnamese, Malays and Nonyas in Malaysia and Indonesians commonly use banana leaf for wrapping fish, chicken, sausages and meats before grilling and for wrapping glutinous rice and desserts before steaming. In Thailand and Malaysia, banana leaves also serve as natural and elegant plate liners that provide visual appeal to meals. It is an indispensable ingredient in Balinese cooking for wrapping food before roasting or steaming.

In Malaysia, Singapore and Thailand, fish, chicken pieces and rice are wrapped in pandan leaf (or screwpine leaf) before grilling, barbecuing or steaming. The fragrant Nonya's and Malay's fish and rice dishes commonly use pandan leaf. Local condiments and desserts are also served in pandan leaf.

Salam leaf is a favorite wrapper of the Balinese in Indonesia for fish and ground meat cooked with spicy condiments. Indonesians also use the long, narrow, fragrant turmeric leaf to wrap spiced fish or meats. Cabbage or local salad leaves are also used to wrap beef before grilling. In Thailand, glutinous rice with dried shrimp is steamed in lotus leaf which adds a unique flavor to this dish.

In Sarawak, Sabah, Sulawesi and West Sumatra, indigenous populations stuff hollow bamboo segments with glutinous rice and finely chopped meats or poultry seasoned with spices and place them directly in open fires to cook. These bamboo segments provide a slightly smoky aroma and a distinct texture to the finished foods.

Caribbean

In many Caribbean regions, whole fish, corn dough, rice and beans and vegetables are wrapped in corn husks before cooking. In Cuba and Puerto Rico, tamales (masa or corn dough filled with seasoned chicken, beef, fish or cheese), fish, chilies or vegetables are steamed or grilled in corn husks. Jamaicans, Puerto Ricans and Dominicans create their picadillos by stuffing spicy ground meat into corn husks that are then steamed.

Latin America

Fresh corn husks or dried corn husks are used in Mexico, Central America and South America for steaming, grilling, baking, simmering or roasting tamales, vegetables, moles, meat, fish, recados and other seasoning pastes. Before use, the dried corn husks are soaked in hot water until they soften. In Michoacán, Mexico, masa with fillings are wrapped in fresh corn leaves (or hojas de milpa) before they are steamed to make tamales.

Maguey leaf from the Agave plant (not a cactus) is a common wrapper in the Yucatan and other regions of Mexico. It has a thin, transparent skin called mixiote that is like a pliable parchment paper, which wraps seasoned goat (barbacoa), pork and chicken before steaming.

Banana leaf is used to wrap and slowly roast pork, venison and other meats. In the Yucatan, banana leaf is used to wrap marinated whole pig that is cooked in a pit to make the local speciality called cochinita pibil. Banana leaf is sold fresh or frozen. Before use, it is steamed for about 20 minutes to make it pliable and soft.

Hoja santa (also called acuyo/holy leaf/hierba santa) is a large green leaf in which corn dough with seasoned fish, moles and meats are wrapped before they are steamed or roasted. The leaf gives an anise-like flavor to the cooked foods.

SEAFOOD FLAVORINGS

In days before refrigeration, and even today in many Asian, European and Caribbean cultures, fish and other seafoods are dried, salted or brined. This is an economical way to preserve them as well as to add flavors and textures to foods. Many indigenous cultures combined them with spices to "spike up" a meatless meal. The ancient Greeks used a fish seasoning called garon, and the Romans used fish and its entrails with salt, called garum.

Seafood flavorings are essential to the cuisines of many countries, particularly in Asia, to enhance main entrees and side dishes. Seafood flavorings come in whole, ground, sauce or paste forms. They are obtained from dried, smoked, salted or fermented fish, oysters, shellfish, scallops and squid. Their preparation varies in different regions of the world. The paste is the solid portion of the dried and or fermented product, and the sauce is the liquid portion.

The fish used as flavorings are usually the small fish, such as anchovies or whitebait (also called ikan bilis in Malaysia and Indonesia or karuvaddu in

Sri Lanka, South India and Malaysia). In the Caribbean and Mediterranean, the larger codfish is also used. Brazilians use whole dried shrimp or codfish (bacalhau) to flavor many dishes. In Europe, anchovies and pickled herring are used as flavorings and appetizers.

SEAFOOD FLAVORINGS IN ETHNIC CUISINES

European Seafood Flavorings

Small bony fish or tiny shrimp, along with the bones and heads, are used to make fish stocks that are the basis of many fish soups, stews and sauces. The French and Italians use cured anchovies with salt, capers, olive oil and spices to flavor many of their soups, stews, salads and marinades.

In southern France, a popular spread called anchoiade is a salted paste made with anchovy fillets, black pepper, olive oil, garlic and vinegar. In Piedmont, Italy, anchovies with garlic, butter and olive oil are made into a dip for raw vegetables called bagna cauda. Anchovies are also added to tomato sauces that are then used as spreads. Liquamen, a fish sauce that is a by-product from anchovy manufacture, is used for intensifying the flavor of seafood dishes in southern Europe. In Holland and Belgium, anchovies are marinated in wine vinegar with onions and lemon rind and used as a condiment. The Swedish use pickled herring with sugar, vinegar, allspice, onions, bay leaf and dill to flavor many of their cream sauces.

Salt-dried cod or bacalao, from the Basque region of France and Spain and from Italy is a popular flavoring for sauces, paellas and vegetables. In Liguria, Italy, salt cod or baccala is softened in water and cooked with parsley, capers, dried mushrooms, pine nuts and vegetables for sauces or is combined with flour to make fritters. It is also soaked in milk and cooked slowly in olive oil, spices and vegetables to make a codfish stew called stoccafissa (or stocchefisso).

African Seafood Flavorings

In West Africa, dried and smoked fish or shrimp are used to flavor many dishes. Some examples include dried, smoked shrimp used in the Gulf of Benin, dried, smoked mollusk and guedge, (a dried, smoked fish), both used in Senegalese stews and sauces. In Egypt and Sudan, a seasoning sauce called faseekh is made by fermenting small fish with salt.

Asian Seafood Flavorings

East Asian Seafood Flavorings

Fish sauce and pastes or dried whole seafood, called yu-jiang and yu-lu in China, or shotturn, dashi or gyomiso in Japan, are used abundantly to season cooking sauces or soups. Oyster sauce, a popular cooking sauce, contains oyster juice (from cooked, dried oysters), fermented soybeans, sugar, starch and other ingredients.

Southeast Asian Seafood Flavorings

Fermented and dried shrimp are a staple flavoring and food in Southeast Asia. Southeast Asians use fermented or salted, dried seafood to add zest and enhance rice, noodles, vegetables and seafood dishes and condiments. Ikan bilis is used as a seasoning as well as a crunchy side dish and appetizer in Malaysia and Singapore. In Vietnam, Indonesia, Philippines and Malaysia, a meal does not taste right without salted or fermented seafood condiments. Kecap ikan, a fish sauce, is commonly used in many Indonesian and Malaysian dishes. The Vietnamese nuoc mam, Thai nam pla, Malaysian belacan or Filipino patis are essential ingredients for enhancing many sauces, curries, salads, marinades and condiments. These fish or shrimp flavorings are unacceptable by themselves, but when cooked and added to other dishes, they round off other flavors or add zest to sauces, soups or salads. They come with different flavor intensities and consistencies as viscous pastes or as thin liquids. They can be light colored with mild flavors or dark colored with heavy, pungent flavors.

Seafood flavorings come as dried whole pieces or ground, as pastes and as sauces. Southeast Asians use the pastes and sauces as condiments or seasonings, much like the French use roux or wine, and the Chinese and Japanese use soy sauce or rice wine. They flavor broths, soups, stir-fried noodles and sauces or are used as marinades. They become part of Japanese dashis and Chinese stir-fries, as well as pad thai, nuoc cham or sambal udang. Other dishes using these flavorings include: Vietnamese pho bo (beef noodle) soup, Nonya laksa (fragrant curry noodles), Laotian laab (minced beef salad), Malay sambal (fiery sauce), Burmese balachaung (condiment), Thai nam prik gaeng ped (red curry paste) and Filipino chicken binakol. The Cristangs (of mixed Portuguese and local ancestry) from the Malacca region of Malaysia use a popular condiment called chinchaluk, made from salted and fermented krill (a type of shrimp) with chile peppers, shallots and lime juice.

Commercial fish sauce is most commonly made from fish but can also be made from shrimp. It adds a subtle flavor, enhances other ingredients or mellows stronger flavors. It is added as a freshly prepared table condiment, as a dipping sauce for spring rolls, as a marinade, or toward the end of cooking. Though the sauce has a fishy aroma and salty taste, the fishy aroma disappears during cooking and adds a delicate overall flavor to the product. It pairs well with chile peppers, lemongrass, lime juice, sugar, garlic, coconut milk, cilantro, spearmint, basil and galangal.

Fish sauce (nam pla) is the main flavoring in Thai cooking. It is a light brown extract made from salted and fermented small fish. The fish are dried in the sun for a few hours then layered in barrels with salt and water and kept submerged. These barrels are exposed to the sun, whereby the mixture ferments for about three months. Then, the liqueur is drawn off slowly, put in a ceramic urn and aged another month before bottling. The first extraction from the fermented fish is used in a fresh batch of dried fish and fermented for another three months. This method produces the saltier, darker and stronger fishy note that is so desired in local cooking.

Dried whole shrimp or prawns (boiled and dried), called kung haeng in Thai, udang kering in Malaysian or tom kho in Vietnamese, are commonly used for vegetables, fried rice, noodles, stir-fries, condiments, soups and sauces. They are soaked in warm water for about five minutes and then used unpeeled or peeled. These dried shrimp are used whole, chopped, sliced or pounded. They can be deep-fried and served as an appetizer with a dip or added to vegetables, sambals, rice and noodles. They are pinkish brown in color and perk up blander ingredients such as noodles, rices, tofu, cassava, potatoes and cabbage. They can be ground to a coarse powder and sprinkled over salads or stir-fried for sauces and soups.

Shrimp paste is a moist greyish to a hard, crumbly brownish black block. It is a shrimpy, salty and pungent seasoning and is generally sold as a purplish pink to beige paste. The aroma is intense and overpowering, but when cooked, its raw taste and aroma disappear and it develops a more fragrant flavor. Shrimp paste is generally cooked with other ingredients, but if it is used in uncooked foods, it is toasted before being added to the food. This toasting creates a more acceptable flavor in the finished product. It does not dominate the flavor of the dish but enhances the other flavors.

Shrimp paste adds pungency and a fragrant aroma to condiments, sambals, soups, salads and stews. The heavier dark to black shrimp paste, called petis or hay koh, is a molasses-like seasoning. It is a favorite of the Nonya and Muslim cooks in Malaysia, Singapore and Indonesia, who use it in noodle soups, stewed noodles and spicy salads called rojak.

Some of the more popular fish and shrimp flavorings include the following:

(1) Small fish/anchovy sauces and pastes: patis (Filipino), sauejot (Korean), nam pla (Thai), nuoc mam (Vietnamese) and mamcho (Laotian). Also, heavier fish sauces include bagoong Balayan (Filipino), ngapi (Myanmar) and liquamen (southern European).
(2) Shrimp paste: bagoong alamang (Filipino), trasi (Indonesian), belacan (Malaysian), kapi (Thai) and mam ruoc (Vietnamese). Heavy, darker pastes include petis (Indonesian) and hay koh (Singaporean).
(3) Whole dried fish or shrimp: ebi (Indonesian), dried bonito flakes/katsuobushi (Japanese); whole salted fish/ikan bilis (Malaysian, Indonesian), salted fish/pla haeng (Thai), dried shrimp/udang kering (Malaysian), salted

shrimp/kung haeng (Thai); salted shrimp/tom kho (Vietnamese), maldive fish and nettali karuvaddu (Sri Lanka).

These dried and or fermented sauces and pastes are rich in protein, B vitamins, calcium and omega 3 fatty acids.

South Asian Seafood Flavorings

Bombay Duck, a Goan condiment from southwest India, is a dried, gelatinous, spicy fish paste used to perk up sauces or condiments. It is also eaten as a snack. In Sri Lanka, pounded and dried maldive fish are popularly used in a variety of sambols (local condiments) combined with green ginger, chile peppers, garlic, lemongrass, cinnamon, onions, coconut and lime juice. Sambols are prepared in different ways and are commonly eaten with local breads such as thosai, puttu or even bread and butter, especially for breakfast. Dried small fish called nethali karuvadu are combined with green chile peppers, shallots, sugar and vegetables for sauces, curries and as condiments.

Caribbean Seafood Flavorings

In the Caribbean, salt cod fish (saltfish or bacalao) is a common flavoring for stews, soups and fritters called bacalaitos or accra. It is shredded and added to vegetables, such as ackee, made into souffles and pies and flavors rice and beans, eggs and many root vegetables. It is mixed with habanero peppers, onions, tomato and olive oil to create buljol, which is served for breakfast or brunch, in Trinidad. In Puerto Rico, bacalao is combined with sweet chile peppers, tomatoes, pimientos and garlic to flavor mashed potatoes, yautia and cabbage.

Brazilian Seafood Flavoring

Salted codfish (bacalhau) and dried whole shrimp are common flavorings in many stews and sauces. A refogado is made with salted codfish or dried shrimp, onions, tomatoes, chives, parsley, chile peppers and sweet bell peppers and is used to flavor potatoes, spinach, coconut milk, rice, omelettes, squash and fish.

FRUIT FLAVORINGS

Fruit flavorings come from the edible fleshy pericarp, sometimes the receptacle and the skin of the fruit. Fruits are used fresh, dried, pickled, pureed, ground or chopped. Whole fruits are chopped, sliced or grated in salads, stews and curries. With some fruits, only the rind is used. Pureed, they provide flavor, consistency and color to many dishes and condiments. The flavor profile varies between ripe and unripe fruit and also changes when the fruit is dried or cooked.

Nuts are fruits that will be discussed separately. Tomato, avocado, breadfruit and cucumber are fruits but will be placed with vegetables because they function as vegetables.

Traditionally, North Americans and Europeans have used ripe or dried subtropical or temperate fruits to provide characterizing fruity flavors or sweetness to beverages, desserts and confectionaries. Fruits that are typical of North American and European cuisines include berries (cranberry, blueberry, strawberry), citrus, grapes, stoned fruits (apricot, cherry), plums, melons and pomes (apples, quinces, pears).

Fruit flavorings can create new flavors, balance flavors or tone down pungent flavors. They provide natural colors and visual appeal and can also become natural acidulants. They can be used to flavor or color vinegars, cooking oils and sauces. Fruits complement many spices, nuts, fish flavorings and chile peppers.

Fruits provide tartness, sweetness, fruitiness, consistency and color to dishes. Sweetness in fruits is due to sugars, mainly fructose and glucose, and varies with their ripeness. Carambola (or star fruit) can be mildly sweet when ripe and sour when unripe. Because of their tartness, fruits can be natural acidulants. The acidity is due to organic acids such as malic, citric or tartaric. Lemon, lime, sour orange, tamarind, carambola, kalamansi and pomegranates are used to provide acidity and astringency to foods and can be tapped as "natural" souring agents for foods and beverages to replace citric or malic acids. At the same time, they will add other enhancing notes to the food—sweet, fruity, astringent or floral. Fruits can be further explored as natural coloring agents, antioxidants and antimicrobials in beverage and food applications.

Mediterranean, Caribbean and Asian cuisines combine fresh, dried or pureed fruits to complement their savory and hot flavors and to create visual appeal. Tropical fruits also provide the variety that consumers seek. Fruits are added fresh or cooked to main dishes, side dishes and condiments or serve as attractive serving vessels for condiments, salads and all other foods.

Mango, banana, fig, lime, dates, papaya, pineapple, guava, rambutan, star fruit, guarana and lychee (or litchee), are many examples of tropical fruits. Many unique fruits such as kokum, pomegranate (anardhana), green mango, belimbing wuluh, green papaya, yuzu orange, sour jackfruit, olive and tamarind are commonly used in Asian, Caribbean and Latin American cuisines and are also becoming known in the U.S. These tropical fruits, such as mango, acerola or melon, differ in their flavors, colors and textures, depending upon their regional origins. For example, mango can have a pulpy or stringy texture and can be sweet and fragrant or sweet and tart.

Today, North Americans are adding fruits to savor dishes and to create unique salsas, soups or salads. Fruits complement many savory ingredients. Apples and apricots go well with pork, cherries and plums with duck and star anise, cranberries with turkey and nutmeg, gooseberries with mutton and coriander, orange peel with chicken, duck and basil, pineapple with pork, ham and lemongrass, dried or pickled plums with beef and shiso, pomegranate with legumes and bay leaf, prunes with pork and ginger, mango with lamb curry and chile peppers, coconut with pungent laksas, kokum with fish and chutneys, olive with garlic and tomato or kalamansi with steak and pork.

Consumers are also seeking "natural" ways of staying healthy. Fruits are great sources of vitamins A, C, K and E, minerals and other important nutrients.

FRUITS IN ETHNIC CUISINES

Fruits, when ripe or unripe (green), give flavor and color to many ethnic savory dishes. Mango is used fresh or dried (unripe) in Indian cooking and provides tartness and flavor to curries and chutneys. Mandarin orange peels give sweet and citrusy notes to Chinese stir-fries and sauces. Papaya, carambola and guava add unique fruity notes to hot Caribbean sauces and beverages. Grapefruit, sapote and tomatilla spike up Mexican condiments and salsas, while passion fruit and jackfruit provide consistency to Indonesian sambals and gulais.

Southeast Asians, Indians, Caribbeans or North Americans are fond of the flavor of unripe (or green fruits) and sour flavors and use them to flavor curries, soups and salads, including green papaya (Thailand, Vietnam), green mango (India, Caribbean, Southeast Asia), green jackfruit (Vietnam, Indonesia), plantain (Caribbean, Latin American), plum (Japan, China), ackee (Caribbean), belimbing wuluh (Malaysia, Indonesia), green carambola (Caribbean, Southeast Asia, Florida), and green apple (U.S., Europe).

Unripe carambola is used as a flavoring in many Southeast Asian and Caribbean foods. Belimbing or bilimbi, smaller varieties of carambola, are added to soups, pickles and marinades in Malaysia, Indonesia and Singapore and curries of South India and Sri Lanka. Kamia, a relative of star fruit is popularly used as a souring agent for marinades, stews, salads and soups in the Philippines.

Dried fruits such as amchur, prunes, figs, dates and raisins and dried persimmons and pomegranate add flavor to many Mexican, Asian and Mediterranean dishes. Prunes, dates and figs are popular flavorings in Middle Eastern rices, meats and desserts. In Mexico, prunes and raisins are blended to add flavor and texture to desserts, moles, meat picadillos, flans and rice puddings. Fruits are also candied and used in many dishes of the Mediterranean.

Table 23 describes various fruits, their flavors and other qualities they add to foods and the ethnic cuisines in which they are typically found.

Specific Fruits as Ingredients

Below is a detailed description of some of the more significant fruits for food applications.

Mango

Mango is indigenous to India, and the word "mango" is derived from the Tamil word "mangai." There are many types of mangoes that vary in size, shape, flavor, texture and color. They are grown in India, Southeast Asia, the Caribbean, the Pacific Islands, Africa and Latin America.

Mangoes' colors range from a yellowish red to yellow to a yellowish green. The inner flesh is generally yellow. Fresh mango has a sweet to slightly tart

TABLE 23. Global Fruits and Their Flavoring Properties.

Fruit Flavorings	Flavoring Property	Ethnic Preference
Star fruit/carambola	Sour, fruity	Caribbean, Mexico, Southeast Asia
Belimbang, Bilimbi	Sour, fruity	South India, Southeast Asia, Sri Lanka
Kalamansi/ponzu	Sour, fruity	Philippines, Japan
Mango, Guava	Sour, floral, sweet	India, Caribbean, East Asia, Southeast Asia
Rambutan	Fruity sweet	Southeast Asia
Guarana	Caffeine, astringent	South America
Persimmon	Astringent, sweet	East Asia, Southeast Asia
Pomegranate	Fruity, sour	Mediterranean, South Mexico, India
Plum	Sour, fruity	Japan, China
Atemoya	Creamy	Mexico, Caribbean
Cherimoya	Creamy	South America, Mexico
Cranberry	Sour, color, acidic	U.S.
Langsat/duku	Sweet, sour	Southeast Asia
Fig	Sweet, fruity	Mediterranean, U.S.
Jujube	Sweet, astringent, acidic	China
Acerola/West Indian cherry	Acidic, fruity	South America, Caribbean
Papaya (tenderizer)	Sweet, floral, papain	Southeast Asia, Caribbean
Pineapple	Sweet, fruity	Southeast Asia, Caribbean
Orange (Seville, Mandarin)	Sweet, sour, fruity, bitter	Latin American, Asian, Mediterranean

taste, depending on the variety. Indian varieties are much sweeter and aromatic and less fibrous than other varieties.

Mango has a high level of beta-carotene (an average mango has 8000 IU), vitamin C, dietary fiber and potassium. Traditionally, it was used to help the body fight infections.

In the U.S. and Europe, mango is generally made into jams, sherbets and ice cream. In Asia, fresh mango is sliced and eaten after a meal, added to salads or made into desserts or juice beverages. In the Caribbean, it is made into beverages or added to sauces.

Mango pairs well with coriander, chile pepper, mint, ginger, tamarind and kaffir lime leaf and is generally added to poultry, lamb and vegetable dishes.

The dried green mango, called amchur in Hindi, is sliced and ground into a powder. The powder has very little aroma but gives tartness to foods. It has a sour, astringent taste and is especially popular with vegetables and pickles. It is also used in chutneys and in marinades for the barbecued meats (tandooris) of North India. Amchur tenderizes the meats and chicken cooked in the tandoors, thus making them succulent and juicy. Its flavor is due to citric acid and terpenes such as ocimene, myrcene, limonene, with aldehydes and esters. The dried pieces are pickled and become a popular condiment in every Indian home. It is also an essential component of a green spice blend called chat masala that is commonly used in North India.

Green mango is also eaten with fermented fish paste or is mixed with salads in the Philippines and Bali. In Malaysia, Indonesia and Singapore, it is mixed with chilies, brown sugar and other fruits and vegetables for a spicy salad called rojak. In the Caribbean, it is added to curries and chutneys, especially in Trinidad and other islands where there is a substantial Indian population.

Pomegranate

Pomegranate is native to Persia. They are large beige to red skinned fruits with numerous pulpy seeds that have a sweet-sour taste. It is these seeds that are used in culinary applications. Pomegranate is called anardana in Hindi, anar in Farsi, rumman in Arabic, melogranate in Italian, mattalam in Malayalam or granada in Spanish. Pomegranate juice, its seeds and paste made from the seeds (pekmez) are essential flavorings in Turkish, Iranian and other Middle Eastern regions. They are used as marinades and in sauces and desserts. The dried seeds are sprinkled over salads, hummus and desserts. They give a sweet, sour, fruity taste to meats and vegetables.

Dried pomegranate is darkish red in color and adds sourness to cakes, desserts, chutneys and curries. In North India, pomegranate seed juice is used to marinate meats and is added to desserts. In Punjab, pomegranate seeds are dried, crushed or ground and used as a garnish and souring agent in chutneys, pakora fillings, paratha stuffing and curries. They are commonly used in the vegetarian cooking of Gujerat, which is noted for its spicy sweet flavors. The vegetarian Jain community of Gujerat uses pomegranate seeds as toppings on fiery legume curries.

In Mexico, pomegranate seeds are used as garnishes for a white sauce topping on stuffed poblanos. This is a renowned dish of the Puebla region called chile en nogada and the red color of pomegranate seeds gives visual appeal and provides a bittersweet flavor to the product.

Papaya

Papaya or pawpaw is a juicy, deep yellow to orange fleshy fruit with centrally located black seeds. The ripe fruit has a sweet taste and enhances many spices, such as chile peppers, onions, ground mustard, cardamom, nutmeg, mint or cinnamon. The juice of the unripe fruit is used as a meat tenderizer in many Southeast Asian regions. In the Caribbean, papaya is added with habaneros, mustard and vinegar to create a hot fruity sauce. A variety of green papaya, called du du xanh, is peeled and shredded to give a crunchy texture to spicy salads of Thailand and Vietnam.

Papaya is high in vitamin C, beta-carotene (about five times that of orange) and fiber. Traditionally, it was eaten to help resist infection, aid digestion and help prevent hardening of the arteries.

Orange

Oranges are native to India but spread to China and were brought to the Mediterranean by the Arabs. There are many types of oranges commonly used to flavor savory foods—naranga china (sweet), naranga agria (Seville, sour/bitter orange), Mandarin orange, bergamot orange and tangerine. Sweet oranges, such as Spanish, Mediterranean, blood or navel types, have a firm, tight skin. They are common eating oranges and are used in desserts, beverages and baked products. The sour and bitter types of oranges are commonly used to flavor meats, seafood and salads. Seville oranges, or narangia agria, are used for marmalades, cakes, stir-fries and condiments of Asia and the Mediterranean and is a popular flavoring in Latin American marinades, ceviches and stews. Mojos in Cuba, adobos in the Dominican Republic, ceviches in Ecuador and recados in Yucatan contain this sour, bitter orange juice.

Mandarins and tangerines are loose-skinned oranges that have less acidity than the sweet orange. Bergamot orange is a variety of bitter orange.

The orange peel or rind is strongly aromatic with a bittersweet taste. It has about 2.5% essential oils, mainly limonene, citral, bitter glycosides and rutin. The highly scented flowers of oranges are extracted for their essential oil, called neroli oil.

Orange peel, fresh or dried, whether sliced, grated or ground, is an important flavoring in many European and Asian cuisines. In France, it is added to bouquet garni, fish and meat dishes. It adds a fruity, bitter flavor to the many hot, sweet and sour Szechwan dishes of China. It pairs well with star anise, ginger, soy sauce, fennel, mustard, brown sugar and rosemary.

Dried orange peel is a common ingredient in many Chinese-based sauces for chicken and beef that require long simmering or stir-frying.

Yuzu orange, called som na (in Thai), cidra (in Spanish) and turung (in Arabic) is commonly used in Japanese, Southeast Asian and Mediterranean cuisines. The rind of yuzu orange is an essential ingredient in the Japanese spice mixture, shichimi togarashi. Kumquats are tiny orange-like fruits but are not related to oranges. The rind and peels are used in Chinese, Southeast Asian and Mediterranean cooking and also as a flavoring in many European liqueurs.

Oranges are high in vitamin C, beta-carotene, potassium and flavanoids. They are used in traditional medicine to protect against cancer, lower LDL and prevent colds.

Lemon and Lime

Lemon is derived from the Persian word, limun, or Arabic word, lemun. Lime and lemon are fundamental ingredients in Southeast Asian cooking. In many regions, lime is often confused with lemon, and they have similar names. Lime juice and lemon juice have long been used to flavor many cuisines. Lemon is called limon (Spanish), lemun (Arabic), limun (Farsi), zitrone (German), ning meng (Chinese), citron (French), limau (Malaysian), elimicham (Tamil) and manao leung (Thai).

Lemons are rich in essential oils (6%) that are mainly composed of limonene (90%) and citral (5%), with some glycosides and coumarin. Lemon is the principal souring agent in the U.S. and in Europe. Its peel is aromatic and bitter and is also used as a flavoring for fish soups, stews, couscous, beverages and desserts. In India and Iran, lemon peel is ground and used to flavor rice.

Lime is called limon agria (Spanish), lai meng (Chinese), limette (German), limau nipis (Malaysian) and manao (Thai). Limes are green to greenish yellow with thin, tight skins and are generally smaller than lemons. Typically, lime has a sour and slightly bitter juice and, depending on their origins, can have a lemony flavor.

Kalamansi, also called limau kesturi or ponzu, is a less acidic and more fragrant small lime that is used in noodles, soups, sambals and desserts and as marinades in Southeast Asian and Japanese cooking.

Kaffir lime does not have juice, but its peel and leaves are popular in the curries and sauces of Malaysia, Indonesia and Thailand.

Other Tropical Fruits

Salty pickled sour plum (umeboshi) gives a fruity aroma to plain rice which is taken for breakfast in Japan. It aids digestion of a meal. Plum sauce, made from salted plums, chilies and vinegar, is a popular dip that is added to many Cantonese poultry and pork dishes.

The small, pale green astringent belimbing wuluh (relative of star fruit) is an acidic fruit that provides tang to the fish dishes, laksas and sambals of Malaysia, Indonesia and Singapore. In the Philippines, it is pickled and added to soups. The juice of the sour kalamansi (called jeruk cina in Indonesia) is used to marinate fish and meats of these regions. It also adds aromatic citrus notes to noodles, soups and desserts.

Jackfruit or nangka (Indonesian and Malaysian) is used green or unripe to flavor stews, sambals or curries. It has a banana- and cantaloupe-like flavor.

The soft seeds of the Mahlab cherry (a black cherry), either ground or whole, provide a delicate aroma and bitterness to Turkish and Middle Eastern sweets, biscuits, breads and stews.

Olives

Called oliva (Italian), zaytun (Arabic), aceituna (Spanish), zeytin (Turkish) or saidun (Tamil), olives are indigenous to Greece, Italy, Spain, northern Africa and other regions of southern Europe and western Asia. One of the oldest known fruits, they were valued by the Egyptians, Romans and Greeks, who decorated Olympic winners with olive branches.

Olive is a fleshy fruit (a drupe), and there are many types that vary in size, color, oil content and flavor. Examples include sevillano from California, manzanillos from Spain, kalamata from Greece, liguaria from Italy and nicoise from France.

Olives are picked at various stages of maturity. They are tiny, hard and green when they begin to ripen, become purple before they are fully ripened, and are plump and black when fully ripened. Each type goes through different curing methods after a lye or water wash. Brine is used for the green (unripe) olive, while salt is used for the fully ripe olive.

Olives are sold whole and pickled (in brine), crushed as a paste, fermented or extracted as olive oil. Raw olives are very bitter but lose their bitterness and develop an aroma after several months of curing. Green olives are picked and treated with lye solution before pickling, which removes the bitterness. In Greece, the fully ripened or black olives develop a strong flavor when they are treated with salt, which also removes the bitterness. Pickled olives are black or green depending on whether they were picked ripe or unripe. Leafy spices are added to pickled olives to enhance their flavor and create new flavors.

Olives contain a glycoside, called oleuropein, a phenol that causes the bitterness in the unripe fruit, and that decreases during ripening. The oil (25–55%, depending on the variety) is from the mesocarp. The oil mainly contains oleic acid, linoleic acid and palmitic acid, with some eicosenoic acid and palmitoleic acid. The green coloring of the olive is due to chlorophyll.

Olives are an essential ingredient in Mediterranean cuisines, especially green olives that are used in cold (or uncooked) dishes such as salads and in sauces, breads and spreads. They are even eaten as a snack. Olives are commonly served as tapas or mezzes accompanied by cheese, breads, nuts or tomatoes. Black olives are used in many cooked dishes of Italy, included as a pizza topping and in tomato sauces. In Provence, France, black olives are an essential ingredient of tapenade, a spread with garlic, anchovies and capers.

In Morocco and Tunisia, olives are cooked with lemons, chicken and spices in tagines. Olives add slightly sharp, tart and bitter notes to salt cod, potatoes, chicken and eggplant. In Spain, the pits are removed, and the green olives are stuffed with sweet red pepper, capers, anchovies or almonds. In Mexico, olives are chopped and combined with ground meat and spices for sweet, savory picadillos. In the Veracruz region, they give a sour note to the popular red snapper dish.

Olive oil comes as heavy and fruity or light with a delicate perfumed note, depending upon its stages of extraction or pressing of the oil. These different types of oils are used for specific applications and are often added at the end of cooking to add flavor.

Olive oil is used for frying, cooking and as a flavoring. It adds flavor to hummus, zhoug, aioli and pasta or is used as a dip for breads and salads. It goes well with garlic, tomato, sesame beans, many leafy spices (basil, thyme, oregano, rosemary, tarragon and parsley) and green chile peppers,

Kokum/Kodampoli

Kokum fruit, scientific name *Garcinia indica,* is a popular flavoring in India, especially in the Kerala, Gujerat and Maharashtra regions. In Kerala, it is known as kodampoli or referred to as "fish tamarind" and is used for many fish dishes instead of the tamarind. It has a pinkish purple or deep purple skin that is sticky to the touch when ripe. It is available as a whole fruit, dried rind or skin or as a paste. It has a fruity, sour, slightly smoky and slightly salty taste. Its sourness is due to malic and tartaric acids.

Kerala is noted for its hot and sour flavors, and kokum is used to provide the sourness to fish, vegetables, coconut-based curries and many chutneys. It is generally discarded before the dish is served.

Garcinia globulosa (or called asam kandis in Indonesia) is a small, thin-skinned fruit that is used as a substitute for tamarind in Indonesia, especially in Sumatra. The skin of the fruit, which is bitter, is dried into a hard black piece that is used in many of the local dishes. It is believed to relieve skin allergies and sunstroke.

Coconut

Coconut's scientific name is *Cocos nucifera*. Called "shripal" in Sanskrit or "fruit of the Gods," it is a holy fruit for the Hindus and is used in Hindu religious ceremonies, weddings and prayers. It is broken as an offering to the Gods, and its creamy firm kernel (meat) is eaten as a blessing. Hindus believe that the three eyes on the coconut represent the Hindu trinity of Lord Shiva, Vishnu and Brahma. It is also called kelapa (Malay), kokos (German), coco (French), polgaha (Singhalese), thenga (Tamil), nalikeram (Malayalam), nariel (Hindi), coco fruto (Spanish) or maprow (Thai).

Coconut provides many flavorings—milk, water, vinegar and a fermented beverage called "toddy" or "arrak." The inner fleshy pulp (endosperm) is scooped and eaten fresh when the coconut is green or unripe. When the coconut becomes ripe, the endosperm or meat is grated and used fresh or dried (called kopra or copra) to thicken sauces and to provide texture or flavor foods. The dried coconut is sold as desiccated and comes in three grades—fine, medium and coarse.

Coconut juice is extracted with water from the grated coconut or endosperm and is called coconut milk or santan in Indonesia and Malaysia. It is sweet, milky white and nutty and is a popular flavoring in South India, Southeast Asia and the Caribbean.

Coconut milk is available canned, dried or as a paste concentrate that can be reconstituted. The copra can be sold desiccated, toasted, creamed (canned sweetened or unsweetened) or as extract (concentrated form) and contains 60% to 70% oil, mainly lauric (40–55%), myristic (15–20%), caprylic, capric and palmitic acids, with small amounts of oleic acid. Coconut oil is the most popular frying medium in South India. The coconutty flavor is due to delta-lactones, especially gamma-deca lactone, while the toasted coconut flavor is due to maltole and cyclotene.

Coconut flavoring is indigenous to Southeast Asian, Sri Lankan, Caribbean and South Indian cooking. It is an essential ingredient in the local curries, marinades, soups, stews, sauces, breads and sweets. Coconut's sweet flavor rounds off other flavors and tones down any heat.

Coconut cream is a must in many dishes of Kerala, especially in the local fish curries. The dessicated coconut gives texture and visual appeal to many dishes and steamed breads. Avial, a popular local vegetable dish, contains toasted, desiccated coconut and dried chilies, kari leaf, asafetida, turmeric, toasted lentils and other spices.

In Sri Lanka, coconut milk is used with toasted spices to mellow and give body to hot curries. Coconut milk is also used in the steamed, fermented breads called "hoppers," which are made from rice flour, coconut milk and yeast and are served for breakfast.

In Indonesia and Malaysia, coconut milk and grated coconut are essential flavorings in rendang (which is beef simmered in coconut milk), curries and vegetables. They are also popular in desserts and cakes such as wajik, dodol, black rice pudding and dadar (pancakes with a sweet coconut filling).

In Thailand, coconut milk is an essential ingredient in curries, including red, green, yellow, Mussaman or Panang. Coconut pairs well with fish sauce, belacan, turmeric, lemongrass, tomato, galangal, basil, eggplant, potatoes, chocolate, fruits, vanilla, pandanus, glutinous rice and gula melaka.

Palm Sugar

Palm sugar is made from the sugary sap of the sugar palms of Southeast Asia and India and the flowers of the coconut palm. Palm sugar (called gula melaka, gula merah, nam taan peep or gula jawa) flavors the sweet and savory dishes of Thailand, Malaysia, Singapore and Indonesia. It is sold as golden to light brown compressed cylindrical packs or thick pastes. It has a caramel, maple-sugar-like flavor. The darker types are stronger in flavor. It is used commonly in desserts and puddings in Thailand (with water chestnuts, pumpkin custard and bananas), in Nonya cooking (with adjuki beans, coconut milk and pandan leaf essence) and in Malay desserts (with jackfruit, banana and glutinous rice). Thai sweet dumplings called kanom tom daeng are served with a sauce of palm sugar, coconut milk and caramelized sugar. Gula melaka is also the name of a Malaysian dessert that contains palm sugar, coconut milk, pearl sago and pandan leaf.

Jaggery, used commonly in Indian desserts, comes from another palm called palmyrah. Jaggery is light brown to dark brown, has less sweetness than cane sugar and has an aromatic flavor and caramel-like taste. The Mexican piloncillo or panela are sold as large, flat, compressed cakes and small or large cones. To use these hard sugars, they must be grated, chopped with a knife or softened in water, the latter method being the best.

Since palm sugars are unrefined sugars, they contain minerals and vitamins as compared to refined sugars.

VEGETABLE FLAVORINGS

In the west, vegetables traditionally have been boiled or served raw and have been eaten mostly because of their nutritive value. In many regions around the world, such as India or Southeast Asia, certain cultures survived only on vegetarian meals, either for economic or religious reasons. These cultures learned to season vegetables to give them great flavors. Asian preparation techniques, such as sautéing, grilling, steaming, baking, braising or currying, have created many unique flavors and textures in vegetable dishes. Many vegetables contain volatile oils that give sauces and beverages pungent aromas and savory sweet or bitter tastes.

Today, with consumer's interest in a healthy lifestyle, vegetable extracts and purees can be substituted for meat and chicken stocks or pastes for flavoring soups, sauces or stews. They are a good source of fiber, vitamins, especially C and A, minerals and antioxidants.

The cruciferous vegetables such as broccoli, cabbages (green savoy, red, Chinese cabbage (or bok choy), collard greens, mustard greens, kale, kohlrabi and cauliflower are high in vitamins, monoterpenes, thiols and indoles. Broccoli contains high levels of calcium, pantothenic acid, vitamin A, thiols and indoles and sulfur compounds that have antimicrobial activity. Cauliflower has high levels of iron, boron, thiols, indoles and sulfur compounds. Carrots contain a high level of carotenoids and antioxidants.

Spinach contains iron, carotenoids and chlorophyll and is an especially good source of lutein. Bitter melon has polypeptide P (an insulin-like compound that can replace insulin); eggplant is high in bioflavanoids; plantains have high folic acid, vitamin C and potassium; and pumpkin has folic acid, vitamin A and vitamin C. Mushrooms, especially Asian types, are rich in zinc, pantothenic acid and folic acid.

VEGETABLES IN ETHNIC CUISINES

Table 24 lists various vegetables, their sensory qualities and their ethnic preferences.

TABLE 24. Global Vegetables and Their Flavoring Properties.

Vegetable	Flavoring Quality	Ethnic Preference
Beet	Color, sweet	Mediterranean, U.S.
Bitter melon/kareli	Bitter	India, Southeast Asia
Broccoli	Color	U.S., Mediterranean
Cabbage	Texture, color	Mediterranean, Asia, U.S.
Carrot	Color	U.S., Asia, South America, Mediterranean
Collard greens	Color	Southern U.S., Africa
Eggplant	Color, texture	Asia, Mediterranean
Plantain	Mealy texture	Caribbean, Latin America
Pumpkin	Texture, color	Caribbean, Asia, Latin America
Spinach	Color	U.S., Europe, Asia
Tomato	Acid, color	Global
Shiitake mushroom	Meaty, texture	East and Southeast, Asia, U.S.
Chayote/cho cho/christophene	Texture	Caribbean, Latin America, Southeast Asia

SPECIFIC VEGETABLES AS INGREDIENTS

Some vegetables that add significant flavorings are described in detail below.

Tomato

Tomatoes are a global flavoring. Tomatoes are eaten raw or cooked and are used chopped, pureed, as paste, juiced or whole. They provide sweetness, acidity and color to many ethnic foods, such as salsas, sambals, chutneys, curries, pasta sauces and pizzas.

The volatile oils in raw tomato include 3-hexenal, 3-methylbutanal, beta-ionone, 3-hexenol and eugenol. Canned tomatoes primarily have dimethyl sulfide. Tomatoes are high in lycophene.

Southeast Asians add tomatoes to sambals to balance chile peppers and fermented shrimp notes. The Chinese add them to characterize their sweet and sour sauces. Indians add tomatoes to chutneys and curries to balance the spiciness, and Mexicans add them to their salsas to give sweet notes. Sometimes, Southeast Asians add ketchup as a source of tomato flavoring in their sweet-sour sauces, sambals and curries.

Tomatilla or Mexican green tomato is not a true tomato but comes from the cherry family. It has a sweet-sour taste, with a fresh hay-like aroma. A calyx surrounds the tomatilla and as it grows, the calyx also grows and becomes a brownish and papery form that encloses the tomatilla. Tomatilla is green to pale yellow in color and is eaten when unripe. It was a staple flavoring in Mayan and Aztec cooking, and today, is used in Mexican, Central American and South American green salsas (or salsa verdes). In the U.S., it is popular in Southwestern dishes.

Mushrooms

Mushrooms are great flavorings for many foods, especially to provide meaty, earthy profiles to vegetarian cooking. Mushrooms are fungi and come fresh or dried in many varieties, each with its own unique flavor. In ancient times, the Egyptians only served mushrooms to the pharaohs, and the Romans and Greeks saw them as food fit for their Gods. In Asia, they were eaten as a tonic for longevity. The French began cultivating mushrooms in the seventeenth century and called them mousseron, which means "growing on moist moss."

Mushrooms come wild or cultivated. Each mushroom has its own type of flavor. The common cultivated field mushrooms are the button mushrooms, which have very little flavor when raw, but which develop a rich nutty and earthy flavor when cooked. They have a white, tan or brown color. They provide texture to salads, soups or curries. Cup mushrooms, which have stronger flavors, are good for stuffing, while the more intense flavored flat mushrooms are used in sauces, soups or stews. Cremini mushrooms, which are tan to dark brown, are a variety of button mushrooms that has a richer flavor and a firmer texture.

Portobellos or brown mushrooms are bigger than the creminis and have more intense and steak-like flavor. They have thick, flat, deep-brown caps. They are popular grilled or roasted.

The cultivated wild mushrooms include shiitake, which has a meaty flavor and is used in East Asian dishes, and the oyster mushroom, which has a chewy texture and subtle flavor that it loses when overcooked. Other types include the boletus (called porcini in Italian, due to their resemblance to pigs, and cepe in French), which are usually dried. These dried types are dark brown to reddish brown in color and have a rich, earthy and meaty flavor.

Chanterelles (from Europe), also called girolle, are a "poor man's truffle." They are orange, bright yellow or red, have an apricot-like aroma, and a nutty taste that mellows with long cooking times. Morels, or sponge mushrooms, are conical shaped and spongy looking. The dried types have a stronger, earthier and smokier flavor. They are popular in Europe, especially in French cuisine.

Truffles, called the black diamonds, are the king of wild mushrooms. There are many varieties, such as the black truffle (in Perigord, France) or the white truffle (in North Italy). Truffles are said to have the best flavor, and they are eaten raw or slightly sautéed to intensify their flavor.

There are many Asian mushrooms that provide great textures to soups and stir-fries of East Asia. Shiitake, which is also called the Chinese or Oriental black mushroom, is dark brown to almost black, with an umbrella shape. It has a slightly smoky and meaty flavor, and when cooked, gives a firm chewy texture.

Cloud ears, also called wood fungus, wood ears, black mushrooms or tree ears, are shriveled and greyish brown with an earthy flavor. They are popular in Chinese and Korean cuisine. Enokis, also called golden mushrooms, are creamy white and have a mild flavor with a crisp texture (sold as enokitaki which are clusters of cream-colored stalks with tiny caps). They are popular in Japanese soups, sauces and sushis.

Oyster mushrooms, also called maiitake, are fan shaped and cream colored, with a delicate flavor and chewy texture. Other Asian mushrooms include nameko (with reddish brown cap and a slippery texture), reishi (in Japanese soups), paddy straw (used in Chinese stir-fries), shimeji (aromatic and meaty) and white matsutake (also called Japanese oysters mushrooms) that are marinated, or used as toppings for seafoods and steaks.

Huitlacoche, or corn smut that grows on corn kernels, is eaten as a delicacy in Mexico. It is bluish gray and has a smoky, meaty flavor. It pairs well with chile peppers, chocolate, caramelized onions, vanilla and epazote. It enhances meats and corn-based foods.

Mushrooms can be used whole, stuffed, chopped, sliced, julienned or pureed. They are available fresh or dried. Mushroom extract is from the dried mushroom and is a dark brown viscous paste with a rich meaty flavor. The earthy and meat-like flavor of mushrooms enhances the flavor of soups, stews and sauces. Mushroom pairs well with meats, poultry, potatoes, ginger, olives, garlic, cilantro, cumin, coriander, turmeric and black pepper. Cooking releases the true flavors of mushrooms. Most dried mushrooms have more intense flavors than fresh mushrooms.

Fresh mushrooms contain about 7% sugar, 3–5% protein, 0.4% fats, 88% water and some volatile compounds. Dried mushrooms have no sugars but are high in folic acid, potassium and phosphorus.

The flavor of mushrooms is due to glutamic acid (which is the major component of MSG). Mushrooms have two distinct flavors, when fresh (or uncooked) and when they are cooked or dried. Certain mushrooms can take on the flavor of the food with which they are cooked. Unsaturated alcohols are found in fresh mushrooms. Fresh shiitake has no flavor, but the dried shiitake has a smoky sulfury aroma due to 1-octane-3-ol, 3-octanone, 2-octenol, lenthionine and octanol.

Chinese dried mushrooms, such as cloud ears, have a mild flavor but when reconstituted add a glutinous, chewy texture for soups and stir-fries. Dried morels enhance cream sauces, rice, polenta, eggs or butter. Dried shiitake, a popular ingredient in Chinese and Japanese dishes, has a tough and fibrous texture when reconstituted. Dried boletus goes well with pasta sauce or risottos. Porcinis taste great when grilled, marinated or slowly braised or sautéed with leafy spices, garlic and shallots and added to risottos, vegetables and salads.

Dried mushrooms are soaked in boiling water for 20 minutes, strained and then used. The strained water is sometimes added to soups and stews or in marinades. The mushrooms can be coarsely blended or ground to powder and added with other ingredients to vegetable dishes to give meatier, more in-depth notes. Sautéing and long cooking bring out the flavors of most mushrooms and are popular cooking methods used in Chinese and Japanese cuisines.

Mushroom extracts are added to soy sauce, chile pastes, soups, salad dressings and stews to provide brothy beef-like notes. They can enhance fermented soy blends and are used to coat fish and vegetables to give a meaty brown color. They can enhance mayonnaise, rice products, noodles, couscous, eggplant and tomato. They can be blended with olive oil, nuts and leafy spices to create intense spreads or pestos (Table 25).

Asian mushrooms have been used traditionally to thin blood, lower cholesterol and stimulate the immune system.

TABLE 25. Global Mushrooms and Their Culinary Uses.

Mushroom	Ethnic Preference	Culinary Uses
Chanterelles	Europe	Omelette fillings, with scrambled eggs
Button	U.S. and Europe	Soups, sauces, stir-fries
Shiitake	China and Japan	Stir-fries, soups, sauces
Enoki	Japan	Soups and salads
Cloud ears	China, Korea, Southeast Asia	Stir-fries, soups
Truffles	France	Grilled or smoked meats, toppings for pasta or as spreads
Huitlacoche	Mexico	Spreads, quesadilla fillings, mousse, soups
Morels	France	Stuffing, soups, sautéed in butter

LEGUME FLAVORINGS

Legumes, also called pulses, are the edible seeds and pods of certain plants, including beans, soybeans, lentils, peas and peanuts. Legumes are popular in vegetarian cooking as a source of protein. They are combined with rice, meat or vegetables as main dishes, or used whole, pureed or ground into flour for snacks, soups, stews, desserts, noodles or crepes.

Certain legumes are very flavorful and can add variety to a meal, such as black beans, masoor dal or chickpeas. Legumes can provide different textures and mouthfeel properties depending on cooking techniques. They can give background notes to a sauce and, therefore, replace chicken- or meat-based roux. Other types are economical because they are nonassertive in flavor and can "stretch" a meal. Soybeans are made into fermented pastes and sauces that are commonly used as seasonings in Chinese, Korean, Japanese and Southeast Asian sauces, condiments, soups and stir-fries.

Beans originated in the New World, and today, they are a staple item in Asian, Mediterranean, Caribbean, Latin American and North American cooking. There are many types of beans or lentils that vary in flavor, color, texture, size and shape.

LEGUMES IN ETHNIC CUISINES

Beans are known by different names, depending on the culture or region, include frijoles, habichuelas, granos, pois, kacang or dals. White beans (lima, navy, white kidney, cannellini), pintos, black, red or pink beans, red kidney beans and black-eyed peas are commonly consumed legumes in the U.S.

Black-eyed peas, called cowpeas in the U.S., pigeon peas and gandules in the Caribbean, were brought here by the West Africans. They are now a staple in Creole, Southern regional and Caribbean cooking. Chickpeas or garbanzos are widely used in Mediterranean and Indian cuisines. Channa dal, the split version of kabuli channa, and kala channa, the smaller and darker variety of chickpeas, are favorites of North India. They have meatier and sweeter tastes than the yellow split pea and are used as braised vegetables or in dal soups.

Lentils, lentejas ordals, are the most commonly eaten legumes in India. They come in white, pink yellow or brown colors. The word dal is used for beans, lentils or peas. Lentils are aromatic and delicate in taste when cooked. Texture and consistency differ among lentils. Dals are generally available hulled and split and come with different names and colors. Sometimes the same split dal has a different name from the whole dal. The unhulled lentils look attractive, but take longer to cook and are harder to digest. Once they are hulled (skin removed) and split, they become easier to cook and digest. These forms are popular in India. The split masoor, urad, mung and toovar dals are commonly used in Indian vegetarian curries, seasonings and sauces. They are also pureed or processed into flour for snacks, soups and desserts. The urad and toovar dals are toasted (dry roasted) and combined with spices to create spice mixtures that enhance many sauces, soups and chutneys (Table 26).

SPECIFIC LEGUMES AS INGREDIENTS

Soybean

The soybean, referred to as one of the "grains" in China, is ubiquitously used in Asian cooking. Fermented soybeans (misos, tan or tow jiew, goit chu jang, taucheong and dou bain jiang) and soy sauce (called jiangyou, shoyu, namsiew, kecap or taoyu) made from soybeans are essential flavorings for condiments and many dishes in East Asia and Southeast Asia. East Asians have developed their own unique processes for fermenting soybeans to achieve varying flavoring end products. The yellow or brown, sweeter and less bitter tian mian jiangs used in stir-fried vegetables of northern China, and the more pungent black, bitter la ba jiangs for seafood, pork and beef dishes in Shanghai style as well as Southeast Asian dishes, are the more commonly used pastes. These fermented pastes are smooth or chunky, hot, salty, sharp or sweet depending on the added ingredients and how they are prepared.

Misos are used abundantly in Japan to flavor soups, stir-fries or salads. They have varying flavor profiles and colors depending on other added ingredients and the method of preparation. Miso made from only soybeans is dark red with a salty taste; miso made from soybeans and barley is dark brown and semisweet, while miso made from soybeans and rice is light yellow and sweet. Different misos are used for different applications. Many foods and appetizers use misos to enhance their tastes, such as shumai (or steamed Japanese dumplings) that are dipped into a miso-based sauce.

TABLE 26. Legumes and Their Ethnic Preferences.

Beans	Ethnic Preference
Soybeans	China, North India, Southeast Asia, Korea, Japan
Adjuki beans	Japan, China, Korea, Southeast Asia
Black beans	Cuba, South Mexico, Venezuela, Brazil South, Northeast and Southwest U.S.
Red beans	South, Northeast and Western U.S., Caribbean
Pinto beans	Northern Mexico, Puerto Rico, Southwest U.S.
Red kidney beans/ rajma dal	Caribbean, North India, South and Southwest U.S., Mediterranean
Lima beans	Southern U.S., Peru, Caribbean
Flageolets	Mediterranean, Italy, France, Northeast U.S
Lentils	
Masoor/red/pink (whole or split)	India, Egypt, Spain, Mediterranean
Urad/black gram (whole or split)	South India, Malaysia, Singapore
Toor/toovar/translucent yellow dal	India, Southeast Asia
Moong bean/green gram/ whole or yellow split dal	India, Southeast Asia, Mediterranean
Peas	
Chickpeas/garbanzo/ kabuli channa/ Bengal gram	Middle East, North Africa, India, SE Asia, U.S, Europe
Pigeon peas/arhar dal/gandules	Creole, Southern U.S., Caribbean
Black eyed peas/ chowla dal (split)	U.S., North India, West Africa, Brazil
Yellow, green, brown split peas	Mediterranean, U.S.
Kala channa (whole dark brown)	North India, Southeast Asia

Soy sauce originated in China over 2500 years ago. The fermented or brewed soy sauce that is commonly used in Asian regions has a characteristic umami (brothy or MSG-like) taste. Its unique flavor is due to over 300 chemicals that it contains—amino acids, alcohols, organic acids, esters, sugars and salt and are thin, thick, salty, sweet, garlicky or hot, depending on local preferences and applications. It is used to flavor vegetarian foods that do not use garlic or shallots.

Adjuki bean, which has a sweet taste, is ground for desserts and fillings for buns and pastries. In China and Southeast Asia, it is seasoned with star anise, rice wine, ginger and shiitake mushrooms and is added to pork, vegetables or glutinous rice.

Nutritionally, soybeans are a good source of protein and fiber and have little or no fat. They are also a good source of folic acid and minerals such as iron, calcium and phosphorus. They contain isoflavones that help prevent breast and colon cancer, slow calcium loss from bones, lower serum cholesterol and help alleviate menopausal symptoms.

NUT FLAVORINGS

Nuts are seeds or fruits that consist of kernels and are surrounded by shells. Examples include almonds, candlenuts, cashew nuts, pistachios, hazelnuts, walnuts, macadamia nuts, chestnuts, Brazil nuts (also called cream nuts, butternuts or Para nuts), kola nuts (or cola nuts), gingko nuts (ya-chio), jackfruit nuts, peanuts (that are legumes), coconuts (a drupe), pecans and pili nuts (Java almonds). Peanuts and Brazil nuts are not true nuts, but they are included here because they function as nuts in foods. (Nuts and seeds are included together here because they have similar applications and functions. Sometimes, seeds are classified with nuts and vice versa.)

Nuts have been used since ancient times as food and as an oil. The Romans used almonds, the Aztecs used peanuts and pecans, Asian Indians used cashews and South American Indians favored Brazil nuts. The Arabs, Persians and Turks used nuts to thicken sauces and to garnish many of their dishes.

Nuts provide consistency, flavor and visual appeal. Some can be eaten raw, like pine nuts, sunflower, lotus or pumpkin seeds. Some possess sweet tastes, such as chestnuts or cashews, and can be used in desserts or as a crunchy garnish. Nuts can be roasted, boiled or steamed to provide toasted flavors to soups, stews, curries, ice cream or for cakes, in which nuts such as pistachios, lotus, hazelnuts, jackfruit nuts and almonds are used. Roasting brings out their full flavor. They can be added during cooking, toward the end of cooking or as a topping.

Nuts can be used whole, chopped, sliced, slivered, flaked or ground. Candlenuts, macadamia nuts, peanuts, lotus and Brazil nuts are ground coarse or fine to provide a creamy texture and a rich flavor to sauces, stews, curries and many vegetarian dishes. They go well with chocolate, cinnamon, chilies, saffron, raisins, cardamom, coconut, rice, caramelized onions and citrus.

Nuts can be boiled, steamed, roasted, stir-fried, tumised or curried. In European regions, they are mostly used in baked goods, snacks and confectionaries for textural effects. In many other cultures, nuts are used as flavor and texture contributors for sauces and curries. Nuts can even be used in beverages. The kola nut is indigenous to West Africa and is used to provide the "cola" type of flavoring to many carbonated soft drinks.

Seeds are also used as flavorings in foods. Popular seeds include pine nuts (pignolias/pinons), pumpkin seeds, chia seeds, flaxseeds, lotus seeds, sunflower seeds, poppy seeds, watermelon seeds and sesame seeds.

Nuts and seeds have many health benefits. They are good sources of protein, vitamin E, calcium, magnesium, phosphorus, folic acid, niacin and pyridoxine (vitamin B6). Nuts are an excellent source of arginine. They are high in fat, generally 50–70%. The fat is nearly all monounsaturated (with walnuts having polyunsaturated fats). Almond is a good source of omega-3 fatty acid and linolenic acid (60%). Ginkgo nuts have a high amount of flavonoids and traditionally have been used to help prevent memory loss and improve blood circulation.

Pumpkin seed has high magnesium, zinc and omega-3 fatty acid. Sunflower is high in vitamin D and B complex vitamins and pectin.

NUTS IN ETHNIC CUISINES

Nuts, such as almonds, walnuts and pistachios, are an essential ingredient in Middle Eastern cuisines. They are used in rice dishes, chicken stuffings and pastries. Walnuts are commonly used in Turkish tarator and baklava. Pistachios are indigenous to the Middle East. They have a bright green color and are used in a dessert called locoum (known as Turkish Delight), cakes, sausages and grilled meats. In North India, almonds, cashews and pistachios are commonly used in biryanis, vegetarian dishes, desserts, puddings and ice creams. They are also used to thicken curries.

Roasted peanuts (a legume by classification) provide the dominant flavor in Southeast Asian sauces, such as satay sauce, stir-fried noodles and laksas. Called groundnuts in Africa, they are also a staple ingredient in West African cuisines. They are used to provide flavor and consistency to sauces, stews, soups and curries.

Cashews and peanuts are used in Szechwan and Southeast Asian stir-fries and sauces with chile peppers, sugar and spices. Gingko nuts have a mild, grassy, cheese-like flavor and are grilled or boiled by the Chinese and Japanese. Chestnuts are typically used in sweet dishes, as garnishes and in the soups of southern Europe. Hazelnuts are roasted and used in romesco sauce of Spain and the meat dishes and confectionaries of the Middle East. Nuts are also ground into flour and are sometimes added to make polenta or crepes, as thickeners for soups and stews and in desserts.

Pine nuts, also called pignoli in Italian, or Indian nuts, are indigenous to the Mediterranean and the U.S. They are used in a variety of Italian dishes, pestos, desserts and in Middle Eastern minced meat stuffings and fruit dishes. In Lebanese cuisine, they are toasted and sprinkled over lamb patties. Walnuts are ground and used in pestos of Italy and Provence.

In Mexican cooking, pumpkin seeds, sesame seeds and pine nuts are toasted and or fried and then finely ground to thicken and provide flavor to sauces and moles. Pipian, containing ground pumpkin seeds, is a popular sauce in Oaxaca and other parts of Mexico. Pipian is also mixed with smoked pasilla, chipotles, cinnamon and aniseed and added to corn dough and then steamed in hoja santa leaves to create tamales. Walnuts go well with fruits, salad dressings, stuffings and fillings and are an important ingredient in chile en nogada of Puebla, Mexico. Almonds are also used in the famous moles of Oaxaca and in chocolate beverages.

Kola nuts are ground and used as flavorings in beverages of West Africa (from which the cola-flavored beverages sprouted globally). They were the original ingredient (along with coca leaf extract from Peru) in the Coca-Cola® beverage and are still used as a flavoring in many cola-type beverages. Their taste is slightly bitter and sweet. They have a high level of caffeine and tannins, and, thus, are used as a stimulant in West Africa. They were traditionally chewed by West Africans to relieve fatigue.

In Europe, hazelnuts, almonds and chestnuts are used in desserts, confectionaries and pastries. In the U.S., nuts and seeds are added to give texture to burgers, pastries, rolls and grain-type breads. Pecans, indigenous to the southern U.S., are used in stuffings, cakes, pies and ice cream. Sweet almonds are used in butter, pastries, cakes, liqueurs, and pralines.

Table 27 lists some popular nuts and seeds, their properties and the region in which they are preferred.

TABLE 27. Global Nuts and Seeds and Their Flavoring Properties.

Nuts/Seeds	Flavoring Property	Ethnic Preference
Almond	Texture, sweet	India, Middle East
Walnut	Texture, flavor	Mediterranean
Macadamia	Creamy, sweet	South America, Asia, Middle East
Lotus	Texture	China, Vietnam
Peanut	Creamy, toasted	Southeast Asia, Africa, China
Pecan	Sweet, pulpy	U.S.
Pinon	Texture, buttery	Italy, U.S.
Pumpkin	Flavor, oil, texture	U.S., Europe, Mexico
Sunflower	Nutty, texture, oil	Asia, Europe, U.S., Asia
Watermelon	Nutty, flavor	West Africa, China
Candlenut	Nutty, consistency	Southeast Asia
Kola nut	Bitter, caffeine	West Africa, Brazil, Java
Coconut	Creamy sweet	India, Southeast Asia, Caribbean
Pistachio	Texture	Middle East, India, Iran
Gingko biloba	Nutty	China, Japan, Korea

SPECIFIC NUTS AS INGREDIENTS

Some of the more important nuts and seeds are examined in detail below.

Almonds

The scientific name for almond is *Prunus amygdalus* (*P. amygdalus*), which is the soft interior of a fruit from the Rosaceae family. There are two varieties, a sweet type (*P. a.* var. *dulcis*) and a bitter type (*P. a.* var. *amara*). Almonds are also called badam (Hindi), almendra (Spanish), mandorla (Italian), lawz (Arabic) and amande (French). Indigenous to North Africa, central and western Asia, almonds are now cultivated in Spain, Italy and the U.S. (California).

Almonds have a bland fixed oil of 54% and 18% protein. The deoiled press cake of bitter almonds has 0.5–0.7% volatile oil. Bitter almond contains amygdalin, which breaks down to benzaldehyde and hydrocyanic acid (prussic acid). These are poisonous compounds that are removed through heat before it is used as an oil. The oil of almond is colorless to pale yellow and consists mostly of benzaldehyde. Bitter almond has about 3–5% amygdalin, while sweet almond has only small amounts of amygdalin. Bitter almond is distilled for essence that is used to flavor cookies, cakes and marzipan. Almonds have about 50% fat, mainly oleic acid (80%), linoleic (15%) and palmitic (5%).

Almonds are blanched and toasted before use. They are popular ingredients in Middle Eastern and North Indian cooking. They thicken curries, are fried in ghee for biryanis and pilafs and are mixed with saffron for kheer. Sweet almond is mostly used for desserts and confectionaries. It is toasted until golden brown and sprinkled over chicken in Tunisia and over sauces and condiments in North India. In the west, almonds are mixed with rose water and sugar in a confectionary called marzipan.

Almonds are a good source of tocopherol E, folic acid, niacin, phosphorus, calcium, omega 3 fatty acids and arginine.

Pine Nuts/Pignoli Nuts

Pine nuts are seeds from the pine tree species, of which there are many varieties. The Greeks, Romans, Chinese and Navaho Indians used them for flavoring, as offerings to their Gods or in magical potions. In the U.S., they are also called Indian nuts. Pine nuts are small, thin and white and have a delicate buttery taste and a soft texture. They can be eaten raw or lightly roasted, and they are used in pestos, sauces, curries, desserts, vegetables and as garnishes in Mediterranean, Mexican and Middle Eastern foods. They are high in protein and potassium.

Candlenuts

Candlenuts are also called buah keras (or "hard fruits" in Malaysian or Indonesian) or Indian walnuts. They are finely ground and used to thicken and enrich curries, laksas and sambals of Southeast Asia. They are oily fruits that are also used abundantly in Hawaii and other Pacific Islands. They contain a poisonous residue, which after extraction, is inactivated at high heats and long cooking times.

Candlenuts have a walnut-like flavor and are roasted, cracked open and ground with shallots, garlic or chile peppers for sauces. They are named candlenuts because at one time they were threaded into the rib of a palm leaf and then burned as candles.

Pumpkin Seeds/Pepian/Pepito

Pumpkin seeds can be eaten raw or toasted and are excellent as snacks, on salads, in breads as purees for sauces and cakes or used as a cooking oil. They are pureed and commonly used in Mexican moles, sauces, salads and stews.

Pumpkin oil (dark green) is used in India, Europe and America for sauces, pasta and vegetables. It contains omega 3 and omega 6 fatty acids.

SWEET AND BITTER FLAVORINGS

There are many other flavorings, such as ghee (clarified butter), yogurt, vanilla, cacao, chocolate, coffee and tea, that are used around the world to flavor sweet and savory products and are used as beverages and cooking oils. While some of these ingredients are familiar to North Americans, their use has been substantially limited to beverages and sweet products. As North American exposure to emerging ethnic cuisines grows, we can anticipate finding these ingredients as enhancing flavors of savory applications with much greater frequency.

SWEET AND BITTER FLAVORINGS IN ETHNIC CUISINES

Next, vanilla, cacao and chocolate and coffee will be discussed. These popular ingredients provide characterizing creamy, sweet, bitter or aromatic flavor profiles to many ethnic dishes. For example, Indian, Middle Eastern and African cuisines often utilize the sweet, creamy, buttery flavors of vanilla and ghee. The bitter notes offered by chocolate and coffee provide a balance or background roasted notes to the spicy profiles of Mexican moles, the sweet, pungent notes of Indonesian sambals and the sour tastes of Caribbean colombos.

While not technically flavorings, barbecuing, roasting and other cooking and preparation techniques also add flavors to prepared foods. In addition to the flavoring ingredients, the next section discusses how various cultures use these preparation techniques to add flavor to foods.

SPECIFIC OTHER FLAVORING INGREDIENTS

Vanilla

Vanilla is the second most expensive flavor after saffron. It was called tlilxochitl by the Aztecs, which means "black flower." The Aztecs flavored their chocolatl (cocoa beans-corn-chile pepper-honey-annatto drink) with ground vanilla pods. Their ruler Montezuma offered this drink to the Spanish invader Hernan Cortez who then introduced it to Europe.

Vanilla is derived from the Spanish word, vainilla, meaning, "little pod." It is derived from the fruit of the orchid vine. Synthetic vanilla, made from wood wastes or other beans, is much cheaper but lacks the flavor of true vanilla.

Scientific Name(s): Vanilla planifolia (or *V. fragrans*): French or Bourbon vanilla; *V. pompono:* West Indian variety, also called Antilles vanilla or Guadeloupe vanilla; and *V. tahitiensis:* Tahiti vanilla. Family: Orchidaceae.

Origin and Varieties: vanilla is indigenous to southeast Mexico and is now cultivated in Madagascar, Reunion Island, Indonesia, Tahiti, Comoro Islands, Caribbean islands and Central America. French vanilla beans (from Reunion, Comoro islands and the northwest tip of Madagascar) supply about 75% of the world's demand for vanilla. These beans are called Bourbon vanilla because they were first grown on Isle de Bourbon, which is now known as Reunion Island.

Common Names: vanila (Arabic), vanille (Dutch), vanille (French, German), vanil (Hebrew), vanilla (Hindi, Urdu, Malayalam), paneli (Indonesian), vaniglia (Italian), banira (Japanese), waneela (Malaysian), hsiang ts'ao (Mandarin), vanilje (Norwegian), baunilha (Portuguese), vanil' (Russian), vainilla (Spanish), vanilj (Swedish), vanikkodo (Tamil) and waneela (Thai).

Form: vanilla is the cured dried fruit or seedpod of the creeping orchid. Vanilla bean is a dark brown, narrow, long and waxy wrinkled fruit. Vanilla comes as whole beans, ground and as extracts. Fresh vanilla pods do not have any flavor because vanillin, which is responsible for its flavor, is bound as a glycoside and needs to be cleaved by enzymatic action. Vanilla is expensive because it is pollinated and harvested by hand.

Powdered vanilla is obtained by mixing the ground beans with sugar or by extracting the flavor from the beans and then mixing with sugar. Vanilla flavor is extracted with alcohol.

Vanilla is sometimes adulterated with other ingredients, such as tonka beans. Synthetic vanilla is made from wood or pulp waste and lacks true vanilla flavor.

Properties: the flesh of the cured vanilla bean has a perfumed, tobacco-like aroma. Pure vanilla has a mellow, rich flavor. Vanilla extract, prepared by steeping the macerated beans in alcohol, has a delicate sweet aroma and a mellow aftertaste. Various grades of vanilla beans are blended to obtain vanilla for ice cream and other products. Synthetic vanilla (from pulp waste) or imitation vanilla (from clove oils) has a heavy aroma and a grassy and bitter aftertaste.

Generally, vanilla has a sweet, delicate, rich and creamy flavor. It enhances the flavor of other ingredients such as fruits, chocolate, butter, chile peppers or spices. Different vanillas exhibit varying flavor profiles: Bourbon vanilla, now grown on Reunion Island and Madagascar, has smooth, rich and creamy flavors. The Reunion type is sweeter and more delicate in aroma than Madagascar vanilla. Indonesian (Java) vanilla is smokier, woody and has stronger notes. The Tahiti type is more floral, fruity and perfumed. Mexican vanilla has a creamy, delicate and mellow aroma, similar to Madagascan vanilla.

Chemical Components: fresh vanilla has no taste. When picked, the green pods have no vanillin because the vanillin is bound to a sugar molecule. Vanilla is released through enzymatic reaction during the curing process to give aroma and taste. The curing process differs, depending on the region, blanching for Bourbon vanilla or steaming for Mexican vanilla, thus resulting in different quality beans. Enzymatic action of beta-glucosidase on the precursor, glucovanillin, gives vanillin and sugar. Vanillin appears as a white crystalline covering on the surface of the beans. It is developed through curing (sweating, fermenting and drying) the beans. The fermented bean contains about 2% vanillin.

Other components are glucovanillin, vanillic acid, anisic acid, aldehyde, *p*-hydroxy-benzaldehyde, *p*-hydroxy methyl ether, phenols, alcohols, lactones, acids and esters. Vanilla has about 25% sugars, 15% fat, 6% minerals and 30% cellulose. Natural vanilla extract includes hydrocarbons, aldehydes, acids, carbonyls, lactones, furans, alcohols, esters and phenols.

Oleoresin vanilla is derived by evaporation under vacuum of the vanilla extract. The oleoresin is diluted with solvents to give one-, two- or 10-fold strengths.

Vanilla contains calcium and phosphorus.

How Prepared and Consumed: Europeans introduced the world to vanilla in sweet dishes such as cakes, cookies, puddings, custards, sweet sauces, sweets, yogurts and ice cream. Vanilla is also used in soft drinks, liqueurs and chocolates. It pairs well with sugar, milk, chilies, almonds, chocolate, fruits, tea, coffee, cardamom, cinnamon, anise, ginger, shallots, saffron, mint, parsley, eggs and wine.

Pre-Columbian Indians used vanilla to flavor chocolate and savory foods such as corn, annatto and chile peppers. It can be used to round off stronger flavors in salsas, chutneys and curries and can provide sweetness to lemon chicken, baked fish or mole negro. In West Africa, it is used to flavor stews.

Spice Blends: corn chowder blend, mole negro blend, sambal blend and ice cream blend.

Therapeutic Uses and Folklore: in ancient times, vanilla was used as a stimulant and an aphrodisiac. It was sometimes used as a digestive stimulant, antidote to poisons and as a treatment for nervous disorders.

Chocolate and Cocoa

The word cocoa comes from the Greek word, theobromo cacao, meaning "food of the Gods." The Aztecs first drank chocolatl (chocolate), mixing cacauatl (cacao beans) with chile peppers, corn kernels, annatto and sugar. Aztecs toasted the cacao beans, crushed them between stones and formed the paste into cakes, which they diluted with water and spiced with annatto and chile peppers. This mixture was beaten and stirred slowly over fire until it became a frothy liquid. The Aztec King Montezuma offered this chocolate drink to the Spanish explorer Cortez who brought its manufacture to Spain. Both cocoa powder (for beverages) and chocolate are made from the cacao beans.

Scientific Name(s): Theobromo cacao. Family: Sterculiaceae

Origin and Varieties: there are two main varieties of cacao beans, Criollo and Forastero, both cultivated in South America. Cacao is indigenous to South America and Mexico. The commercial Venezuelan types are Caracas, Carupano and Porto Cabello; and Criollo types include Arriba, Machala, Bahia and Guayaguil. Their flavors and aromas vary from mild and sweet to bitter. There are other varieties from Mexico, West Africa, Sri Lanka, Java, Malaysia and the Caribbean (Trinidad and Jamaica).

Common Names: chocolat (French), cschokolade (German), shokolad (Hebrew), cioccolata (Italian), chocolate (Spanish, Portuguese), chocolade (Dutch), choklad (Swedish), sjokolade (Afrikaans), chakeleti (Swahili), suklaa (Finnish), sjiokolade (Norwegian), csokolade (Hungarian), cokolada (Czech), coklat (Malaysian, Indonesian) and chokolet (Thai).

Form: the ripe pods are 7–12″ long, reddish dark brown to purple, depending on the variety. They are filled with a pale pink to whitish soft pulp that contains the seeds. The Criollo varieties are sweet with less bitterness, while the Forastero variety is very bitter.

Properties: cacao seeds are bitter because of the tannins. After fermentation, they become more reddish and less bitter. When cacao is dried, it becomes less astringent, and bitterness almost disappears. Roasting develops the bean's aroma and flavor.

Cacao beans contain an outer shell and an inner kernel, called the nib. The nib is ground into a fine paste called chocolate liquor, while the fat is separated and made into cocoa butter. The chocolate liquor is blended to give the desired, fat free cocoa powder or chocolate. Chocolate products are white/milk chocolate (with whole milk and sugar), sweet chocolate and semisweet or bitter chocolate.

Chemical Components: cocoa contains caffeine (230 mg/100 g) and theobromine (2057 mg/100 g), tannic acid and oxalic acid. Cocoa butter contains 59.7% saturated fat, 32.9% monounsaturated fat and 3% polyunsaturated fat. Chocolate is rich in phenylethylamine, which is also produced by the brain when one is "in love." Cocoa powder contains tannins (which give cocoa its color and flavor), purine, theobromine, caffeine and starch.

How Prepared and Consumed: chocolate and cocoa powder are used in confectionaries, baked goods and beverages. Europeans have traditionally used cocoa powder and chocolate in beverages and sweet products, but it is now being used more frequently in savory applications.

Chocolate is an essential ingredient in the popular mole negro from Oaxaca, Mexico. Chocolate pairs well with chile peppers, cinnamon, annatto, curry, fennel, ginger, clove, nutmeg, lemongrass, mints, pandan leaf and many flower essences. It can add a new dimension with its chocolatey bitter notes to sofritos, recados, fish soups, corn chowders or meat roasts. In Spain and Italy, it is used with garlic, onion, tomato and parsley in meat and fish dishes.

Spice Blends: mole negro blend, sofrito blend, recado negro blend, chicken curry blend and corn chowder blend.

Therapeutic Uses and Folklore: the Aztecs and Mayans used cacao beans as currency. Chocolate was forbidden to the ladies during those ancient times, but today in Europe, the U.S. and all over the world, it has become the symbol of "amor."

Coffee

Coffee comes from the Arabic word, "qahwah," and began in ancient times as a food rather than a beverage. It was first used by the Galla tribe in Ethiopia around 575 to 850 A.D, who mixed dried coffee berries with fat to make food balls. Later, in 1000 A.D., wine was made from the dried beans, including the skin. Coffee was introduced to Turkey in the thirteenth century, and from then on, the Arabs roasted and ground coffee to produce their favorite beverage. It was spread by Muslims to Italy, the Netherlands, France and England. The first coffeehouse was opened in Italy in 1645 and in England in 1650. Coffee came to the U.S. in 1668, and the first coffeehouse opened in New York in 1696.

Scientific Name(s): there are many types of coffees, the two main varieties being *Coffee arabica* and *Coffee robusta*. Family: Rubiaceae.

Origin and Varieties: differences in coffee flavors depend on the type of beans, soil, altitude and climate conditions. *Coffee arabica* beans grown at higher altitudes, are the most highly valued because of their delicate flavor.

C. robusta is less delicate with an earthy flavor but has higher caffeine than the former. Other types, such as *C. liberica* and *C. stenophylla,* are less fragrant than the former two. Coffee is grown in India, Indonesia, Brazil, Uganda, Ethiopia, Mexico and Columbia.

Common Names: the names of coffee originate from the country, region or quality grade of coffee the beans come from, such as Java, Costa Rican, Kona or Kenya AA, Arabian coffee, mocha (Yemen coffee, also the name for coffee with added chocolate). It is also called café (French, Portuguese, Spanish), kaffee (German, Danish, Swedish, Norwegian), caffe (Italian), koffie (Dutch, Afrikaan), kahvi (Finnish), kave (Hungarian), kahawa (Swahili), ko pyi (Korean), kopi (Malaysian, Indonesian) and kopee (Tamil).

Form: the coffee bean or fruit contains two seeds with adjacent sides that are flat and contained in the fleshy pulp. They are separated by a central groove. The seeds are greenish yellow in color and are separated from the rest of the pulpy material. Coffee's color changes as it ages. Coffee is ground to different degrees of fineness depending on the method of brewing.

Coffee is sold as whole roasted beans, coarsely or finely ground, decaffeinated (97% caffeine free), as blends or as flavored coffees. Instant (soluble) coffee is roasted coffee extracted with water under pressure and spray dried or freeze dried and finely ground. Roasted coffee ranges from light to very dark colors and varies in flavor. Examples are American roast (usually medium color), French roast (very dark brown) and Italian roast (almost black). Flavors such as almond, cinnamon or chicory or chocolate are added after roasting to give flavored or regional coffees.

Properties: green coffee does not have any aroma. The chemical composition of coffee beans changes as it matures and ages. Green coffee beans are fermented, dried and or washed, thus separating the pulpy material. Then, seeds are roasted at about 400°F, anywhere from about 3 minutes to 17 minutes, during which chemical changes called pyrolysis occur. Chemical substances are released that characterize coffee's taste and aroma. During roasting, the beans also become a brown color, lose moisture and swell (from 50% to 100%). Its moisture level decreases from 7% to 2%, and its acidity also decreases. After roasting, the beans are quickly washed and cooled. The longer coffee is roasted, the darker it becomes and the higher its acidity, with lower tannins. Over-roasting gives rise to bitterness and or burnt and smoky aromas.

When the beans are roasted to their peak flavor, the beans are cooled quickly (by air cooling, water quenching or a combination of both) to stop the roasting process.

Chemical Components: coffee contains chlorogenic acid, caffeine, cafetannic acid, tannin, caffeic acid, trigonelline, starch pentosans, cellulose, hemicellulose, sucrose, proteins and oil. Roasting decreases chlorogenic acid, tannin and some trigonelline.

How Prepared and Consumed: globally, coffee is used principally as a beverage. It is also used to flavor desserts and ice creams. Sugar is usually added to decrease its bitterness. Spices have been used with coffee to create regional coffee favorites. In Saudi Arabia, dark-roasted coffee, mixed with ground cardamom, called gahira or gahwa, is served to guests as a symbol of hospitality. In the U.S., cinnamon and chicory coffees are popular. Fondues and other sauces can be flavored with coffee blended with spices or fruits.

In certain cultures, coffee's bitterness balances savory and other taste profiles. Coffee can enhance or give roasted notes to savory foods, such as red salsas, dark curries, sambals with shrimp paste and even roast turkey or chicken. It tends to pair well with tamarind, kaffir lime leaf, lime, soy sauce, galangal, cardamom, allspice or mango.

Spice Blends: curry blends, sambal blends and roast chicken blend.

Therapeutic Uses and Folklore: traditionally, Ethiopians and Middle Easterners chewed on coffee to increase energy or appetite. Today, it serves the same purpose.

PREPARATION AND COOKING TECHNIQUES IN ETHNIC CUISINES

Many spices and ingredients are common to different ethnic cuisines, yet their flavors assume distinct flavor profiles and colors in these different foods. Why? Because these ingredients are prepared and cooked in different ways. How spices are cooked (refer to Chapter 3), including toasting coriander seeds, roasting chilies or popping mustard seeds in hot oil, produces distinct flavors and colors.

Cooking techniques also remove raw notes, intensify flavors and sometimes modify colors. For example, dry roasting cumin seeds mellows some of their bitterness and gives them a different characteristic flavor profile. Boiling or roasting ground cumin also creates different flavors. With certain bitter spices, such as fenugreek, clove or mace, roasting in oil tends to take out some of their bitterness.

TABLE 28. Types of Cooking Techniques.

Dry Heat	Grill, broil, roast, saute/stir-fry, barbecue, pan fry and deep fry, bake
Moist Heat	Steam, boil, poach, simmer, pressure cook
Dry and Moist Combination	Braise, stew

TABLE 29. Specific Global Cooking and Preparation Techniques.

Grill, roast, bake, broil, smoke, barbecue: tandoor, bakar, pibil, barbacoa, jerk
Deep fry, stir-fry: tumis, balti, bargar, bhoona, kwali caldero, tarkar
Steam: couscousiere
Braise: korma, daubiere, pokkum
Stew: kallia, refogado, claypot, tagine
Pickle, ferment: escabeche, chutney

North Americans traditionally added spices in boiling water, simmering stews or pickling solutions. These preparation methods did not necessarily give spices their optimum flavor characteristics. Today, North Americans are learning some newer techniques that can be used with spices to obtain more enhanced and intense flavor profiles. Thus, we see a growing demand for smoked chile peppers, roasted spices or grilled tomatoes.

Cooking techniques fall into three basic categories: dry heat (air, fat), moist heat (water, steam or other liquids except fat), and a combination of the two (Tables 28 and 29). Globally, these three techniques are called by many names. The equipment the food is cooked in creates distinct flavors, colors and textures. The cooking utensil's size, shape, and most importantly, the material from which it is constructed, affect the food's ultimate taste.

Typical materials of cooking utensils include stainless steel, cast iron, enameled aluminum, anodized copper and ceramics, (clay, Crock-Pot).

The order in which ingredients are added to a recipe is crucial in creating a well-balanced flavor profile. For example, in making Southeast Asian sauces, garlic is sautéed first, then ginger and onions are added, followed by spices and chilies and tomatoes. Spices that add bitterness, such as fenugreek or clove, are generally added toward the end of cooking.

As discussed earlier, corn husk, banana leaf and pandan leaf are used to wrap chicken, fish, rice or meat which are then steamed, smoked, grilled or barbecued. These wrappers enhance flavor and provide moistness, color, texture and visual appeal to foods. They are discussed earlier in this chapter. Though commercially we may not be able to present food in these wrappers, we can simulate the flavors of these cooked wrappers.

Photo 1. Spices, chile peppers and flavorings.

Photo 2. Spices in a mortar/molcajete/batu lesung.

Photo 3. Spices on a tray.

Photo 4. Top right, clockwise: red cayenne peppers, bunga kantan/torch ginger bud used in Malay curries and sauces in Malaysia, small lime (limau kasturi), eggplant.

Photo 5. Fresh chile peppers, from top left, first row: New Mexican, habaneros; second row: jalapenos, serranos, cayennes; third row: cubanelle, Thai peppers, bird peppers/cili padi.

Photo 6. Dried chile peppers, from top left, first row: mulato, Tabasco, New Mexican; second row: cascabel, chipotle, pasilla; third row: chiltepin, cili kering/fried cayenne; fourth row; pequin, ancho, chile de arbol.

Photo 7. From left: fenugreek seed/methi; right: fenugreek leaf/sag methi.

Photo 8. From left: coriander seed; right: coriander leaf/cilantro.

Photo 9. Spices in masala dani/spice box, from left to right, clockwise: green cardamon, fenugreek, cumin, fennel, black peppercorn, coriander; center: ground turmeric.

Photo 10. Kari/curry leaves/karivepillai.

Photo 11. Clockwise from left: brown cardamon, green cardamon, white cardamon.

Photo 12. Clockwise from top left: nigella/kalonji, black cumin, cumin, ajowan, anise; center: fennel.

Photo 13. Clockwise left to right: broken mace blades/aril, whole nutmeg, ground nutmeg.

Photo 14. Turmeric: whole, ground.

Photo 15. "Wet" market in northeast Malaysia.

CHAPTER 6

Emerging Spice Blends and Seasonings

INTRODUCTION

NORTH American tastes are becoming culturally diverse. Consumers, whether traditional or adventurous, are looking for something new and flavorful. Some consumers want familiar foods but with new twists. Other consumers are not afraid to try something totally different, so unique ingredients and bolder flavors appeal to them. Both types of consumers want more flavor, good visual appeal and distinct textures. They have little time to prepare meals and want products that are convenient. Spice blends and seasoning blends can help create the foods that these consumers desire.

Asian curries or teriyaki sauce, Latin American adobos or moles and Mediterranean harissas or pestos can provide desired flavors to meats, vegetables, rice or pasta. Many ethnic seasonings have spices such as ginger, cilantro, cinnamon or sweet basil that are recognizable to mainstream consumers and can safely "spice up" traditional foods. Others contain unique spices such as fenugreek, black cumin, galangal or recao leaf that will appeal to the adventurous consumer.

Spice blends and seasonings are mixtures of spices and/or other ingredients, such as vinegar, pomegranate juice, soy sauce, wine or fish sauce. Their possible combinations are almost endless and are left to the imagination of the creator or spice blender.

When combining different ingredients to create a seasoning blend, the food designer must understand that not all generic ethnic blends are the same. That is, not all salsas, dukkahs, masalas, pestos or sambals taste the same. Among different cultures or regions, variations in flavor (sourness, sweetness or heat), color or texture (creamy or thin) can occur.

For example, a basic curry blend from India predominantly contains cumin, turmeric, coriander and ground red pepper, with a little of other spices such as cinnamon, clove or cardamom. However, curry blends vary depending on the countries or cultures of origin, the applications in which the blends are used, how their constituent ingredients are treated and the order in which they are added to the applications. The Punjabis in the north of India eat milder curries with cardamom, anise, yogurt, nuts and raisins, while Tamils in the south eat very hot, spicy curries with mustard seeds, tamarind and curry leaves. Other variations in curry blends occur wherever they are enjoyed. Thai curry blends are pungent, fiery, hot and sweet, with cilantro, galangal, lemongrass and hot chile peppers. The fragrant Nonya curries of Malaysia contain pandan leaf, spearmint, candlenuts, dried shrimp pastes and coconut milk. On the Caribbean island of Guadeloupe, curries have ground mustard, black pepper, habaneros, vinegar and chives.

The food creator must also understand spices and flavorings in all of their varieties, because spices are the fundamental building blocks of these seasoning blends. There is more than one type of basil, sesame seed, chile pepper or citrus flavor. Some cultures or regions have specific preferences for particular varieties. By understanding the similarities and differences among the spice varieties, we can create authentic spice blends or other appropriate fusion blends. Additionally, spices and seasoning blends interact in a system based on other ingredients (such as starch, fat and proteins), pH and processing techniques. The method of preparation of the ingredients (braising, roasting or tumising) and how the foods are presented in a meal contribute to the flavor and texture of the foods.

Seasoning blends in western countries are usually a simple combination of spices, which are generally mild. In Asia, Latin America and the Caribbean, they are more complex in their combinations, and they are hotter and more intense. Malaysia, Singapore, Indonesia, Thailand, India and Sri Lanka traditionally have the most varied, intense and complex spice blends in the world. Some are already well-known in the west. Others are less familiar but are growing in popularity.

Seasoning blends come as freshly made paste, liquid, dried, whole or ground forms that are added to cooked or uncooked products. They can be added "as is" or roasted, fried or smoked. Some blends have fresh pureed (or blended) spices, while others have dried whole or ground spices. The wet mixtures are ground with or added to water, oil, vinegar or coconut milk. These are usually refrigerated after use. Seasoning blends can be created for snacks, soups, marinades, rice mixes, pasta sauces, desserts and beverages. The emerging trend for spice blends will be to provide intense aromas and a perception of freshness. In this regard, the wet and the dried forms that retain their aromas and freshness will increase in popularity.

Spice blends can be authentic or fusion. Generational differences and adaptation to the U.S. by immigrant communities need to be considered when developing seasonings. The first generation will have eating patterns and buying behaviors that are different from the second, third and later generations who have adjusted to a new culture and language. For the latter, fusion blends can be developed to give a "taste of home" combined with North American flavors. Spice blends can also be created for regional U.S. preferences as discussed in Chapter 2.

In this chapter, we will look at some of the popular global seasoning blends categorized according to region. The focus will be on Latin American, Asian, Mediterranean and Caribbean seasonings because they are the hot trends in the U.S. African seasonings will also be discussed briefly to give the creator some idea of its regional flavors.

In addition, the chapter will examine popular regional North American blends and two types of seasoning blends that are popular globally, green blends and chile blends.

Northern European seasoning blends will not be discussed in detail because they are not emerging "new flavors" in North America. Many of these seasonings are found in traditional U.S. cooking. Several popular French, Italian and Spanish spice blends are discussed under the Mediterranean blends.

As in the U.S., European cities are becoming more culturally diverse and, as a consequence, their traditional flavors are slowly changing. They are fusing ingredients from newer immigrants, with unique seasonings emerging from the old favorites. For example, in England, Indian and Asian flavors have become mainstream. Curry to an Englishman is like salsa to a North American. Shepherd's pie, lamb roasts and other traditional English dishes now often contain Asian Indian or Thai flavors. Tandoori, nasi goreng, biryani and chicken tikka masala are becoming mainstream items in the U.K. They are also infiltrating French, German and many other European cuisines. Tex-Mex flavors are also becoming popular in Europe. The discussion of emerging seasoning blends contained in this chapter will help the food product designer also understand emerging European flavors.

As noted above, the subsections of this chapter introduce the foods and ingredients of Latin America, Asia, the Mediterranean, the Caribbean and Africa. The chapter also describes the many regional variations of these complex cuisines.

Included in each subsection are examples of blends that are commonly sold as prepared spice mixtures or seasonings, such as five spice blend, curry blends, adobos and pestos. Also described are the spices and other ingredients that provide each region's characteristic flavors. They can be captured and made into spice blends or seasonings that provide the flavor profiles for many products. Some spice blends in this section are dried, some are liquids and others are wet pastes. Whether pastes, liquids or dry mixtures, they can be tailor made for specific applications once the characterizing spice blend is created.

LATIN AMERICAN SPICE BLENDS

The Latin American category includes the Spanish-speaking peoples and indigenous Indian populations of Mexico, the Caribbean islands, Central and South America and Portuguese-speaking Brazil. In this book, seasonings from Spain will appear under the Mediterranean category.

Some seasonings have mass appeal throughout the Latin American market. Seasonings such as achiote, sazon and sofrito add traditional flavors and colors to foods from many Latin American regions. However, flavor preferences vary among Latin American countries, based on the country of origin and other sociocultural differences. For example, sofritos are used to flavor beans, rice, fish and stews in many Latin American regions. A basic sofrito blend is made from chopped onions, garlic, tomato and green bell peppers, but it varies depending upon its origin. The Puerto Rican sofrito is pungent with culantro, the Yucatan sofrito is peppery with habanero peppers and the Cuban is mild with parsley. Likewise, adobo is a popular all-purpose seasoning that is used to marinate meat, chicken or fish. It also comes in different flavors, adobo con limon for Puerto Ricans, adobo con naranja agria for the Dominicans and adobo con pimienta for those who desire a little heat.

Not all Latin American food tastes hot. Latin American food is milder where there is greater European influence and spicier and hotter where it is more influenced by the indigenous population. Cubans, Puerto Ricans and Dominicans use oregano, tomato, garlic and black pepper with mild chilies to flavor their foods. In these Caribbean regions, flavors tend to be generally milder. In contrast, hot chile peppers are used extensively in the Andes areas of South America and in the Yucatan and Oaxacan regions of Mexico.

Latin American foods, especially Mexican foods, give intense flavors through roasting and smoking of chile peppers, onion, garlic, tomatoes and other vegetables. In the U.S., our taste in food has been greatly influenced by Latin American flavors, particularly Mexican flavors because they are the largest Latin American group in the U.S. Mexican food has been one of the three most popular ethnic foods for the last 15 years. With the increased demand for authentic ethnic cuisines, many consumers are now seeking the more authentic regional Mexican profiles rather than simply northern Mexican or Tex-Mex foods.

Authentic Mexican foods differ in their flavors and heat levels. In northern Mexico, the food is milder, but further south, toward Oaxaca and the Yucatan regions, spicier and hotter foods predominate with the influences of the Caribbean and the Mexican Mayans. The Oaxaca region has the most diverse cuisine of Mexico. It has unique moles, which vary in color, flavor and texture.

The next largest Hispanic groups in the U.S. are Puerto Ricans and Cubans. Together with Mexicans, these three groups comprise three-quarters of the total U.S. Hispanic population. The other Hispanic groups that constitute the remaining one-quarter of the growing Hispanic consumer market are Central and South Americans. Their seasonings are becoming increasingly popular.

Salsas are very popular in the U.S., and we can expect more varied salsa flavors using ingredients from different Hispanic cultures: chimichurri from Argentina, salsa de aji from Peru, pico de gallo nortena from Mexico, pebre from Chile or salsa Criollo from Puerto Rico. These sweet, spicy hot, tangy sauces are served fresh (uncooked) or cooked, with tomatoes, spices, chilies or avocados.

Similarly, empanadas, pasteles and saltenas, which are used as snacks in all of Latin America, are becoming more popular. Like other Latin American foods, their filling flavors vary based on the ingredients and how they are seasoned, with chile peppers, cilantro, fish, garlic, eggs, meat or potatoes.

Let's look into the regions of Latin America and see how they are different and how they influence the overall Latin American flavor.

REGIONAL CUISINES OF LATIN AMERICA

Mexico

Mexican cooking has been influenced by many cultures, through trade or conquest. The Aztecs and Mayans with chile peppers, corn, tomato and chocolate; the Spanish with wheat, rice, citrus and coriander; the French with their cooking techniques and breads; the Chinese with sweet and sour sauce and ginger; the Lebanese with cumin and pastries; and the Germans with their cheeses and beer.

Mexican flavors vary from region to region. In the north, called el Norte (Monterrey region), the food is influenced by the Mennonites (German) and North American cowboys. The food is mild with cheeses, beef, caldillo (beef stews), burritos, guacamoles, frijoles charros, tacos and New Mexican chiles. In the North Pacific region (Jalisco, Sinaloa, Colima regions), seafood, tomatoes, poblanos, pozoles, ceviches and pico de gallo predominate. The South Pacific region, which includes Oaxaca, is influenced by the indigenous groups (Toltecs, Zapotecs and others). Moles, especially mole negro, birria (mutton or goat stews), hierba sante, salsa verde, chipotle, chile de arbol, jalapenos, guajillo, ancho, ginger, corn husk, tortilla, tamale, and pipian stews are popular. Chiapas has milder flavors than the Oaxaca region.

In the Yucatan region, the foods are traditional (from Mayan influence) as well as fusion, influenced by the Caribbean and Florida as well as Asia and Europe. Yucatan flavors include recados, habaneros, sapote, mango, achiotes, epazote, cochinata pibil, tamales, rice and beans, banana leaf, chiltomates, misantle paste and chile xcatik. The Veracruz region (Gulf area) is influenced by Spanish and Creole cooking. It has olives, mariscos, paella, coconut, sofritos, frijoles, tomato sauces and al ajillo.

The area around Mexico D.F. provide old and new flavors such as nopales, squash blossoms, moles, recados, huitlacoche, mango, fish tamale, barbacoa, tlacoyos, taco pastor, tortilla soup and guacamole. The Bajio region, which includes Michoacán, Queretero, San Miguel and Guanajuato, is influenced by Spanish and indigenous Mexicans. It has cactus, cheese, enchiladas, anchos, moles, pinto beans, corn tamales, thyme, cumin, avocado, beef, pork, fish and jalapenos. Central Mexico has Spanish, Iberian, Muslim and indigenous Mexican influences. Its foods include mole poblano, chile en nogada, ancho, mulato, pasillas, chipotle, gusanos, maguey, taco pastor, chalupas and the famous mixiotes.

Caribbean

The Caribbean islands of Cuba, Puerto Rico and the Dominican Republic are influenced by the Taino Indians, Arawaks, Caribs, Spanish and Africans. Cuban flavors can be mild or hot depending on the region and feature pork, seafood, black beans, mojos, parsley, cachucha, annatto, plantain, congri (rice and beans), squash, tamale, ajaico (vegetable stew) and citrus. Puerto Rican and Dominican foods are mild to slightly spicy. They feature adobos, sofritos, achiotes, culantro, cilantro, oregano, paprika, aji dulce, picadillos, mofongo and paellas.

Central America

The flavors of Central America differ depending on what cultures influenced their cooking (Mayans, Spanish, Germans, Italians, North Americans or Africans) and what ingredients are grown in the region, whether rice, corn or chile peppers. Central Americans who live along the east coast use a more Caribbean style of cooking, which is very different from the cooking of those who live inland. This is the Criollo or Creole cooking and is influenced by the various cultures from the Caribbean and Africa, with rice and beans, coconut, seafood, bananas and curries. Guatemalan cooking has strong indigenous Indian influence and some influence from Mexican cooking.

Smoky tasting tortillas form the staple meal with grilled (parillas) or roasted meats (asados). They are served with spicy condiments of cilantro, onions and chile peppers. Rice and beans and plantains, pacaya, chile rellenos, ceviches, recados, bistec, stews (caldos), chuletas (pork), tamales, frijoles, refritos (boiled mashed beans refried in lard) and enteros (boiled beans with onions) are commonly eaten.

South America

South American foods are also varied. Certain regions that had more European influence (Germans, Italians or Spanish) have milder flavors, while those that had Inca and other indigenous Indian influences have hotter and more pungent flavors. Colombian and Venezuelan foods are mild with corn, rice, arepas, tortillas, chicken, seafood, casseroles, creamy sauces, plantains, apio (celery root), pabellon criolla, black beans and empenadas. Brazil has many cultural influences—indigenous Indians, Portuguese, French, Lebanese, Chinese, Japanese, Africans, Dutch and Italians. Their flavors vary from hot to mild. Where Indians have a strong influence, foods are spicy and hot. Brazilian flavors include fish sauces, peanuts, malagueta pepper, rice and beans, annatto, dende oil, dried shrimp, cassareep, coconut, couscous, lemon, vatapas, feijoada completa, empenadas and hearts of palm. Argentina and Uruguay have similar flavors with grilled steaks (matambre, churrascos and parillas) served with chimmichurri, salsas (parsley based), empenadas, humitas, corn and fruits.

Paraguay is very much influenced by the indigenous population (Guarani Indians) and Europeans. It has soups, beef, cheese, oregano, onion, parsley, poblano and olive oil. Bolivia has picante flavors and uses rocotos, ajis, saltena, potatoes and eggs. Chile uses less meat and more hot chile peppers in foods. Its foods include fruits, potatoes, squash, shrimp chowders, fish stews, garlic, aji molido, pebre, cilantro and empenadas. Chile's national dish is porotos granados, which has seafood, beans and corn. It is served with hot pebre sauce. Peru's staples are corn and potatoes, which were the staples of the Incas. Potatoes are seasoned in a variety of ways with cheese, milk, chile peppers, onions, peanuts, jalapenos or olives. Ecuador is well known for its hot, sour and slightly bitter ceviches. Potatoes, corn, cheese, naranga agria, ajis, paprika, fish, cilantro, plantains and peanut sauce are commonly eaten.

Following, find the basic Latin American spice and seasoning blends that need to be known in order to develop authentic Latino and increasingly popular Latin-influenced foods.

POPULAR LATIN AMERICAN SPICE BLENDS

Adobos

Description

Adobo is a popular, all-purpose seasoning that has a savory, garlicky flavor. It is an indispensable item in Puerto Rican, Cuban and Dominican kitchens. The earliest adobos were salt and vinegar mixtures used to preserve meat. The word adobo comes from the Spanish, meaning pickling sauce of olives, vinegar and spices, or from adobado, which means pork pickled in olives, vinegar or wine and spices. Today, the dry spice blend, adobo, is used with vinegar and olive oil to marinate beef, pork, poultry and fish, or to zip up rice, stews and sauces. Commercially, it comes as a dry, ground mix.

Ingredients

Adobo consists of salt, garlic, black pepper, oregano and turmeric. Optional ingredients include onions, olive oil, lime juice, sour orange juice or vinegar, which are added to create marinades for meat, fish or chicken.

The basic adobo has mass appeal among Latin Americans, but specific ingredients, such as onion, seville orange juice, lemon or lime juice, parsley, chipotle or cumin are added to appeal to specific Hispanic groups. Cubans use garlic, cumin and sour orange juice; Puerto Ricans prefer vinegar and oregano; Mexicans enjoy a fiery hot adobo made with habanero or chipotle peppers and sour orange juice; Dominicans use sour orange juice, while Nicaraguans prefer it with annatto. Adobo without black pepper is prepared for Hispanics who do not want any heat. Accordingly, the adobos in Table 30 can be tailored for each Latin American region.

Applications

Adobo provides flavor to flank steaks, chicken, pork, fish and shellfish before grilling, deep-frying or sautéing. In Latin America, this preparation is so essential that the verb "adobar" (to adobo) is commonly used during food preparation. Adobo can be used as a base seasoning for stews, sauces, sour cream, chicken stock, vinaigrette, baked potatoes, steamed vegetables and as a rub on chicken, fish or meats before they are barbecued or roasted.

Mole Blends

Description

Moles are the "royal" sauces of regional Mexican cuisine. The word mole comes from the Nahuatl word molli, which means sauce or mixture. They were originally created for the rulers of pre-Hispanic Mexico using chocolate (made from cacao), an ingredient reserved for royalty. Nowadays, these

TABLE 30. Types of Latin American Adobos.

Basic adobo	Garlic, oregano, black pepper, turmeric
Puerto Rican	Adobo con limon (adobo with lemon)
Cuban	Adobo con cumin (adobo with cumin)
Dominican	Adobo con naranga agria (adobo with sour orange)
Mexican	Adobo con chipotle (adobo with chipotle)
Nicaraguan	Adobo con annatto (adobo with annatto)

smooth, rich sauces, with their sweet, nutty, roasted and slightly bitter flavors are used with chicken and seafood dishes.

The most popular moles, and perhaps the best, come from the Oaxaca and Puebla regions. Oaxaca is known as the "Land of Seven Moles." However, every Oaxacan seems to have his or her own idea of what the seven moles are. In fact, there seems to be at least eight types that are popular in Oaxaca: mole negro, mole rojo, mole verde, mole coloradito, mole amarillo, mole de almendra, mole manteles and mole chichilo. The uncontested king of moles is Oaxaca's mole negro. Mole poblano is the most popular mole from the Puebla region of Mexico.

Ingredients

There are hundreds of mole recipes in Mexico. In fact, each family seems to have their own variation, depending upon their preferred ingredients and cooking techniques. That is why moles seem to come in so many different colors: black, brick red, yellow and green. However, most moles contain dried chile peppers, roasted tomatoes, pureed nuts and seeds, spices, herbs and chocolate. However, not all moles have chocolate. The chile peppers provide the mole's characteristic color and flavor.

It takes hours to prepare a good mole. Preparation and cooking techniques, such as grinding, pureeing, grilling, browning, roasting and toasting are essential for intensifying ingredient flavors and obtaining the right consistency. Most ingredients are roasted on a comal (a large, lime-coated ceramic pan). After roasting, the ingredients are either hand ground on a heavy, granite metate or commercially ground at the local market in a machine called a molino. Finally, once the ingredients are prepared, the sauce mixture is cooked with lard in a cazuela, or molepot, and is stirred continuously.

Mole negro is a complex, fiery black sauce made from numerous ingredients: toasted and ground dried chile peppers (chilhuacle negro, pasilla, mulato, guajillo and chipotle), blackened chile seeds and stems, almonds, tomatilla, spices (garlic, onions, sesame seeds, black peppercorns, canela, cloves, thyme, marjoram, Mexican oregano) and chocolate.

Mole poblano has several types of dried chilies (ancho, mulato, pasilla, chipotle), with anise, clove, canela, black pepper, fried garlic, tomatillas, sesame seeds, ground almonds, peanuts and bitter chocolate. Table 31 lists the ingredients of nine mole blends. However, there are many interpretations of these nine blends and their ingredients.

Applications

Moles are generally eaten with white rice or tortillas, which are used to scoop the mole. The thick sauce is poured over chicken, turkey, seafood, pork or beef. Mole sauce can also be used as a filling for empanadas and tamales or poured over enchiladas.

TABLE 31. Types of Latin American Moles.

Blend	Ingredients
Mole negro	Chipotle, chilhuacle, negro, pasilla, mulatto, guajillo, chile seeds and stems, almonds, onions, tomatillas, garlic, sesame seeds, black peppercorns, canela, clove, thyme, Mexican oregano, chocolate
Mole poblano	Ancho, mulatto, pasilla, chipotle, anise, clove, canela, black pepper, sesame seeds, almonds, tomatillas, garlic, peanuts, chocolate
Mole rojo	Red chilhuacle chile, ancho, pasilla and peanuts
Mole coloradito	Ancho, pasilla, guajillo, canela, raisins, sesame seeds, almonds, chocolate and fried bread
Mole verde	Green jalapeno, epazote, cilantro, parsley hoja santa, tomatillas and masa
Mole amarillo	Guajillo, chiluacle amarillo, ancho, cloves, cumin, chepil, zucchini, chayote and tomatillas
Mole manches manteles	Anchos, pasillas, chilcostle chiles, sesame seeds, fresh pineapple, ripe plaintains, tiny apples
Mole de almendra	Almonds, ancho, cloves, sesame seeds and fried bread
Mole chichilo	Chihuacle negro, pasilla, guajillo, blackened chile seeds, roasted avocado leaves and masa

A few popular moles are not sauces at all. In pre-Hispanic time, moles were more like soups and were served as one-bowl meals. Some of these ancient moles such as mole de maiz quebrado (corn mole), bean mole and shrimp mole are still consumed today. A popular one, mole de olla, is a hearty soup full of beef, onions, vegetables, cilantro, lime and epazote, with added rice and corn tortillas.

Mojo Blends

Description

Mojo is to Cuban cuisine as vinaigrette is to French cuisine. Also called Cuban garlic citrus sauce, it is tart, tangy and garlicky. It can be used to marinade meats or as a condiment to add zest to meals. It is Cuba's national table condiment.

Ingredients

Mojo's ingredients include garlic, olive oil, sour orange juice (narangia agria) and black pepper. Ground cumin is optional. In Cuba, naranga agria, a green, bumpy orange that tastes like lime juice, is the acid used. In other

Caribbean regions, including Puerto Rico and the Dominican Republic, vinegar is substituted for the naranga agria. In the Miami area, lime juice or other fruit juices, such as pineapple, carambola and grapefruit, are generally substituted for naranga agria.

Application

Mojo is a vinaigrette-like, Cuban sauce used as a marinade, especially for poultry and meat. Mojo is also served on Cuban sandwiches, grilled seafood and meats, boiled yuca and many other foods. Rosemary, cilantro or mint can be added to create new flavors.

Salsa Blends

Description

Salsa means sauce in Spanish, and it comes in many forms and flavors. Some Latin American sauces are called salsas, while others have specific names, but all are added to provide zest or heat to foods. Latin American regions have many versions of salsas that range in flavor from sweet and mild to hot, pungent and slightly tangy. They vary depending upon ingredients and preparation techniques used.

Salsas can be fresh or uncooked (salsa cruda/salsa fresca/salsa de molcajete) or cooked, based on roasted chilies, tomatoes and other ingredients. Salsas can be smooth or coarsely textured, thick or thin. Salsa verde (green salsa) is uncooked and is tart and hot. Salsa rojo (red salsa) is cooked and is sweet and spicy.

Popular salsas include red (rojo) and green (verde) salsas of Mexico, xni pec from the Yucatan, pico de Gallo from northern Mexico, salsa di aji from Peru, pebre from Chile, salsa criollo from Puerto Rico and chimmichurri from Argentina.

Ingredients

When North Americans refer to salsa, they generally mean a condiment of tomatoes, onions, chilies and bell peppers. Spices such as garlic, cumin, oregano and cilantro are also added. Whole tomatoes, diced tomatoes and tomato pastes form the basis of commercial salsas. Lime, lemon, sour orange juice or vinegar are often added to provide acidity. Depending on regional preferences, tomato, bell peppers, habaneros, ajis, jalapenos, corn or fruits are added.

The basic red salsa is made from red roasted tomatoes, red bell peppers and roasted red chile peppers. Green salsa is made from tomatillas, cilantro and green chile peppers with green leafy spices. Nowadays, the emerging trend in the U.S. is to incorporate fruit flavors with salsa blends to create new flavors.

Salsa fresca, salsa cruda or fresh salsa uses basic salsa seasonings with chopped jalapenos, serranos or habaneros, tomatoes, white onions and cilantro. These salsas are eaten raw or are cooked briefly in lard or oil. Cooked salsas generally use dried, roasted chilies, such as chipotle or ancho, with roasted tomatoes and garlic.

Some examples of Latin salsa blends and their ingredients are given in Table 32.

Applications

Salsas are used as dips for snack chips or tortillas or poured over eggs, fajitas and enchiladas. Salsa crudas (uncooked salsas) appear on Mexican tables like salt and pepper. They are used as condiments to add zest to rice and beans, meats, fish, eggs and beans and serve as toppings/garnishes on enchiladas, tacos, quesadillas and antojitos. Puerto Ricans serve salsa ajilimojili with meats, tostones and boiled vegetables. South Americans serve chimmichurri sauce as a dip or condiment for grilled meats or churrascos.

Today, salsas are emerging as toppings for grilled foods, baked fish, fillings for crepes and wraps and as exciting condiments to add extra flavor to foods. They are also being used as marinades, salad dressings and even as a topping for baked potatoes.

TABLE 32. Types of Latin American Salsas.

Salsa Blend	Ingredients
Salsa rojo	Chipotle, red jalapeno, red bell peppers, garlic, tomato, cilantro
Salsa verde	Green jalapeno/ serrano, cilantro, tomatilla, garlic, onion, lime juice
Pico de gallo	Red serranos/jalapenos, cilantro, oregano, onions, tomato, lime, orange, jicama, melon, cucumber, olive oil, shredded coconut
Chimichurri	Garlic, parsley, cilantro, vinegar, black pepper, oregano
Salsa de aji	Aji, onion, tomato, cilantro
Molho malagueta	Malagueata pepper, olive oil, vinegar, onion
Ajilimojili	Garlic, sweet chile peppers, black pepper, olive oil, lemon
Pebre	Onion, cilantro, parsley, green hot peppers, lemon juice, olive oil
Aji molido	Aji, olive oil, garlic, vinegar
Salsa criolla	Onion, tomato, garlic, black pepper, parsley, olive oil, vinegar
Xni pec	Tomato, onion, cilantro, habaneros, sour orange juice

Recado Blends

Description

Recados are ground spice mixtures or wet pastes that are essential to traditional Yucatecan, Central American and Puerto Rican cooking. They come in different flavors and colors, depending upon their ingredients (Table 33). Most widely used is the recado rojo (sometimes called achiote). Others are recado negro (or chilmole), recado cumin de bistek, recado de adobo and recaitos. They can be mild, pungent, smoky and hot in flavor and are red, black or olive colored. The chilmole is black and pungent, while the recaito is greenish. Recados are used to zip up sauces, stews, meats, seafood and beans. They can be served over cooked poultry or rubbed on meats, chicken and pork before grilling or barbecuing.

Ingredients

Recados' ingredients include garlic, chile peppers, oregano, onions, annatto and black peppercorns. Other ingredients are cumin, allspice, cilantro, canela, culantro or cloves depending on regional preferences. There are many variations, even in the Yucatan.

Applications

Most Yucatecan dishes use recados as a base flavoring for many dishes. Turkey is roasted in recado negro and fish is grilled in recado rojo. To create the famous pibils of the Yucatan, recado is rubbed on pork or chicken that is then wrapped in banana leaves and pit roasted. Chilmole is used to season meatballs or ground meat.

TABLE 33. Types of Latin American Recados.

Recado Blend	Ingredients
Recado rojo	Ground annatto seeds (achiote), garlic, peppercorn, onion, cilantro, oregano, allspice and sour orange juice
Recado de bistek	Oregano (olive colored), canela, garlic, onion, black peppercorns, cumin, cloves, chile de arbol and sour orange juice
Recado de negro (also called chilmole)	Annatto paste, toasted chile xcatik, allspice, roasted onions, roasted garlic, cumin and burned tortillas
Recado de adobo	Chilies, oregano, garlic, clove, cumin, cilantro, saffron, canela, Seville orange

In Puerto Rico, an olive green spice mixture called recaito is used as a base seasoning in the local sofritos. Recaito contains onion, cilantro, recao leaves, garlic and aji dulce.

Sofrito Blends

Description

Sofrito is an aromatic mixture of spices used to flavor meats, rice, beans, stews and fish. It originated as tomato sauce from the Catalan region of Spain. Now, it is an essential seasoning in Puerto Rican, Spanish, Cuban and Mexican cooking. The word sofrito comes from the Spanish word sofrier, which means lightly cooked. Sofrito blends range in color from bright orange to red, and their flavors are mild or pungent. Preferences and ingredients vary with country of origin (Table 34).

Ingredients

Typical sofrito ingredients are green and red bell peppers, onions, annatto, garlic, oregano, tomatoes, cilantro or parsley. Puerto Rican sofrito is pungent, and its ingredients include recaito leaves, culantro stems and roots, aji dulce (also called rocotillos or sweet chile peppers), annatto, garlic, onions, oregano, cilantro and tomato sauce. In Puerto Rico, it is added with ham, tocino (salt pork) or alcapaardo (salted capers, olives and pimentos). Cuban sofrito is mild, with garlic, onion, cachucha (or aji dulce) and cured ham. Spanish sofrito is sweet, with bell peppers, onions, tomatoes and olive oil. The Yucatecan sofrito is peppery and spicy with habaneros, roasted garlic, black pepper and cumin.

Applications

Sofrito is used as a starting base seasoning or added toward the end of cooking to give a finishing touch to many Latin American and Spanish dishes.

TABLE 34. Types of Latin American Sofritos.

Sofrito Blend	Ingredients
Basic sofrito	Garlic, tomato, sweet bell peppers, mild chile pepper
Spanish sofrito	Sweet, mild with garlic, red or green bell peppers, onions, olive oil
Puerto Rican sofrito	Pungent, slightly spicy with recao leaves, culantro root and stem, aji dulce, tocino, alcaparrado
Cuban sofrito	Mild with garlic, onion, cachucha, cured ham, parsley
Mexican/Yucatan sofrito	Peppery hot with roasted garlic, black pepper, cumin, habaneros

Cubans flavor soups, stews and casseroles with sofrito or use it as a topping sauce over grilled chicken, beef and pork. In Puerto Rico, sofrito is lightly sautéed in lard or olive oil and added to stewed beans, braised chicken or fish. In Spain, sofrito is made with toasted almonds and is an essential ingredient in a seafood stew called zarzuela de pescado.

When sofrito is made with annatto, the annatto is first heated in oil or lard until it becomes red, then the other ingredients are added to complete the sofrito blend. Annatto gives a wonderful orange yellow hue to the foods in which the sofrito blend is used.

ASIAN SPICE BLENDS

The abundance and variety of Asian spices and spice blends provide a "fresh" and intense taste found in no other cuisine. Asian foods have spice and seasoning blends that mainstream consumers are comfortable with, such as hoisin, sesame paste and teriyaki sauce. They also have stronger, more exotic flavors that excite adventurous consumers, such as sambals, curry powders and kimchis. Authentic Asian blends have intense flavors and textures, great visual appeal and variety that consumers want.

The growing demand for foods that promote a healthy lifestyle is also fueling consumer interest in the Asian way of eating. Asian cuisines have strong flavor profiles derived from seasonings blends that are nutritious. Many blends also contain ingredients such as lentils and soy products that consumers perceive as natural and healthy.

North Americans are combining Asian food concepts, including ingredients, cooking techniques and presentation styles, with traditional foods to create new American foods, including Pan Asian, Asian Latino, New Californian and Pacific Rim.

Some Asian ingredients and seasoning blends, such as ginger, soy sauce, fish sauce, garlic, star anise and curry may have mass appeal throughout Asia, but Asians come from many ethnic backgrounds—Indians, including Pakistanis, Chinese, Japanese, Koreans, Thais and Vietnamese—and each Asian region has its own specific seasoning preferences. There are also differences in flavor profiles within each region. For example, in Southeast Asia, Malaysian foods are distinct from Thai or Vietnamese foods. Even within Malaysia, flavor differences occur because of its diverse cultures. The Malay foods are different from the Chinese, Indian or Nonya foods.

Religion also adds to the variety and unique differences in seasonings found in Asian foods. Hindus, Muslims, Buddhists, Christians and Jews of Asia have specific preferences and prohibitions for ingredients. Certain ingredients and preparation techniques are specific to different religious groups. Rice wine, widely used in Chinese and Japanese cooking is Haram (not Halal), so regions of Muslim predominance, such as Malaysia, Indonesia and Pakistan, do not use it in their cooking. As a consequence, there are endless varieties of seasoning blends in Asia for use with fish, chicken, beef or vegetables.

Asian concepts of meals and food preparation differ from North American and European cooking in several basic ways. In traditional western meals, there is a tendency to segregate tastes. Potatoes have one distinct flavor, meats another flavor and sweet flavors are left for dessert. In contrast, Asian cuisines, including Chinese, Indian and Southeast Asian foods, create balance and harmony in a meal by combining tastes and textures, often even within a single dish.

For example, Indian cooking categorizes foods into six tastes: sweet, sour, salty, spicy, bitter and astringent. The proper, well-balanced Indian meal contains all six tastes. This principle explains the complex spice combinations and depth of flavor in Indian foods. Similarly, in Chinese cooking, there are also six different tastes: sweet, salty, bitter, sour, spicy and umami. The proper balance of these different flavors with appropriate textures and colors creates good taste in Chinese cooking.

Asian cooking principles go beyond the balancing of tastes. Foods can be hot, cold, moist, dry, heavy or light. Every meal should be well-balanced between these sensations to promote digestion and well-being (see Chapter 3).

Asian presentation techniques are designed to carry out this philosophy of eating. For example, Asian foods redefine the concept of an entree. In traditional western meals, the entree typically consists of a large piece of unseasoned or lightly seasoned meat or fish. In contrast, Asian meals feature smaller servings of meat or fish that are more seasoned and sauced, cut thin and served with steaming white rice. In Malaysia and Indonesia, the word "nasi" means cooked rice but is also the word for "meal." Many Asians believe that rice nourishes body and soul. Without rice, food is only a snack, not a meal. Thus, rice is the centerpiece of every Asian meal, accompanied by a variety of crunchy stir-fries, pungent condiments and aromatic curries. Rice is also a comfort food for Asians. Rice porridge with shredded chicken and pickled vegetables, called congee, is often taken when one is sick.

In Southeast Asian cuisines, the concept of a "main entrée" and side dishes sometimes blend together to form a "one-pot" or "one-dish" meal. A one-pot meal may be a stew, soup, sauced noodles or fried rice containing onions, tofu, chilies, leftover meat, vegetables or anything available in the pantry, served with a spicy condiment. However, the addition of these ingredients is not simply random. They are chosen and added to create a balanced "system" of taste and texture. A variety of seasoned rices and noodles are commonly eaten all over Southeast Asia, for lunch, dinner, breakfast and even as "pick me ups."

Side dishes and condiments contribute variety to an Asian meal. The hot, sour and crunchy flavors of side dishes and condiments, whether chutneys, curries, sambals or dals, enhance or provide balance to the overall flavor and texture of the main entree. Condiments are an integral part of Asian meals, whether teriyaki, sambal trasi, mango pickle or goit-chu jang, that perk up a dish and provide added appeal to individual tastes on a meal table. One of the most notable aspects of Thai and Vietnamese cooking is the emphasis on condiments based on fish sauce, garlic, tamarind juice, leafy spices and chile peppers.

Aromatic hot sauce to an Asian is like ketchup to a North American or salsa to a Mexican. Every region has its own hot sauce or condiment, fresh, pickled or cooked. Southeast Asia is noted for its pungent fruity, sweet, sour and hot sauces. There are a variety of hot sauces—sambal olek, chile garlic, Siracha, sambal trassi and nuoc cham, which contain vinegar, tomato, brown sugar, shallots, tamarind, coconut milk, garlic, lemongrass, turmeric or black pepper.

Artfully presented garnishes and toppings also provide a final note of contrasting textures, flavors and colors. Malaysian and Indonesian seasoned rices and noodles are never served without nutty roasted peanuts, caramelized crispy shallots or crunchy bean sprouts. Fresh leafy spices are important Southeast Asian garnishes that provide fresh aromas and many textural variations.

Preparation techniques of spice and seasoning blends are significant in Asian cuisines to provide distinct flavors, colors and textures to Asian foods. Chile peppers, coriander seeds or sesame seeds are smoked, pickled or toasted before being added to spice blends to give more intense flavors. Steaming or grilling meats or rice in pandan or papaya leaves gives unique flavors and visual appeal to Asian foods.

In Asian cuisines, texture is an important element, especially in vegetable entrees. Consumers perceive crispy, crunchy vegetables as fresh and natural. In traditional American cooking, vegetables are often boiled soft and are unappetizing. In contrast, Cantonese steaming and stir-frying techniques using very light seasonings bring out the vegetables' fresh colors, flavors and textures.

Appearance is also an important factor in Asian cooking. Thai and Japanese foods add sculptured fruits, chilies, flowers and vegetables to rice dishes, noodles, grilled chicken and fish.

Spice blends give Asian foods their strong flavors and heat profiles and can create hot, sweet, sour, savory and aromatic sensations all in one meal.

The spices commonly used in Asian cooking are ginger, cumin, cassia, coriander, star anise, galangal, chile peppers, coriander leaf, basils, spearmint, turmeric, clove and garlic. A variety of fresh leafy spices, including mint leaf, cilantro, basil leaf, lemongrass and scallions, is important in flavoring and garnishing foods and drinks. Seasonings such as hoisin, masalas, sambals and panchphoron create magic with fish, steamed chicken, stews, soups or sauces.

Certain spice blends have a "crossover" appeal among different Asian ethnic groups, such as curry, sweet and sour or teriyaki. However, with other blends, such as nam pla, taucheo or chat masala, regional preferences exist.

Also, not all generic spice blends have the same flavor. They vary depending on where they originate, what ingredients are added and what end product they go into.

Asians have mastered the art of preparing spice blends—as dry roasted, toasted in oil or braised, all of which enhance the flavor of the finished product by removing bitter notes, intensifying flavors or adding fragrant aromas. Different preparation techniques suit specific applications and create a multitude of flavors and colors in Asian cuisines.

In the next section, some of the Asian spice blends from different regions will be discussed to provide an understanding of the similarities and the distinct differences in creating authentic Asian foods or Americanized Asian flavors.

REGIONAL CUISINES OF ASIA

South Asian

South Asians have mastered the art of combining different spices to create numerous blends with unique depths of flavors. A basic bouillon seasoning used in India and Pakistan is an example of regional variations in South Asia. Yakhni or akni is the South Asian seasoning equivalent to the bouillon cubes of the western world. It is made from mutton or lamb with basic spices such as onions, garlic, ginger, green and red cayennes, bay leaf and parsley. A North Indian version is made from onion, coriander seed, cloves, green ginger, fennel seed and cayenne. Black cumin, mustard seed, fenugreek leaf, cardamom, cinnamon, kari leaf and parsley are added to create regional tastes. This blend is used as a base flavoring for many Indian dishes, such as curries, sauces, snacks, soups and grilled foods.

Pakistani foods have Arabic, Iranian, Turkish, North Indian and Afghan influences. They are noted for their festive biryanis, wheat-based breads called naan, marinated mutton and lamb kebabs. Pork and alcohol are rarely consumed because of Pakistan's Muslim population.

Nepal and Tibet are influenced by Chinese, Indians and Europeans, and they use plenty of dried and fermented foods derived from buffalo meat, mutton, vegetables, fish, fruits and legumes. Corn, rice, millet and wheat are eaten, along with soups, chutneys and dumplings. The seasonings are mild, and curries are sweeter.

Bangladesh, formerly called East Pakistan, has flavors similar to India and Pakistan. Because the population is mostly Muslim, pork is prohibited. Rice, fish, legumes, chutneys and vegetables flavored with mustard, poppy seeds and nigella are popular foods.

Indian foods have tremendous variety because India is influenced by many cultures and religions. The major religious influences are Hindus, Muslims,

Buddhists, Jains, Sikhs and Parsis from Iran. Indian cuisines have great aromas and in-depth taste profiles that are derived from a complex combination of spices and preparation techniques of spices. Rice and wheat are popular ingredients, depending on regions. Other commonly used ingredients are legumes, vegetables, fruits, nuts, fish, mutton, lamb, dairy products, coconut and numerous spices. In the north, milder and yogurt-based curries are popular, while the south has fiery tamarind and coconut-based curries.

There is no one type of Indian spice blend or flavor. Indian cuisine possesses distinct flavors that vary, depending on regional and cultural preferences. In the northern regions (around Punjab and Kashmir with Mogul, Pakistani and Afghan influences), ghee, nigella seeds, fenugreek leaves, almonds, dill seed, mint, bay leaf, pomegranate, lotus seeds, lemon juice, fennel seeds, saffron, rose petals and garam masala flavor parathas, lamb kebabs, pilafs, tandooris and kormas. In the south (Kerala and Tamil Nadu, influenced by Hindus, Muslim Keralites/Moplas, Arabs, Christians and Jews), kari leaf, tamarind, kodampoli, turmeric, hot chilies, coconut, fenugreek seeds, black pepper and mustard seeds flavor fish curries, sambars, urad dal, sambars and sour dosais. In the southwest (Goa), the Portuguese-influenced vinegary vindaloos and tomato-based flavors are popular. In the east (Bengal, with British influence), mustard oil, mustard seed, nigella, chilies, raisins and panchphoron flavor chutneys, kedgeree and crabs. And, in the northwest, around Gujerat and Bombay (with Parsi, Jain, Muslim/Bohris and Hindu influence), dried apricots, green chile, coriander leaf, black cumin, tamarind, asafetida, heavy cream and kala masala flavor sweet and sour sauces, spicy sour curries, bhel poori, channa dal and aromatic biryanis.

Sri Lankan cooking is influenced by Hindus (Tamils), Buddhists (Singhalese), Portuguese, Arabs, British, Malaysians and Dutch. It has fiery hot curries of fish, beef, chicken and vegetables, sambols, chutneys, savory rices and fermented breads. Their cuisine is flavored by darkly roasted spices, hot chile peppers, coconut milk, dried fish, kari leaf, tamarind, green jackfruit and mango.

Southeast Asian

The countries of Southeast Asia are geographically close and have common ingredients and similar preparation techniques. Yet, each region has its own unique flavors. This region has flavors that are a fusion of many ingredients and cooking techniques. Ginger, chile peppers, fish sauce, shrimp paste, tubers, leafy greens, noodles and legumes are widely used in all of these regions. Each region prepares these commonly enjoyed ingredients differently, thus creating unique variations.

For centuries, Southeast Asia has been the crossroad for trade and religious exchange between the East and West. Conquerors and traders have had a significant impact on their languages, religions and foods. Some of the cultures that have influenced Southeast Asia include Chinese, Indians, Sri Lankans, Arabs, Jews, Portuguese, Dutch, English, French, Spanish and Americans. The Malays use their pungent shrimp pastes, sambals and coconut milk; the Indians introduced curries, fenugreek and dals; the Chinese introduced soy sauce, taucheos and tofu; the Spanish introduced tomatoes, beans and adobos; the Portuguese introduced their vinegar and mustard seeds; and the Arabs introduced kebabs, breads and cinnamon.

As a result, the cuisines of Southeast Asia are some of the most varied in the world. When we taste Southeast Asian foods, we experience a burst of differing flavors and textures at one meal. Simple or complex combinations of many well-balanced sensory perceptions can be tasted at the same time, including hot, sweet, bitter, sour, crunchy, astringent or salty. Their foods are fragrant, peppery, spicy, sweet and sour and have wonderful textures and colors.

A typical Southeast Asian meal consists of white rice, surrounded by one or two meat, poultry or fish dishes, a vegetable dish and several condiments. It is finished with fresh cut fruits. Soups are typically seen in Chinese meals.

One-dish meals are a local favorite for lunches, quick-fix dinners, as snacks and pick-me-ups. They are either stir-fries or soup dishes that consist of rice or noodles, meats or seafood, vegetables and an accompanying condiment.

In Thailand, foods are influenced by the soups and noodles of China, curries from India and Sri Lanka and pungent fish and coconut sauces from North Malaysia. Flavors vary among its regions. Mild and hot flavors predominate in the mountainous regions of the north, with wild game, mushrooms, steamed glutinous rice and chile-lime dips. Fiery, tangy salads (laab) are popular in the northeast, with minced pork and chicken, lime, chile peppers and pungent shrimp paste. Spicy, sweet coconut-based curries, satays and lemongrass soup characterize the foods of the south. In the central regions, hybrid flavors with elegant presentations of rice, noodles, seafood and desserts are enjoyed.

Foods in North Vietnam have more Chinese flavors, including soy sauce, ginger and black pepper. The south and central regions are influenced by Cambodia and Thailand and by Indians, so fish sauce, chilies, fruits and curries predominate.

Indonesian flavors in cooking vary widely among the different islands, with influences from Malaysia, Singapore, China and the Arab world. Most islands have a Muslim background, except for Bali, which has Hindu influence. Sumatra has fiery hot rendangs and spicy root vegetables, Java has stir-fried noodles and hot salads and Bali has pungent curries and fish sambals.

Foods from Malaysia and Singapore have similar flavors, with pungent Malay sambals and tamarind-laced fish, Chinese stir-fried noodles and soups, Nonya chile-based laksas, Indian curries and dals, Cristiang (Portuguese influenced) hot condiments and vegetables and European cutlets and salads. They have unique regional and cultural flavor variations derived from fermented bean pastes, shrimp pastes, wrappers, sour fruits, leafy vegetables, coconut milk, nuts and a variety of spices.

Kampuchean (or Cambodian) foods have a strong Chinese and Indian influence with rice, fish, peanuts, ginger, black and white peppers as essential flavorings. Laotian foods have flavors similar to Cambodian and Thai foods as well as origins from China. Sticky rice, beans, tubers, pork, chicken, chilies, coriander, mint, coconut milk, shrimp and fish sauce are important ingredients. Philippine foods have Malay, Chinese, Spanish and North American influences with pungent fish pastes, coconut, adobos and fruit juices. Foods from Myanmar (or Burma) are influenced by neighboring India, China, Cambodia and Thailand. Rice, noodles, duck, chicken and fish curries, fermented fish pastes, garlic, turmeric, tomatoes and onions are popular ingredients.

East Asian

East Asian flavors such as Japanese and Chinese are more familiar to North Americans. Chinese foods have some influences from Japanese, Mongolians and Indians, depending on the regions. In South China, people eat mainly rice or noodles, while the northerners eat wheat, millet and sorghum. The most popular meats are pork, beef and lamb, the latter two being preferred by the Muslims. Vegetables are eaten in abundance.

Regional cooking of China is becoming better known in the U.S. Cantonese foods have little seasoning and, like French cuisine, rely on the subtle juices of ingredients and preparation techniques. Their seasonings are soy sauce, ginger, garlic and oyster sauce. Beijing (or Peking) cuisine uses pancakes, steamed buns, stir-fried or steamed dishes, pickled vegetables, tofu, scallions, roasted or barbecued pork and lamb. Shanghai cuisine on the east coast has sweeter flavors and uses plenty of noodles, sesame seeds, fish, dark red sauces, sesame oil and black bean sauce. Szechwan and Hunan cuisines in the west are hotter, using chile peppers, preserved vegetables, fagara and cloves.

Korean foods have been influenced by neighboring Japan and China. Rice is the staple with noodles, seaweeds, pickled vegetables, pork and kimchis. Essential flavorings include sesame, ginger root, ginseng, gingko nuts, garlic, soy sauce, fish sauce and chile peppers. The south has spicier and saltier flavor profiles than the north. A typical Korean meal, called panchan, consists of steamed white rice with side dishes of kimchi, soup, spinach "namul," toasted seaweed, pickled vegetables, meat or pickled seafood.

Japanese foods have aesthetic appeal and simplicity. They have some influence from China and Korea. Japanese foods are lightly seasoned, focusing more on preparation techniques, textural and visual appeal. If seasonings are used, such as gomasio or shichimi, they are used as garnishes and condiments. Their foods delight the senses and are based on cooking techniques using fresh ingredients.

Their staple is glutinous rice, with a variety of noodles, fish and other seafood, pickled vegetables, mushrooms, seaweeds, soy sauce, wasabi, sansho, ginger, miso, mirin and sesame. Japanese flavors are obtained mainly from the way ingredients are prepared and cooked (such as grilling, steaming, simmering or deep frying).

POPULAR EAST ASIAN SPICE BLENDS

Fermented Soybean Blends

Soybeans have inspired many dishes in China, Japan, Korea and Southeast Asia. Soybeans are frequently fermented to achieve end products that are richer in appearance, flavors, proteins and vitamins. These fermented bean pastes are called jiang (Mandarin), jang (Korea), taotsi (Philippines), taucheo (Malaysia) and miso (Japan).

The sensory properties of fermented soybean pastes vary with fermentation time, aging and added ingredients—chile peppers, seafood, peanuts, sesame, onions or garlic. They can be smooth or chunky, salty, sweet or hot. They are commonly used as seasonings to flavor the meats, seafood, vegetables, soups, dips and dressings of China, Korea and Japan. They are a major ingredient in Szechwan sauces, marinades, hoisin and oyster sauces. They are also important in Singaporean braised meats and stir-fries and Indonesian sambals.

Fermented bean pastes are made to complement or enhance particular dishes. Some of the fermented soybean products are the Japanese and Korean misos, the salty Malaysian taucheo, pungent Korean twoenjang and the salty, sweet, dark black and brown Chinese bean pastes. Miso is a common ingredient in Japanese and Korean soups, sautéed dishes and marinades. They are sweet, nutty, or salty and white, brown or dark red colored. Their flavor and color depend on whether they are made from soybeans alone or with rice or barley, their fermentation time, and other added ingredients. The yellow or brown bean paste is slightly sweet and flavors stir fries of Peking origin. The black, salted and fermented pungent soybean paste is made with garlic and is used to flavor seafood and pork dishes of Shanghai origin. The red, sweet bean past made from adjuki beans is used in pastry and bun fillings and in the puddings of the Guandong (Cantonese) region.

Fermented Bean Seasonings

Fermented bean seasonings include soybean sauce/paste (jiangs, taucheo), sweet bean sauce (timcheo), oyster sauce, misos and twoenjang.

Description

Asians ferment, salt and pickle soybeans and other beans to create sauce blends with intense flavors, colors and textures. These blends are rich in flavor and have high levels of B vitamins and minerals. Each region uses its own distinct aging process and raw materials to create unique, intense flavors. They are red, black, white or yellow and are pungent, spicy, sweet, nutty or salty, depending on the added ingredients (rice, barley, wheat, salt, chile peppers, garlic or sesame oil) and their aging times.

Ingredients

Soybeans and/or rice, barley and buckwheat are the starting ingredients used in creating fermented bean pastes. Salt, sugar, garlic, red chile peppers, seaweed, sesame oil, ginger, rice wine or scallions are some of the typically added ingredients.

Applications

Bean pastes enhance or complement many dishes, such as soups, stir-fries, cooking sauces or marinades. They also thicken and provide consistency to these products. They are mixed with hoisin, sweet and sour or tomato sauce to give a wonderful blend of sweet, pungent and savory tastes to many Chinese dishes.

Miso blend, a must in Japanese cooking, is a fermented product made from soybeans with sea salt and koji, together with other ingredients such as rice, wheat, barley or buckwheat. Aka, hacho, mugi, shiro, shinso, soba or kome are different misos that are mild, salty, intense, sweet or hearty, depending on the raw ingredients and fermentation time. The three most popular misos are the aka miso (all soybean), which is dark reddish black and salty; the shiro miso (soybean and rice), which is sweet and light yellow; and the shinso miso (soybean and barley), which is semisweet and dark brown. Misos' applications vary by region, depending on the raw materials available, climatic conditions, and eating customs. For example, shiro miso is typically used in the Kyoto region of Japan, aka miso in northern Japan and shinshu miso in Nagano region. Spreads, marinades, salad dressings, stir-fries and soups use different types of misos. Meat and seafood go well with aka miso, while shiro miso pairs well with soups and salad dressings.

Soybeans are fermented to create salted whole beans (taucheo), pungent chunky black bean sauce and the slightly sweet pungent brown bean sauce (timcheo). They are used throughout East Asia and Southeast Asia, wherever there is Chinese influence. The sauces are thick, pungent, salty or sweet and provide consistency and flavor to pork, fish or stir-fries. They are also used as spreads for fresh spring rolls or moo shu pork skins.

Black bean sauce is thick with a deep reddish brown color and a rich flavor, whereas the yellow bean sauce is sweeter and less pungent. Taucheo, the salted whole soybean, is used whole or pounded to flavor fish, noodles, vegetables and pork dishes.

Oyster sauce (also called hou yu jiang in Mandarin, nam man hoi in Thai and kicap tiram in Malaysian, Indonesian and Singaporean) is a thick, reddish black seasoning that is commonly used for Cantonese-style cooking, marinades and stir-fries. It contains oyster extract or juice combined with fermented soybeans, sugar, starch and other ingredients. It is added to fish sauce, hoisin and soy sauce to intensify the flavor of stir-fries, pork, fish and meat dishes and is used with spices as a topping sauce for fried tofu. It adds a sweet, meaty and slightly smoky flavor to beef, seafood, tofu and vegetables.

Dwen jang/twoenjang, a fermented soybean paste, is used in making spicy pungent goit chu jang paste, which is a staple flavoring in Korea. Goit chu jang contains red pepper paste and glutinuous rice paste and is an essential ingredient in stews, vegetable dressings, soups and dips. Garlic, sesame oil, vinegar, sugar, soy sauce or sesame seeds are added to goit chu jang to create a hot pungent table condiment.

Chinese Five Spice Blend

Many Asian regions, such as China, Thailand and India, have their own five spice blend that is commonly used to season many dishes or used as a "sprinkle on" seasoning at the meal table. Each one contains different ingredients based on regional preferences. The Chinese five spice is the most well-known five spice blend in the U.S.

Description

Chinese five spice blend is a ground mixture that is popular in Chinese cooking, especially Szechwan, Hunan and Peking dishes. It is also called wu xiang fen in Mandarin and phong pha lo in Thai. It is used in the Chinese-influenced dishes of Vietnam, Thailand, Malaysia and Singapore. It has a pungent, anisey taste with slightly camphorus notes, and its color ranges from tan to gingery brown. Foods get bitter if the spice blend is overused. The Chinese believe that the number five has curative powers, so this blend started as a spice medicine.

Ingredients

Chinese five spice blend is made from star anise, fennel, clove, cassia and Szechwan pepper. Black peppercorns can be substituted for Szechwan pepper. Today, commercially sold blends also contain ginger, licorice, cardamom and/or ground dried tangerine peel.

Applications

Chinese five spice blend is an important flavoring in hoisin sauce and other Chinese marinades for deep-fried, roasted, grilled, braised or barbecued foods, such as pork, duck, chicken and other meats. It is mixed with red color, soy sauce and sesame oil to create the popular Chinese red, roasted pork called char siew. It is mixed with salt and used as a dipping sauce for roasted meats, poultry and raw or deep-fried vegetables. The blend pairs well with rice wine, dry sherry, soy sauce, shallots and garlic.

Japanese Seven Spice Blend/Shichimi Togarashi

Description

Shichimi togarashi is a finely ground mixture of spices that is essential to Japanese cooking. It has a hot spicy taste with a citrus aroma. Its texture is gritty.

Ingredients

It contains sansho, chile peppers, dried orange/tangerine peel flakes, dried nori/seaweed flakes, white and black sesame seeds and white poppy seeds.

Application

In Japan, this seasoning is sprinkled on prepared foods. It is generally used to season grilled meats, soups, noodle and rice bowl meals. It is also used as a popular table condiment.

Teriyaki Blend

Description

Teriyaki adds flavor and color to raw and cooked foods in Japanese cuisine. It is slightly sweet and soy sauce-like and is sold in dried, paste and liquid forms.

Ingredients

There are many versions of teriyaki sauce. The basic teriyaki ingredients are soy sauce, sugar, vinegar and sweet rice wine. Ginger, garlic, chives, sesame oil and fish stocks can be added to give extra edge or different flavor notes to enhance specific applications.

Application

Teriyaki complements broiled and steamed dishes such as steaks, chicken and spinach. It flavors dipping sauces that are used for wontons, dumplings, sashimi, deep-fried and battered seafood or vegetable tempuras. It is commonly used as a Japanese-style marinade for grilled or roast beef, chicken, seafood and vegetables. Pineapples, oranges, lychees, tomatoes, onions and bell peppers are chopped or sliced and added to teriyaki sauce.

Hoisin Sauce Blend

Description

Hoisin sauce is one of most important seasoning in Szechwan, Peking and Cantonese cooking. It is a thick, reddish brown or dark brown sauce with a slightly sweet, spicy and tangy flavor. Traditionally, it is made with wheat flour and sugar. It originated in northern China and is called tianmian jiang in Mandarin, which means "sweet wheat paste." In Cantonese, hoisin means "sea fresh sauce."

Ingredients

Hoisin sauce blend is made from soybeans, wheat, sugar, caramel, vinegar and five spice blend. Sometimes, other ingredients are used, such as ground sesame seeds, red beans or garlic. Plum puree, sesame oil and/or chile peppers are sometimes added to create differing flavor variations in applications.

Applications

Hoisin is used as a dipping sauce, a marinade for ribs, shellfish and duck and as an accompaniment to Peking duck and moo shu pork. It is combined with chilies, garlic, plum sauce and vinegar to create char siew or roast pork (a Cantonese specialty), double cooked pork (a Szechwan specialty) and stir-fry dishes. Hoisin provides a deep, rich flavor and color to stir-fries, roasts and barbecues. It is also used as a table condiment and dipping sauce. Hoisin pairs well with chicken, duck, ribs, shrimp, scallions, sesame oil, rice wine, soy sauce, vegetables and ginger. Hoisin sauce is widely used in Cantonese-style cooking to season and to provide consistency.

POPULAR SOUTH ASIAN SPICE BLENDS

Curry Blends/Kari Podis/Masalas

Description

Historically, curry is believed to have originated in South India. Through trade, colonization and immigration, curry spread to other parts of Asia, Europe, Africa and the Caribbean, with resulting variations, such as rendang, w'et, gaeng or colombo.

While spice mixtures called kari podis and masalas for flavoring sauces have been made for thousands of years in India, the term "curry" or curry powder is not an Indian word, but it is an English term for a commercialized spiced sauce or spice mixture from India. There are many versions of how the word "curry" originated. Some say that the English coined the term curry from the word "kaari" (a vegetable-based sauce) or "kaaree" (a meat-based sauce) in the South Indian Tamil language, which means, "sauce" or "gravy". Other stories say that the British took the word curry from "kari leaf," a pungent leafy spice used in kari (or sauce) of South India. Another origin of the word curry is derived from "karahi," a wok-like vessel used to cook curries in North India and Pakistan. Yet another source traces curry from the Hindustani word, "turcurri," which was shortened to "turri" and mispronounced as curry by the Westerners.

Curry blends are used not only to flavor sauces but also to flavor stews, snacks or rices. Generally, curries are a mixture of many different spices (whole, ground or crushed) with fruits, vegetables, nuts, seeds and other ingredients. There are hundreds of curry blends in South Asia. Curry blends differ in flavor and color, depending on who creates them, the types and proportions of ingredients used, their preparation and the end application. They are becoming "hot" in the U.S. because curries satisfy consumers' increasing appetite for spicier ingredients and stronger flavor profiles. Whether called curry powder, garam masala, panchphoron, chat masala, vindaloo paste, Madras curry powder, korma paste or sambar podi, all contain mixtures of spices blended for specific applications and regional preferences.

Ingredients

A good curry blend or masala contains a well-balanced proportion of spices. It would be difficult to develop an industry-wide standard for curry blends because of the numerous types of curries made. A basic curry blend from South Asia consists primarily of cumin, coriander, black or red pepper and turmeric. Many versions exist depending on regional preferences, cultural influences, the availability of ingredients and the application—meat, fish, vegetable or poultry. Spices such as cardamom, clove, cinnamon, mustard seeds, fenugreek, kari leaf, mint, coriander leaf and celery seeds are added to create these variations. South Asians toast the whole spices then grind them before using to obtain more fragrant aromas and crunchiness and to remove raw tastes.

Curry blends pair well with garlic, ginger, shallots, coconut milk, yogurt, lime juice, lemon juice, tamarind, lemongrass, mango, tomato, other vegetables and vinegar.

SOUTH INDIA

South Indian curries, or kari, are hot, spicy, coconutty and pungent with some sour notes. They can be dry, saucy, soupy or thin and are generally made with fish, goat, chicken, lentils or vegetables. Essential ingredients in the curry blend are chile peppers, coconut milk, tamarind, turmeric, black pepper, fenugreek and kari leaf. "Popped" mustard seeds are usually added, along with shallots, ginger and kokum. Optional spices are clove, cinnamon and green cardamom.

These curry blends characterize vindaloo, Madras curry, konju meen, sorpotel, keera kootu, meen kolambu, keema bafat, erachi kotu (mutton) and xacuti curry.

NORTH INDIA, PAKISTAN, BANGLADESH

Curry spice mixtures in the north are called masalas (plain masala or garam masalas), which comes from the Arabic language. Curries from North India, Pakistan and Bangladesh are influenced by Mogul and Persian cooking and often have yogurt, dried fruits and nuts to characterize them. Other ingredients are coriander leaves, methi leaves, paprika, ground mustard, saffron and ghee. They are mild, sweet, rich, nutty, fruity and or aromatic, creamy or dry. They are generally made with lamb, chicken, vegetables and legumes.

Some popular curry sauce blends are saag blend, jalfrezi blend, rogan josh blend, korma blend, channa makhani blend, masala murgh blend, matar keema blend, chicken masala blend, Kashmir curry blend and kofta curry blend.

SRI LANKA

In Sri Lanka, most curries are chile-based, pungent and extremely hot. The milder curries contain coconut milk and tamarind. The black curries contain kari leaves, darkly roasted spices such as coriander, cumin, fennel and fenugreek, and other ground spices including cloves, cinnamon and cardamom. Tiny bird's eye peppers provide the heat. Some curries have dried and powdered maldive fish, eggplant, mango and unripe jackfruit.

Some popular Sri Lankan curries are fish pindun, beef smoore, pork padam, frikkadel and duck padre.

Specific Blends

Chat masala is a popular green curry mixture used in North India. It comes as a wet paste or a dry mix, and there are many versions of it. A typical blend uses green chiles, mint leaves, coriander leaves, coriander seeds, cumin seeds, ajwain seeds, black peppercorns, cloves, ginger, asafetida and dried mango powder (amchur). Others contain pomegranate seeds, lime juice, bay leaf, cinnamon and cardamon. Chat masala has a tart, salty and hot flavor. It is commonly used to flavor sauces and snacks.

Garam masala is the most popular masala used in North Indian cooking. It is generally used with curry blends to add zest to sauces. Garam masala means "hot or warm spice mixture" containing spices that create heat in the body. These spices are ground for use in North Indian curries. Its ingredients vary, but, traditionally, garam masala contains the "warmer," more pungent spices such as brown cardamom, black peppercorn, clove, cinnamon, mace and nutmeg. Today, it includes the cooler spices, such as fennel, green cardamon and bay or cinnamon leaf. Based on regional preferences, cassia, fennel or celery seeds are added. It usually lacks turmeric.

Garam masala is dry roasted and used to enhance many curry dishes. It contributes to the aroma of a finished sauce and is usually sprinkled on just before serving or toward the end of cooking to avoid bitterness caused by prolonged cooking. It is generally added to meat dishes, less in poultry and biryanis (layered meats and rice) and rarely in vegetable or fish dishes.

Korma is a braised, dry curry from North India. It is brown, aromatic, and mild, with a creamy consistency. *Korma blend* has almonds, cardamom, cumin, coriander, turmeric, yogurt and onion. It goes well with lamb, chicken and certain vegetables. In Muslim-style biryanis, korma sauce is layered between cooked basmati rice and is baked to give biryanis their characteristic aromatic flavor.

Madras curry blend is a hot pungent curry from Tamil Nadu in south India. It contains roasted dry spices such as dried ground red chilies, coriander seeds, cumin, mustard seeds, black pepper, fenugreek, turmeric, asafetida and kari leaves. It can also contain ground, toasted lentils and rice. This spice blend is added to chicken, fish, mutton or vegetables.

Kashmiri masala is mild and aromatic with green cardamon as the predominating spice. Other spices are similar to the garam masala. It is used for baked dishes. *Xacuti curry blend* is from Goa and contains cumin, coriander, fenugreek, black peppercorn, chile peppers and shredded coconut, which are roasted until a dark color is obtained.

Char masala, a four spice mixture used in Afghanistan, contains any combination of cumin, coriander, cinnamon and cloves, green cardamon or black pepper.

Kala masala, a black spice mixture popular in Sri Lanka, consists of roasted and ground clove, cinnamon, cumin, coriander, red chile peppers and black pepper. Maharashtra, in north India, has kala masala in many versions, a typical one having cumin, coriander, cinnamon, cardamom, sesame seeds, chile peppers and shredded toasted coconut.

Samanthi podi is another popular dark brown spice mixture in Kerala. It contains dried toasted coconut, kari leaf, tamarind, onion and ground red chile. Toasted and ground lentils are optional. It flavors condiments and chutneys that are served with local breads such as puttu, thosai or idli, especially at breakfast.

Tandoori blend, an aromatic, spicy marinade, was created by the Moguls of North India who baked food in a "tandoor" or clay-type oven heated with charcoals. Prior to baking, chicken, lamb, shrimp or fish are marinated in a blend of yogurt, lemon juice and ground spices, such as ground chile pepper, ginger, cumin, coriander, clove, cinnamon, cardamom, nutmeg, mace, black peppercorn and bay leaf.

Panchphoron is a five-spice mixture from Bengal that contains fennel, mustard, cumin, fenugreek seeds and kalonji. It generally complements fish, dals and vegetable dishes.

Applications

Commercial curry blends come as powders, pastes, sauces or oils that offer convenience and consistency for a recipe. They also are available hot, medium or mild depending on the amount of chile pepper added. While the U.S curry blends contain ingredients that have mass appeal, such as cumin and turmeric, they often lack the flavor-enhancing effect required for a specific application. To create the authentic products that consumers demand, curries must include specific or specially prepared ingredients that will enhance the flavor, texture and color profiles of fish, meat, poultry or vegetables. It is also important to maintain the aroma and taste of freshly ground spices. Commercial curry blends are sold as powders, masalas, curry pastes, curry sauces or curry oils.

When creating a sauce or curry, ginger, garlic, onions, potatoes, vegetables, nuts or seeds provide the "body." The consistency of a finished curry varies and can be thick or thin, dry or wet. Color is also an important sensory attribute of curry. Color is contributed by turmeric, saffron, paprika, green leafy spices (cilantro, basil, mint), certain vegetables, nuts or chile peppers.

Ingredient preparation and cooking techniques create numerous flavor, texture and color profiles in a curry sauce blend. By using ground or whole spices and cooking techniques, such as dry roasting, toasting, braising or sautéing, different flavor dimensions can be achieved.

The order and time when ingredients are added in the cooking process are also crucial in obtaining the flavor desired. Some spices are more volatile than others. Cumin, coriander and fennel are dry roasted and added first. The more volatile spices, such as mace, cinnamon and cardamom, are added later to retain their aromas.

Sambar Podi, Rasam Podi, Dal Podi

Description

South India vegetarian spice mixtures often contain "dals" (lentils) and are used in sauces, soups and curries and to enhance chutneys, dips and snacks. These spicy lentil mixtures include sambar podi, dal podi and rasam podi. They are added directly into cooking "as is" or they are dry roasted or "popped" whole in hot oil to render them fragrant, then ground into the "podis" or mixtures.

Ingredients

Indian "dals" or lentils are fried or dry roasted, then added whole or ground with spices and other flavorings. The podis usually contain urad dal, channa dal or toor/toovar dal. Many other lentils or beans are also used in combination with spices to add zest or texture to foods.

Sambar podi contains channa dal (smaller chickpea/Bengal gram), toovar dal and urad dal (black lentil/black gram) that are mixed with spices, such as coriander seeds, cumin seeds, mustard seeds, fenugreek seeds, black peppercorns, asafetida, kari leaves and ground red chiles. The spices are roasted for a few minutes, then cooled, ground and used for stews and soups.

Rasam podi contains toor dal, channa dal, coriander, cayenne, black peppercorns, cumin, turmeric and kari leaves. This blend is added with tamarind juice to make a spicy, sour soup.

Dal podi contains toor dal, channa dal, urad dal, turmeric, cumin, coriander, asafetida, cayenne, black peppercorn and onions. It is used in stews, sauces and dips.

Applications

In Indian cooking, dals are used in many different applications, and an understanding of their usage will help create appropriate lentil spice mixtures. Channa dal, popular in all of India, is used raw in chutneys, roasted whole as a spicy snack, ground for sweets or whole with vegetables and chutneys. Urad dal is widely used in South India where it is fermented with rice and mixed with spices and curry leaves to create dosai, steamed idli and snacks such as vadai or pappadum. Toovar dal exhibits a thick and more gelatinous consistency and is combined with channa dal, spices and chilies to create sambars. Kala channa, a smaller, darker chickpea, is popularly used in the Punjab region and is braised for use in tangy, spicy cumin and coriander-based sauces. The leafy spices are usually added crushed or added whole to the ground spices.

These podis are added to chutneys, pickles, soups and sauces of south India. They are usually eaten with local breads or rice.

Pickle and Chutney Blends

Description

Pickles and chutneys are essential accompaniments to Indian meals, snacks, soups, breads and appetizers. Chutneys originated in India and were introduced into England and Europe by British colonials. Chutneys or chatnis, which translates to "licking good" in Sanskrit, and pickles, are essential condiments and side dishes in every Indian meal. They add zest and texture to the main meal with their hot, salty, sour and sweet flavors. They consist of spices, fruits, vegetables, vinegar and sugar. There are unlimited variations of these condiments, depending on regional preference as well as on the type of food it is eaten with—meat, fish, rice, bread or vegetables.

Ingredients

The main ingredients used in pickles and chutneys are mango, lime, coconut, apples, chile peppers, ginger, garlic, mustard oil, mustard seeds, turmeric, sesame seeds, mint, cilantro and tamarind. These ingredients are usually fermented or pickled to achieve intense, characterizing flavors. They can also be freshly prepared (uncooked) or cooked.

Pickles contain discrete pieces of mangoes, green chile peppers or limes, whereas chutneys contain smooth or coarsely textured ingredients such as coconut, mint or tomato. There are many variations of chutneys and pickles, and some are included below.

The British chutneys tend to be sweet and tart, incorporating fresh or dried fruits such as citrus, pineapples, cranberries, apples, plums, figs, peaches, apricots and dates. They are served alongside hot curries and cold foods. Today, fresh vegetables such as corn, squash, bell peppers, cucumbers, carrots and green tomatoes are added. These chutneys can be compared to the freshly made or cooked salsas of Latin America.

Mint/pudina chutney is fragrant, uncooked, smooth-textured chutney that goes well with roasted/grilled meats, tandooris and fried foods. It contains mint, ginger, coriander leaves, black pepper, green chile peppers with vinegar, lemon juice and sugar. It is also commonly served as dips with samosas, pakoras and other appetizers.

Coconut/nariel chutney is an uncooked, coarsely textured, and spicy sweet specialty condiment of South India. It usually accompanies fermented breads, such as dosai or idli, and fish and chicken. It has toasted coconut flakes, roasted cumin seeds, "popped" mustard seeds, garlic, ginger, green chilies and roasted curry leaves.

Tomato/thakali chutney is an uncooked or cooked, sweet, sour and spicy condiment, commonly used in South and North India. It has roasted tomatoes, onions, ginger, garlic, roasted, ground cumin seeds, black mustard seeds, cayenne, coriander leaves and lemon juice. It complements seafood dishes, barbecues and fried foods.

Mango pickle is a spicy, tart pickle with green/ripe mango with mustard seeds, mustard oil, turmeric, ginger and garlic.

Green chile acar is a pungent spicy pickle with green cayenne, mustard oil, ginger, sesame oil, ground mustard, fenugreek seed and turmeric.

Mixed pickle/sabzi achar is a spicy, sour and hot mixed vegetable pickle commonly eaten all over India with any meal. It contains garam masala, onions, ginger, cayenne, turmeric, black mustard seeds, nigella, fennel, mustard oil, carrots, green mangoes, green chilies and lotus root.

Applications

Pickles containing mango, lime, shredded carrot or chilies with spices and yogurt are generally used to provide bite to vegetarian meals as well as to balance the flavors in curries or dals. They are used for breakfast, lunch, snacks or dinners. Chutneys usually accompany snacks and tone down fiery sauces. They are also used as dips for breads, such as idli, thosai or chappatis. They can be spicy, sour or slightly sweet and include mango, mint, tomato, cauliflower, white radish/mooli, carrot, eggplant, onion, coconut and tamarind.

POPULAR SOUTHEAST ASIAN SPICE BLENDS

Sambal Blends

Description

Sambals are the hot sauces of Southeast Asia and are used the same way Tabasco sauce is used in the U.S. or salsa is used in Mexico. They are essential table condiments used to perk up rices, breads, mild curries, soups and noodles. Sambals are also used as base sauces for cooking shrimp, fish, chicken, meat and vegetables and for spicing fruit and vegetable salads.

Sambals are hot, sweet and very fragrant. The fragrant aroma is developed when the sambal sauce blend is "tumised" or sautéed slowly with constant stirring until the oil seeps out from the mixture. Indonesia and Malaysia have many varieties of sambals with their meals. Some examples include sambal belacan (Malaysia, Singapore), sambal trasi (Indonesia), sambal olek (Southeast Asia), sambal badjak (Indonesia), cucumber and pineapple sambal (Malaysia), sambal tomat and sambal matah (Bali) and nam prik (Thailand).

Ingredients

The basis for most sambals is chile peppers, shallots, tamarind, brown sugar and lime juice. It can be as simple as shallots with cili boh (pureed chilies) and lime juice or more complex with other added ingredients, such as lemongrass, nutmeg, shrimp paste, palm sugar, fish paste, ginger, galangal, kaffir lime leaves, coconut milk, candlenuts and tomatoes.

Each ethnic group adds its special flavoring to these sambals, such as ginger and soy sauce (for less heat) by the Chinese; bird peppers, belacan and coconut milk by the Malays; candlenuts, mint or "bunga kantan" by the Nonyas; or a little curry powder and coriander leaf by the Indians.

Applications

Southeast Asians add sambals to almost anything, even as spreads to sandwiches and as fillings to buns and pastries. A typical breakfast in Malaysia and Singapore, called "nasi lemak," is a coconut milk and lemongrass-based rice dish that is topped or served with sambals made from prawns or ikan bilis (an anchovy-like fish).

Fruits and vegetables, such as pineapple, mango, cucumber, green papaya, petis (bitter bean), belimbing wuluh and carrot are added to sambals. Herb salads, laksas, rices or noodles that are enjoyed by Malaysians and Singaporeans are served with fiery pungent sambals.

Malays add fish, prawns, ikan bilis, cucumber, cuttlefish, tofu, mango, bengkuang (jicama), tomatoes, peanuts or belimbing in their sambals and serve them as side dishes with their "nasi" or boiled white rice. Every meal has a choice of one or more sambals to suit the accompanying dish. Nonyas, from Malaysia, generally serve sambals with laksas, fried rice or fish dishes.

Curry Blends

Description

Indians have influenced Southeast Asia, and as a result, curry blends have become an important flavoring in Southeast Asian cooking. Like sambals, they are used in sauces (wet or dry), stir-fried noodles, soups and condiments. Each region has developed its own distinct curry blends.

Ingredients and Applications

BURMA/MYANMAR

Their dishes are influenced by the Chinese, Indians, Thais and Cambodians. Their curry blends are mild or medium hot as compared to South Indian curries. Burma's curry blends include basic spices such as onion, garlic, ginger, turmeric, fish sauce and chile peppers, with other added vegetables, tamarind and the bark of the banana tree. Their curries are oily, with sesame and peanut oils. Curry blends are used in duck, pork, fish, chicken and vegetable dishes.

Examples are kyethar peinathar hin, bairather sepiyan, mohinga, Chettiar or monsoon curry and pazun hin.

CAMBODIA/KAMPUCHEA

Cambodians dishes have strong Chinese and Indian influences. The essential ingredients are fish sauce, lemongrass, galangal, garlic, coconut milk and chile peppers. The lemongrass provides their curry blends with a fibrous texture, and the coconut milk gives them a creamy consistency. Curry blends are used with meats, such as pork and water buffalo. Bean sprouts, potatoes, peanuts and eggs are other added ingredients.

Examples include lemongrass curry, egg and potato curry, Cambogee and aamoh.

LAOS

Laos is greatly influenced by southern Chinese and northern Thai cooking. Their flavors are similar to Cambodian cooking and have influences from India. The southern region of Laos uses fish sauce, coconut milk, mint and chile peppers as essential ingredients to spice up their sauces. The Hmong cuisine of Laos, influenced by Chinese cooking, uses soy sauce instead of fish sauce to flavor their foods. Laotian curry blends' essential ingredients are ginger, shallots, chile peppers, coconut milk, yams, taro and sweet potatoes. Curry blends are used with pork and chicken, catfish and eggs but rarely with vegetables, which are usually stir-fried. Laotian curries are very smooth and starchy.

Examples are catfish curry, kalee peel and mok pla.

VIETNAM

Vietnamese curries are influenced by Chinese, Indians and French. North Vietnam cuisine is influenced more by the Chinese. South Vietnam has been influenced by Indian traders, and so curries, called caris, are also found here. These curry blends include lemongrass, mint, coconut milk, vinegar, sugar, turmeric, garlic, chile peppers, lime, fish sauce and coriander leaf. They are milder than Thai curries and can be slightly sour and sweet with a strong turmeric color. Curry blends are added to frog legs, mung bean noodles, taro and sweet potatoes.

Some examples include Saigon cari, cari ga and Echnau cari.

THAILAND

Thai curry blends are hot, pungent, sweet, sour and aromatic and some of them have a strong, coconutty flavor. Thai curries, called gaengs or kaengs, are hot and fiery with plenty of fresh or dried chile peppers that include cayenne or bird peppers. Thai red curry is pungent and intense with coriander root, shrimp paste, galangal, and dried, red chilies; green curry has coriander leaves, green chilies, lemongrass and green eggplant; yellow curry includes fresh turmeric and yellow chilies; Mussaman curry, which is influenced by Indian Muslims, contains more spices, including clove, nutmeg, cinnamon and cardamom; and Panang, which has influences from Penang (in North Malaysia), has coconut milk shallots, peanuts and tamarind. Thai curry blends pair well with pork, seafood and chicken. Kaffir lime leaves, Thai basil, lemongrass, fish sauce, shrimp paste, coconut milk, bamboo shoots, eggplant and green beans are some essential ingredients in their local curries. Thai curry pastes are added to pork, chicken or beef and are used to perk up soups, dips, salads and stir-fries.

Red (nam prik gaeng ped), green (kreung gaeng kiow wahn), yellow, Mussaman, Panang and spicy basil are some of the popular Thai curry blends.

MALAYSIA/SINGAPORE

Malaysian and Singaporean curry blends are influenced by South Indians, Malays, Nonyas, Chinese and the Portuguese. Essentially, their curry blends contain Indian spicing with added Malay and Nonya ingredients. They can be dry or soupy. They are generally hot, pungent and very aromatic. Curries are very popular as side dishes to rice, as dips for flat breads (naan, chappati, dosai, roti canai), to flavor rice (called biryanis) or as bases for soups, pickles, sandwich fillings and spicy salads.

In Singapore and Malaysian curries, lemongrass, star anise, galangal, pandan leaf, ginger flower, lime, coconut, tamarind, candlenuts, mint, coriander leaf, turmeric, shallots, shrimp paste or sambals are some of the ingredients that are added for variety.

The Nonyas add tamarind, fish sauce, turmeric, candlenuts and lemongrass to make pungent, tart and fiery laksa curries. The Malays add bird peppers, mint, belacan, galangal and coconut milk to create hot and aromatic fish curries. Indians use many spices, including kari leaves, yogurt, tomatoes, garam masala, ginger, coconut milk and tamarind to make vindaloos, mutton curry, sothis, sambars, tandooris, biryanis or keemas. Their curries vary depending on their South Asian ancestry: South or North Indian, Bengali, Keralan or Sri Lankan.

Singapore, which has a stronger Chinese influence, adds soy sauce, fish balls, seafood and local vegetables to create their famous fish head curry, dry mutton curry, masak lemak and rendangs.

Examples are rendang blend, laksa blends, kari ayam blend, debil kari blend, kapitan curry blend, kurma blend, fish head curry blend, dry mutton curry blend, asam pedas blend, keema blend, mughlai blend and sayur lodeh blend.

INDONESIA

Arabs, Indians, Chinese, Dutch and Portuguese influenced Indonesian curry blends. They are similar to Malay curries, but vary regionally. The Minangkabaus in East Sumatra eat fiery curries with beef and coconut milk; the Javanese enjoy sweet and hot curries with kecap manis and ginger; and the Balinese, with a heavy Hindu influence, use more aromatic spices with the local sambals. In Bali, curry pastes are rubbed on fish, chicken, pork or beef, wrapped in banana leaves and then grilled or baked.

Essential ingredients in curries are galangal, coconut milk, tamarind, trasi, salam leaf and chile peppers. Other ingredients are eggplant, okra, potatoes, manioc and lime leaves. Fried shallots and sliced boiled eggs are common toppings for curries. Some Indonesian curry sauce blends include beef rendangs, gulai kambing, terong kari, ubi kentang kari and gulai putih.

MEDITERRANEAN SPICE BLENDS

Mediterranean foods are tasty, trendy and part of a healthful lifestyle. They use plenty of fruits, vegetables, nuts, olive oil, roasted flavors and spices that provide great variety and visual appeal. Mediterranean blends range from the familiar Italian pasta blends to the more exotic North African harissas. Mediterranean foods take their unique flavors from many regions and encompass all of the varied foods of southern European, northern African and the Middle Eastern countries that surround the Mediterranean.

Certain ingredients are commonly used throughout the Mediterranean. Olive oil, garlic, parsley, beans, lamb, dried fruits and pimientos have mass appeal in all of the regions, but each region has its own unique flavors. Northern Mediterranean cuisine (Provence, Italy, Basque regions) is characterized by sweet leafy spices (basil, thyme, rosemary), tomatoes and wine; the west

(Spain) by bacalao, sofrito and saffron; the east (Middle East) by mint, sesame, pomegranate juice and yogurt; and the south (northern Africa) by coriander, cayenne pepper, pippali and harissa.

Mediterranean foods will continue to grow in popularity because of their light sauces and healthy ingredients and cooking techniques. The Middle East, Turkey and northern Africa (Ethiopia, Egypt, Morocco and Tunisia) have stronger and spicier notes that will excite an adventurous consumer.

Religion adds to the variety and unique flavor differences found in Mediterranean foods. Muslims, Christians, Jews and Arabs have specific preferences and prohibitions for ingredients such as lamb, fish, beef or pork. Certain ingredients and preparation techniques are specific to different religious groups. Wine, widely used in southern French and Tuscan cooking, is not a Halal item, so regions of Muslim predominance do not use it in their cooking. Likewise, Muslims or Jews do not use pork flavoring.

There are no intricate flavor combinations with Mediterranean cooking. The cooking is simple and rooted in the soil. Antipasti, mezzes and tapas using a variety of vegetables, fruits, breads and olive oil are a way of life with Mediterraneans. The natural flavors and textures of the ingredients are kept intact using light sauces. The variety comes from the different ingredients. Spiced and preserved meats and fish, crystallized and dried fruits, pickled vegetables and olives are some of the other unique flavors.

Seafood plays an integral part in the Mediterranean diet with many varieties of fish which can be used fresh, dried, grilled, pureed, as stocks for a main entree or to flavor other dishes. Bouillabaisse, zarzuela de pescado, bourride, soupe de poisson and tagine del hout are some of the popular fish dishes. Lamb and mutton are seasoned and grilled as kebabs in the Islamic regions, where eating of pork is forbidden. Pork and veal are popular in Spain, France and Italy.

Cheese such as feta, mozzarella, romano, gruyere, domiatti, manchego, ricotta, pecorino and Parmigiana (parmesan) are common in Mediterranean meals and are used for salads, with vegetables, in desserts or as toppings for main entrees. They are typically derived from sheep, buffalo or goat's milk. Yogurt is used abundantly in Middle Eastern and North African meals. It is served plain or with fruits and is used as a cooking medium, a thickener or a beverage. It is sauced over fried eggplant, mixed into cucumber soup and served alongside pollou.

Rice is served plain accompanied by stew or condiments, or as a one-dish meal (paella, pilaf or risotto) using arborio, patna or Spanish rice. Each Mediterranean region has its own preference for rices, whether sticky, firm, soft, separate, fluffy or creamed.

Couscous, a popular dish from Morocco and other regions of North Africa, is precooked pasta that takes on the flavors of other ingredients with which it is prepared. It is served as a side dish or a one-bowl meal and made into wonderful desserts with fruits and spices. It can be eaten with stews, fried chicken, meat kebabs, fish or seasoned vegetables. The presentation and preparation of couscous varies depending upon the country. In traditional Moroccan cooking, it is steamed (covered or uncovered) in the aromatic vapors of a meaty stew or broth.

Mediterraneans enjoy a variety of flat breads that are fermented, baked or fried and are used to soak up sauces, scoop up food or as a base to spread olive oil, anchovy paste or tapenade. The long French baguette, Italian focaccia, North African baladi and injera and Middle Eastern pita, barbari and tanoor are eaten with spicy condiments, yogurt dips, lentil salad, spicy lamb stew or clarified butter.

Broad beans, chickpeas, fava beans, flageolets, Borlotti beans, black-eyed peas, green and yellow split peas, red lentils and red kidney beans are the peasant foods of this region. They are prepared and seasoned in a variety of ways and used in salads, falafel, casseroles, antipastos, risottos, spreads, stews, soups and as purees. They are combined with vegetables, rice, wheat and meat to provide a complete meal, such as pasta e fagioli, cassoulet, ful medamus and cholent. Chickpea flour is used in a hearty vegetable soup for breaking the daily fast during Ramadan, a holy time of the year for Muslims. In the Middle East, navy beans, great northern, chickpeas and black-eyed peas provide body and texture to dishes of rice, noodles, stews and salads. In Ethiopia, lentils are seasoned with berbere seasoning and eaten with injera. Falafel, a street food in Israel, Egypt and Lebanon is made with white fava beans or chickpeas with spices and chile peppers and is served at breakfast, lunch or dinner. Hummus, made from pureed chickpeas and sesame, and bissara, a garlicky herb puree made from fava beans, are popular foods eaten in the Middle East.

Olives are ubiquitous in Mediterranean cooking, whether pickled, braised, roasted, pureed, to season foods, as garnishes or as a cooking oil. There are many varieties with different colors, shapes and sizes. Black kalamata, iconian and nafplion olives from Greece, Spanish manzanilla and Spanish Queen and the Italian green and Moroccan types have become regular items on our grocery shelves. Olives are served as appetizers with cheese, nuts, breads and chorizos. They are pickled and seasoned with lemon, vinegar, wine, garlic, fresh leafy spices or cayenne before they can be eaten. They can also be pitted or stuffed with anchovies, almonds or capers. Tapenade, an aromatic olive and caper spread, salad nicoise, pizzas, tagines and puttanesca sauce are many of the familiar items with olives.

Nuts, such as walnuts, almonds or pine nuts add crunchiness to poached fish, lamb patties and salads, thicken soups and are ground into flour for desserts, spreads and sauces or are steamed with couscous. They are compatible with many leafy spices, tomatoes, raisins, butter, lemon rind and chile peppers.

Fruits are eaten fresh, preserved in syrups or added to ice cream, candy, tarts and sorbets. There are a variety of fresh fruits, such as citrus, grapes, apples, figs, nectarines, peaches, pears, melons, pomegranates, persimmons, quinces and strawberries. Dried fruits such as figs, prunes, dates, apricots and raisins are mixed with nuts, raisins, spices and herbs and cooked with fish, pork or duck and used in stuffings for lamb.

Mediterranean foods offer a variety of vegetables, such as aubergine, artichoke, asparagus, savoy cabbage, cardoon, zucchini, courgette, fennel, kohlrabi, leeks, mushrooms, pumpkin, potatoes or radicchio. They are usually eaten raw, grilled, sautéed, deep fried, marinated, baked or steamed. Cucumber and yogurt salad, served as a mezze, is a popular item in eastern Mediterranean cooking and has many variations using garlic, mint or dill. Eggplant is popular in soups and dips, relishes or in Middle Eastern ratatouille. It is fried in oil, grilled, pickled, stuffed and baked, braised or pureed and seasoned with garlic, lemon juice, cayenne, paprika, basil, mint, oregano or yogurt.

There is great diversity not only with ingredients, but also with cooking styles, which vary among the Arabs, North Africans, Turkish and southern Europeans. Marination, grilling, pickling and braising are preparation techniques that create healthy, tasty side dishes with variety and excitement. Grilling techniques using aromatic woods are popular around the northern Mediterranean. Grilling creates wonderful aromas and flavors with vegetables, meats and fish.

Mediterranean foods are aromatic with fresh leafy spices such as mint, thyme, parsley, dill, basil, tarragon, chervil; peppery with cayenne, black pepper, clove and ginger; savory with coriander, cinnamon and caraway; nutty with almonds, pine nuts, pistachios and walnuts; and fruity with citrus, figs, pomegranates and nectarines.

There are numerous seasonings used with fish, meats, couscous, salads and appetizers. Some of the popular seasoning blends are pesto, sofrito, tapenade, harissa, picada, rouille, romesco, salsa verde, aioli, tarator and the many different tomato sauces. Garlic and olive oil are popular items used with a variety of ingredients such as cayenne pepper, parsley, cumin, paprika, eggs, walnuts, mint and bread to create unique marinades, sauces and dips.

Let's look into the regions of the Mediterranean, how they are different and their influence on overall Mediterranean flavor.

REGIONAL CUISINES OF THE MEDITERRANEAN

The northern Mediterranean includes Spain, southern France and the Italian regions.

Spain

Spain had influences from the Moors, Phoenicians, Arabs, Romans, Italians and French. Spanish explorers also brought peppers, chocolate, vanilla, potatoes and beans from the New World. The essential ingredients in Spanish cooking are olive oil, garlic, onions, olives, parsley, tomatoes, sweet red peppers and sweet bell peppers. Other typical flavorings in Spain include saffron, paprika, piquillo peppers, nora peppers, chorizos and mint, which are used in dishes such as paella Valenciana, stuffed piquillos, zarzuela de pescado, gambas al ajillo and gazpacho.

Spanish cuisine has many varying flavors and ingredients, depending on its regions. On the north coast (Galicia and Asturia), seafood, anise, ham and onions characterize its cooking; the north east (Catalonia) with Italian and French influences has various sauces—sofrito, sanfaina (onion, tomato, pepper, aubergine), romesco (dried red chile peppers, garlic, tomatoes, almonds), picada (pureed almonds, crushed garlic, parsley and bread crumbs) and allioli (garlic, olive oil). These sauces are popular with grilled meats and fish and are used as base sauces for rices and stews. Valencia has a Moorish influence with paellas, citrus and almonds. The Basque region, which neighbors France, has garlic, onions, dairy products and seafood. The south (Andalusia), with Arab, Roman and Phoenician influences, has given Spain its olives, olive oil and fried seafoods. Arabs brought saffron, nutmeg, mint, black pepper, almonds, citrus, honey and sweet and sour sauces to Andalusia. In the central region (Extremadura), the hearty cooking of the local Spanish nomads has contributed roast lamb, sausages, cured meats, Manchego cheese (from sheep milk) and pork with tomato and olive oil.

Codfish, tomatoes, onions, garlic and pimiento characterize the flavors of the Spanish Basque region; romesco, garlic and parsley the Catalonia region; saffron, oranges, almonds, chocolate, lemon peel and rice in Valencia; and olives, cinnamon and Spanish paprika/pimenton are typical of Andalusia.

France

France has many diverse flavors that vary with climate, terrain and regional history. The south and southwest of France, which includes the regions of Provence (with Roman, Greek and Italian influences), Languedoc (with Roman and Arab influences), Corsica and the Basque country, form the northern boundary of the Mediterranean. These regions tend to have lighter sauces than the north and other regions of France. Tomatoes, beans, garlic, olives, olive oil, white wine, anchovies, saffron, clove, sweet and hot peppers, red wine, figs, citrus and bouquet garni are popular flavorings. Many sweet leafy spices such as rosemary, thyme, marjoram, sage, juniper, oregano, fennel and savory provide aromatic flavor to the foods.

In Provence, soupe au pistou, sauce provençale, bourride, brandade, bouillabaisse, daube, rouille, tapenade, pissaldiere, soup de poisson, ratatouille and aioli are some of its popular dishes. In Languedoc, leeks, mutton, lamb, tomatoes, garlic and sweet peppers flavor cassoulet (white beans and meat stew), pot-au-feu and polenta, while ham, mushrooms, garlic, charcuterie and potatoes are popular foods in the French Basque region.

Corsica, with French and Italian influences, has hotter flavored versions of sauces and spicier tomato pastes than Provence. Their characteristic ingredients include hot peppers, pimientos, rosemary, juniper and anchoiade, hot sausages, polenta, game meats, raviolis and coulis.

Italy

The flavors of Italy vary with its terrain, climate and cultural heritage and with the effects of trade and conquest. Italian cooking is influenced by Greeks, Etruscans, French, Saracens, Germans, Austrio-Hungarians, Spanish, Chinese and Arabs. These cultures introduced tarragon, salted cod, rice, cream, butter, cayennes, chowders and sausages. Early explorers also brought tomato, pimiento, corn, vanilla, peanut, chocolate and potato from the New World.

In Italy, antipasto leads a typical meal. It is followed by the first course of rice, pasta or soup. The second course is the main dish of meat, chicken or seafood, accompanied by a vegetable dish (contorno). Condiments and sauces, such as red sauce, green sauce and béchamel sauce, are served with pasta.

Italian flavors vary widely around the country, with tomato an essential ingredient in the central and south regions, butter in the north, olive oil and garlic in most regions and hot peppers in the south. Pasta is eaten in all of the regions, but each region has its own flavors and methods of preparation. Olives (made into olive oil) and grapes (made into balsamic vinegar) are significant ingredients in Italian cooking.

While every Italian region does not border the Mediterranean, all of Italy's regional flavors have influenced Mediterranean cuisine.

In Liguria and coastal Tuscany, pesto is added to pasta, minestone soup and trofie (Genoa dumpling). Stoccafissa (codfish stew), stewed veal and seafood chowder are also popular foods. Leafy spices (sweet basil, thyme and marjoram), citrus, olive oil, potatoes and mussels are common ingredients.

In Lazio (Rome), marsala wine, ricotta, pecorino, porchetta, sage and marjoram flavor carbonara sauce, saltimbocca, veal scaloppine and chicken diavola. Typical foods from Emilia-Romagna, Veneto and South Lombardy are risi e bisi (rice with peas), osso bucco risotto alla Milanese flavored with mortadella, proscuitto, red wine, saffron, gorgonzola and creamy mascarpone, lasagna and tortellini. Piedmont and North Lombardy, with German, Austro-Hungarian and French influences, have simple hearty and rich foods such as potato dumplings, bollito (sauced meat and vegetable dish), stewed red cabbage, carpacio, bagna cauda, spinach dumpling and barley soup flavored with bacon, cheese, bread, white truffle, paprika, gorgonzola, horseradish and poppy seeds.

Inland Tuscany, Umbria and Le Marche have simple, light cooking with soups, seafood stews, bean salads, stuffed pork chops, spit roasted pheasant and lamb. Olive oil, ham, beans, sausages, tomatoes, olives, mushrooms and plenty of meat and vegetables predominate. Onions and garlic are essential flavorings along with fresh leafy spices such as rosemary, sage, bay leaf and fennel. Sardinia (also influenced by Greeks) has robust hearty foods such as stewed and roasted meats and game. Citrus fruits, figs, fennel, myrtle leaves, ham, salted pork, pecorino, ricotto and feta cheese flavor stuffed calamari, tuna or pork dishes.

In the Abruzzi and Apulia regions, meat, cheese and spicy chile peppers (pepperoncini) are added to fish chowders, orecchiete alla puglese and fettucine alla abruzzese. Campagnia (Naples) has spicy flavors with diverse pastas such as spaghetti alla puttanesca or spaghetti alla vongole. Basilicata has Greek-influenced pork and lamb sausages and vegetable dishes. Typical ingredients of these regions include basil, fennel, mozzarella, citrus fruits, vegetables, pork, sausages, onions and clams.

Calabria and Sicily (which have Asian, northern African, Arab, Spanish and Greek influences) use pickled vegetables, citrus fruits, sundried figs, capers, couscous, salt cod, cinnamon, nutmeg, raisin, pepperoncini, cheese, fish, lamb, tomato and eggplant. Caponata, stoccafisso and cabbage soups are popular.

Italians have basic stocks to flavor fish, meats, chicken or duck. Generally, garlic, onion, black peppercorn, bay leaf, fennel, thyme and parsley, tomato paste, carrots and celery are used with chicken bones, fish bones or beef bones.

The Middle East

The Middle East includes Israel and Arab regions (Lebanon, Jordon, Syria, Palestine, Gulf States and Yemen). The Middle East has been influenced by Arab, Iranian, Indian and European cultures and thus has varied ingredients. Generally, garlic, saffron, nuts, yogurt, sumac, fenugreek, parsley and mint are popular flavorings in all of these regions. Some regions prefer olive oil, while others use clarified butter (smen/samneh) from sheep or goat's milk. Other popular ingredients include sesame oil, olive oil, garlic, fenugreek, cinnamon, paprika, cayenne, coriander, cumin, cloves, nutmeg, fresh coriander, mint, dill, cassia, allspice, mahleb cherry, sesame, caraway, almonds, pine nuts, grape and vine leaves. Yogurt is a staple flavoring in all of these regions. Torshi or pickles, hilbeh, pickled lemon, tahini, kebabs, pilafs, garlic dips, kibbeh, kofta, cholent and lentil soups are also popular foods.

A popular seasoning that is essential in Middle Eastern cooking is a stock called yahni, which is made from onion, garlic, cardamom, baharat and clarified butter with lamb or mutton bones. These spice mixtures generally come dried. Zhug, zahtar and baharat spice blends are popular in most regions. Aromatic salt, which is sea salt combined with spices, is used as an essential flavoring in many Middle Eastern dishes.

Throughout these regions, there are ingredients and foods that have a mass appeal, but each region has its own specific flavor preferences. The southern Gulf States like hotter and curry-type flavors. The northern regions of Iraq and

Syria enjoy spicy, sweet and pungent spices such as pomegranate, Aleppo pepper and cumin. The west (Israel and Lebanon) has sesame paste, parsley and sour, milder flavors. The east (Iran) prefers aromatic leafy spices and almonds.

Some typical dishes are muhammara, a Syrian spicy hot dip that contains Aleppo pepper, walnuts, cumin, lemon juice and pomegranate juice; kofta which are meatballs made from lamb, baharat, aromatic salt, mint, cilantro and garlic; hilbeh, ground fenugreek seeds combined with garlic, olive oil, green chilies, tomatoes and lemon juice used as a spread on breads before it is baked; and talia, a garlic, onion and chile pepper seasoning used in Yemen.

Greece, Turkey, Armenia, Iran and Cyprus

Greece, Turkey, Armenia, Iran and Cyprus have quite distinct cultures, yet certain seasonings have mass appeal throughout these countries. Turkey, predominantly Muslim, uses garlic, Aleppo peppers, bay leaf, paprika, mahlab cherry, pomegranate juice or concentrate (pekmez), yogurt, walnuts, turmeric, vinegared chile pepper as essential flavorings for their kebabs, dips, soups, fish or meats.

Greece, influenced by Ottomon Turks and West Asians, uses oregano, garlic, olive oil, almonds, black olives, goat cheese, raisins, yogurt and capers as base flavorings for its simple cooking. Other ingredients include onions, pickled green chilies, Aleppo pepper, Boukovo pepper, vine leaf, bay leaf, sesame, mahleb cherry, cinnamon, anise, clove, parsley, dill and saffron. These flavorings are used in many dishes such as dolmas, tarator soup, skordalia sauce, taramosalata, moussakas, pilafs or trahanas.

Iran, with Arab, Aryan, Greek and Ottoman influences, combines sweet, savory and sour tastes using molasses, honey, lemon, pistachio, yogurt, almonds, parsley, dill, marjoram and dried fruits (figs, prunes, dates, apricots). Typical dishes are koresh (a creamy, sweet-sour lamb sauce with cloves, sumac, turmeric, cinnamon, dried fruits, smen and pickled lemon), aromatic pollou (which contains cardamon, cumin, cassia, saffron, clarified ghee and many leafy spices) and mahi now rooz (stuffed fish with spinach, smen, lime, nuts, raisins, cinnamon, mint and garlic).

North Africa

North Africa includes Tunisia, Morocco, Algeria, northern Egypt, Libya and Ethiopia. It has been influenced by many cultures through trade and conquest, including the French, Moors, Berbers, Arabs, Phoenicians, Sephardic Jews, Romans, Greeks and the Far East.

Most North Africans are Muslims, so pork and alcoholic beverages are taboo. Lamb and mutton are preferred meats, and coffees, spiced teas, fruit- and flower-based drinks are enjoyed.

Tunisian blends have spicy and hot profiles with turmeric, chile mixture, harissa and dried mint. Moroccan food is not as spicy and is famous for its couscous, lemons, paprika and tagine dishes. Egyptians enjoy falafel, white fava beans, zahtar, dukkah, sesame paste, cumin and coriander, while Ethiopians prefer sesame seeds, pippali, berbere and injera. The Berbers contributed the slow-cooked tagine dishes to North African cuisine. Pickled foods called cisbech are very common Algerian appetizers or accompaniments to foods and use spices, such as cayenne pepper, paprika, cumin, coriander or many herbs.

Grilled chickpeas, marinated olives, kibbeh, ful medamis (bean with mint, cumin, garlic and onion) and labneh (fried cheese) from Egypt; sohleb (millet porridge), brochettes and pickled sardines from Algeria; nit'ir qibe (spiced butter), kifto and doro w'ets of Ethiopia; and bistilla and tagines from Morocco are commonly eaten foods in north Africa.

Popular flavorings include saffron, cayenne, mint, rose petals, turmeric, roasted spices, paprika, black pepper, cinnamon, cilantro, nutmeg, cloves and sumac. Typical North African seasonings are tabil, chermoula, la kama, harissa, dukkah, ras-el-hanout, semit and berbere. Chermoula, popular in Morocco, is a paste made from garlic, olive oil, parsley or coriander leaf, red chile peppers, paprika, saffron and onion and is used for marinating fish, meat or poultry. La kama, which consists of ground black pepper, ginger, turmeric, nutmeg and cinnamon is also a popular flavoring in Morocco for stews, soups and salads.

Some of the typical Mediterranean spice blends are discussed below.

POPULAR MEDITERRANEAN SPICE BLENDS

North African Spice Blends

Harissa Blend

DESCRIPTION

Harissa is a fiery chile-based sauce that resembles sambals and is a fiery North African version of ketchup. It accompanies appetizers and many other dishes in North Africa. It has a garlicky, slightly sweet, fresh chile flavor. Tomato and red bell pepper tone the heat down. There are many different versions of harissas, depending on the region, Algeria, Tunisia or Morocco. Its other versions are called felfel sudani in Morocco and dersa in Algeria. The Tunisian version is the hottest of all three. It is often sold as pastes.

INGREDIENTS

Harissa has a deep red color with a smooth, thick consistency. It has dried chile peppers, tomatoes, garlic, ground cumin, ground coriander, vinegar and olive oil. Hotter versions omit the red bell peppers and increase the chile peppers. Other added ingredients are crushed mint leaves, caraway, lemon juice, cilantro or olives. For more intense flavor, bell peppers and chile peppers are roasted.

PREPARATION

Dried chile peppers are soaked in water until soft, and are then blended in a food processor or blender with all of the other ingredients to achieve a smooth texture. Enough olive oil is added to form a smooth, thick paste. Traditionally, a mortar and pestle are used to grind all of the ingredients into a smooth puree.

APPLICATIONS

As a table condiment, harissa enhances the flavor of entrees and side dishes. In Tunisia, it accompanies meat and fish and adds zest to salads and soups. It is scooped over couscous, rice or noodle dishes. Harissa goes well with grilled meats and is used as a rub for barbecues, as a topping for hot dogs or to add zest to stews, soups and sauces. In the Middle East, harissa accompanies mezzeh (called aadou in Tunisia and kemia in Algeria) or finger foods and dips for breads. In Tunisia, harissa is placed over a variety of foods and is even used with bread and butter for breakfast.

Ras-El-Hanout

DESCRIPTION

Ras-el-hanout is a Moroccan classic mixture of 20 or more spices. The word in Moroccan means "top of the shop." The shop owner mixes and creates a blend that is perceived as his or her ultimate creation. These vary in ingredients depending on the shopkeeper. All of these blends contain spices, dried flowers and some aphrodisiacs (cantharides).

INGREDIENTS

A typical blend includes cardamom, cassia, cumin, fenugreek, lovage, mace, cubeb, long pepper, allspice, nutmeg, ginger, rose petals, lavender blossoms, orange blossoms, Spanish fly, dung beetle, belladonna berries, monks pepper, grains of paradise, berries, black pepper, chile pepper, turmeric, nigella, ash rhizome, orris root, lesser galangal, chufa nuts, thyme and rosemary. Whole spices are generally mixed with the rose petals and leafy spices.

APPLICATION

It is used to enhance many dishes in Morocco and other neighboring regions. Ras-el-hanout is used on game meats, rices, couscous, lamb tagines and desserts.

Middle Eastern Spice Blends

Za'Atar Blend

DESCRIPTION

Za'atar is a blend of spices that varies in flavor and is from Turkey, North Africa, Lebanon and Syria. It is also the term for thyme in Arabic. In Morocco, it is referred to as a leafy spice that is a hybrid of oregano, thyme and marjoram. Za'atar spice blend can be mildly tart or very tangy depending on the amount of sumac added.

INGREDIENTS

Za'atar is usually a blend of ground thyme, marjoram, sumac and roasted sesame seeds. Pistachios, hazelnuts, olive oil, mint or garlic are added to create varying flavors. Sesame seeds are dry roasted for about 1 minute until aromatic and are then added to the rest of the ingredients and blended well.

APPLICATION

Za'atar is used as a seasoning for dips, to flavor breads and salads and is eaten with cheese. In Syria, it is mixed with olive oil and is used as a dip for pita bread. Za'atar is also used as a dressing in salads with other leafy spices such as parsley and mint. It goes well with lemon juice, garlic, black pepper, tomatoes and olive oil.

Zhug/Zhoug/S'Ug

DESCRIPTION

Zhug is a spicy, aromatic, rich, green blend of spices used in Middle Eastern cuisines. It can be called the "pesto" of the Middle East. Its flavor varies from country to country. Zhug originated in Yemen where it has a hotter flavor profile.

INGREDIENTS

Ingredients include coriander leaf, green chilies, ground cumin, cardamom and garlic. Jalapenos or serranos, mint, lemon and vinegar can be added to give varying or hotter tastes. The ground spices and garlic are fried in oil until they are aromatic and are then cooled, combined with other ingredients and blended into a paste.

APPLICATION

Zhug is a table condiment and is used to add zest to meats, vegetables, breads and salads or is used as a base sauce in stews and sauces. Zhug can be tossed over salads, used as a rub for barbecues and a spread or filling for sandwiches, bagels, pitas or dumplings or as a sauce over pasta or noodles.

Tahini Blend

DESCRIPTION

Tahini is a base blend or paste that adds zest to many cold and hot dishes of the Middle East, Turkey and Northern Africa, such as hummus, baba ghanoush, salad dressings, creamy sauces, soups, cooked meats and fish. Tahini is an oily, shiny paste with a very strong, roasted sesame flavor. It provides a strong, nutty flavor to soups, salads and sauces.

INGREDIENTS

Tahini is made from sesame seeds, sesame oil, garlic and lemon juice. Depending on regional preferences, ground chile pepper, paprika, cumin or clove, chopped parsley, mint, almonds or walnuts, yogurt or olive or hazelnut oil can be added with tahini. Tahini blends well with yogurt, olives, tomatoes, nuts and pomegranate.

In preparing tahini, sesame seeds are lightly roasted, then ground and mixed with the sesame or olive oil and other ingredients to make a smooth paste. Boiled chickpeas with water and the tahini paste are pureed in a food processor or blender to achieve a smooth or coarsely textured puree.

APPLICATION

Hummus is made from pureed chickpeas (garbanzo) and tahini paste, while baba ghanoush is made from pureed roasted eggplant and tahini. They are favorites throughout the Middle Eastern regions, from Israel to Syria to Egypt. Hummus is a very smooth to coarsely textured, thin or very thick puree. It is creamy, with a slightly sweet, nutty and roasted sesame taste. Depending on the type of sesame seeds used (white or brown) or how long seeds are roasted, hummus has different colors and flavor strengths. Hummus and baba ghanoush are typically eaten as dips with pita bread, as spreads or as appetizers. They accompany shish kebabs, vegetables or fish. They are served chilled or at room temperature, usually with a little paprika oil and chopped parsley.

French Spice Blends

Tapenade Blend

DESCRIPTION

Tapenade is a caper and olive paste from Provence, France. Its name is derived from the French word tapeno for capers. There are many versions of tapenade. Generally, it comes as a smooth and dark paste, with a salty pungent, sharp taste.

INGREDIENTS

Ingredients in tapenade include capers, garlic, black olives, ground black pepper, Dijon mustard, lemon juice and dark rum. Other ingredients are anchovies, bay leaf and herbs. Tomato paste, sugar or mashed cooked vegetables are added to tone down its bold flavor.

To prepare tapenade, olives, capers and/or anchovies are soaked to remove excess salt. They are then rinsed, drained and blended with the rest of the ingredients until a smooth, thick sauce is obtained.

APPLICATION

Tapenade is used as an appetizer or spread on toasted bread and topped with herbs and cheese. It is also served with boiled potatoes and cold fish and can be added to perk up blander foods.

Herbs de Provence

DESCRIPTION

Herbs de Provence is a mixture of fresh or dried French and Italian leafy spices. It has its origins in southern France. The blend has sweet, floral and slightly spicy notes. The Romans had a great influence in Provence, and, thus, the presence of Italian sweet basil and fennel seed.

INGREDIENTS

Herbs de Provence has sweet spices such as savory, thyme, rosemary, fennel seed, oregano and marjoram with lavender. Sweet basil and tarragon are the other spices added.

APPLICATION

Herbs de Provence is used as a rub to add a sweet, herb-like flavor to roast chicken, beef, cornish hen and pork tenderloins. They also perk up stews, baked vegetables and salads.

Rouille Blend

DESCRIPTION

Rouille is a garlicky, spicy mayonnaise-like condiment that is used to accompany fish-based sauces and stews in Provence, France. In French, rouille means "rusty," describing its reddish coloring. It is thick, smooth and spicy.

INGREDIENTS

Rouille's ingredients include garlic, roasted red bell pepper, dried hot red pepper, bread, olive oil, black pepper and fish stock. Sometimes, saffron, thyme, parsley, chervil or sage, paprika, cheese and tomato paste are added. One version of rouille uses mayonnaise as a starting base, then adds spices and other ingredients.

APPLICATION

To create rouille, garlic, peppers and bread pieces (soaked in water and squeezed dry) are blended with stock and water in a food processor or blender. The olive oil is slowly added until the mixture gets a thick consistency. Black pepper and salt are added, followed by the warm fish stock. The rouille mixture flavors fish stews and soups, especially the noted bouillabaisse. Rouille is also served in the French islands of the Caribbean, where it is spread over toasted bread, which is then topped with grated cheese and eaten with fish stew or soup.

Quatre Epices

DESCRIPTION

Quatre epices is a mixture of dried spices that is also called Parisian/Parisienne spice or epices fines. It is the most popular seasoning used in France, other parts of Europe and the French islands in the Caribbean. (In France, quatre epices is also the name for allspice.)

INGREDIENTS

The basic ingredients in quatre epices are white or black peppercorns, nutmeg, ginger, cloves, mace, cinnamon and allspice. Leafy spices such as bay leaf, sage, marjoram and rosemary are also added. All of these spices are then blended and pounded with the leafy spices added at the end.

APPLICATION

Quatre epices is used to flavor French charcuterie, slow-cooked meats and red cabbage dishes.

Bouquet Garni

DESCRIPTION

Bouquet garni is a mixture of leafy spices, especially the sweeter types. It flavors many European dishes, especially those requiring long simmering. Bouquet garni is a French term for a bundle of herbs. The types of herbs used and their proportions vary to complement particular dishes, such as poultry, fish, game meats or lamb.

INGREDIENTS

In a bouquet garni blend, the leafy spices that are added need to complement each other. The classic mix is parsley, thyme and bay leaf. Crushed garlic is sometimes included. In Provencal cooking, dried orange or lemon peel is also used. Other spices are added to complement specific applications. Thyme, tarragon, parsley, rosemary and bay are added for poultry; rosemary, mint, parsley and thyme are added for lamb; thyme, basil, marjoram and bay are added for pork; juniper berries, bay and savory are added for game meats; and sage, lemon verbena, dill and chives are added for fish.

APPLICATION

Fresh leafy spices are tied in bundles with a string, while the dried herbs are tied in a cheesecloth bag. These bundles are removed at the end of cooking or when sufficient flavor has been imparted but before bitterness is added to the application. Bouquet garni is used to flavor soups, sauces, stews, and meat dishes of Europe, especially France.

Spanish Spice Blends

Romesco Blend

DESCRIPTION

Called salsa romesco in Catalonia, Spain, it is a popular table condiment, dip or sauce used to perk up fish, pasta, vegetables and chicken. Its name comes from the mildly hot, dried red peppers that are the main ingredients in this sauce blend. It has a rich, smoky, nutty and spicy taste.

INGREDIENTS

Romesco is made from dried romesco pepper, onions, tomatoes, olive oil, garlic, red wine vinegar, roasted almonds or hazelnuts. Other ingredients that can be added are black pepper, salt, toasted white bread and seafood broth. All ingredients are roasted to give romesco its deep rich flavor. If more heat is desired, dried hot red chile peppers are added. There are many versions of romesco from the Catalan region in Spain.

To prepare romesco sauce, tomatoes, garlic (with the skin on) and onions (no skin) are roasted in olive oil. Then, chile peppers are boiled with vinegar. Tomato and garlic are peeled and deseeded. The vinegared chile peppers are added to the onions and roasted until the onions brown. All ingredients with their juices are blended together until a smooth puree is obtained.

APPLICATION

Romesco accompanies fish stews, other seafood dishes, salads and boiled vegetables. It can be used as a spread for breads, a dip for vegetables or added to pasta sauces.

Sofrito/Soffritto Blend

DESCRIPTION

Sofrito is an aromatic mixture of spices with tomato, sweet bell peppers and/or pimientos that are browned in olive oil and added to soups, sauces or stews. This principal sauce characterizes the Catalan cooking of Spain. The Spanish also introduced sofrito to the Caribbean islands, the Philippines, Italy and Egypt, which have other flavor versions. (Latin American versions are discussed earlier.) Sofrito comes from the Spanish word, "sofrier," which means to lightly fry. Called soffritto in Italy, its taste varies from slightly sweet to spicy and peppery. Soffritto is combined with pancetta (cured ham seasoned with black pepper, nutmeg, cinnamon, cloves and juniper berries) and garlic and onions. All are browned in olive oil and are used as a flavor base for seafoods, soups, vegetables and fricassees.

INGREDIENTS

A basic sofrito blend has tomatoes, onions, garlic, olive oil, pimiento and saffron (Table 35). Other Mediterranean versions can be created with bell peppers, chile peppers, annatto oil, butter, cilantro, celery, carrots, oregano, thyme, parsley, lemon and black pepper.

APPLICATIONS

To make sofrito sauce, onions are gently fried in oil until they become light brown. Then, deseeded tomatoes and/or other ingredients are added and gently stirred until the sauce becomes somewhat thick. Sofrito is used to flavor stews, rices, soups and beans, as a rub for chicken and fish and as a condiment. It is used as the starting base for many dishes and adds zest to dishes such as the famous Catalan seafood stew called zarzuela de pescado, paellas, soups or sauces.

Italian Spice Blends

Pesto Blends

DESCRIPTION

Pesto is a green blend of spices and herbs that is traditionally used in the cooking of Genoa, in Liguria (Italy). The traditional blend, pesto alla Genovese (or basil pesto) is sharp, pungent and fresh tasting. The French version, called pistou (in the French Riviera, adjoining Liguria), is influenced by the Genovese pesto but is much milder in flavor.

TABLE 35. Sofritos of the Mediterranean and Characterizing Ingredients.

Sofrito Blend	Ingredients
Basic sofrito	Garlic, onion, tomato, olive oil
Spanish sofrito	Sweet and mild with sweet bell peppers, onions, pimiento, olive oil and saffron
Italian soffritto	Mild and savory with onions, celery, carrots and parsley pancetta

INGREDIENTS

There are many versions of pesto. The essential ingredient is sweet basil. Pesto alla Genovese contains sweet basil, garlic, olive oil, pine nuts and cheese (pecorino or parmesan). Other versions add sundried tomatoes, parsley or sweet red bell peppers. The more recent versions add walnuts, almonds, green chile peppers, mint or cilantro.

APPLICATIONS

In Italy, basil leaves are deveined and ground in a mortar and pestle with garlic, salt and pine nuts. Olive oil is added drop by drop until the desired consistency is obtained. Finally, cheeses are ground and added. Nowadays, a food processor is used to make pesto. Pesto is typically served with gnocchi, tossed with pasta or stirred into soups, such as minestrone. Today, pestos are used as dips, flavoring for rice or a topping sauce for chicken and seafood.

Pasta Sauce Blends

DESCRIPTION

Italian sauces give their foods character. Endless varieties of these sauces are used for flavoring bland pastas. Each region of Italy has its own characteristic sauce. The heavier tomato-based sauces are more popular in the south, while some creamy sauces are common around Rome and the north. Emerging pasta sauces contain fire-roasted garlic, sundried tomato, sweet and sour flavors, saffron or Burgundy wine flavor. These sauces can be made into seasonings that can contain a tomato base, garlic, onion, vinegar, sugar, olive oil and many spices such as basil, thyme, parsley or sage.

INGREDIENTS

A basic cooked pasta sauce blend has tomatoes, olive oil, garlic, black pepper and basil. Sugar and vinegar are added for a sweeter or more acidic preference. Other sauces contain ingredients that are generally preferred on a regional level, such as cayenne pepper, capers, anchovies, mushrooms, cream, meats, clams, squid, spinach, carrots, peas or pancetta.

APPLICATIONS

These sauces are generally applied over differently shaped and sized pasta, such as spaghetti, penne, fusilli, orecchiete, tagliatelle, angel hair, fettucine, linguine or pappardelle. They are added fresh or cooked. Most pasta sauces are cooked, but there are freshly made uncooked sauces that are most frequently eaten during the summer.

Generally, in making a pasta sauce, garlic is sautéed first in olive oil, then tomatoes are added and finally, the other ingredients are added. Cheese is usually added before serving. The seasonings are simmered. These sauce blends can come as pastes or dried mixtures that are then added with vinegar, wine or cream to give varying flavors for a finished bottled sauce.

EXAMPLES OF CLASSIC PASTA SAUCE BLENDS

Aglio e oilo, the most basic pasta sauce blend, contains olive oil, garlic and parmesan cheese. Other ingredients such as parsley and red pepper flakes are optional.

Bolognese sauce blend is a creamy meat sauce from Emilia-Romagna, containing ground beef, meat broth, sausage, tomatoes (for color and flavor), olive oil, butter, white wine, porcini mushroom, onion, garlic, parsley, sage, bay leaf, celery, carrots and Parmigiano reggiano (parmesan cheese).

Puttanesca sauce blend is called the "style of the prostitute," because Italians say that the sauce is spicy, like the ladies of the night. Another says that these ladies attracted their clients with the sauce's enticing aromas. A third mentions that the sauce was quickly prepared as a fast food and eaten in between clients. It is an aromatic, pungent sauce that contains olive oil, tomatoes, garlic, black olives, anchovies, capers, parsley, fresh oregano, crushed red peppers and black peppers.

Marinara sauce, popular in the south, is a sweet aromatic sauce that contains tomatoes, basil, oregano, garlic, black pepper and olive oil. Olives and capers are optional ingredients.

Arrabbiata sauce is a very hot, spicy, deep reddish orange sauce, from the far south, that has tomatoes, olive oil, red cayenne pepper, pancetta, garlic, basil and pecorino romano. It is usually tossed over penne.

Alfredo sauce is creamy with butter, heavy cream, Parmigiano reggiano, garlic, nutmeg and black pepper.

Carbonara sauce is a mild and creamy sauce that contains pecorino romano (or Parmigiano reggiano), white wine, pancetta, olive oil, butter, eggs, parsley and black pepper.

Vongole sauce contains clams, olive oil, white wine, garlic, parsley and red pepper flakes. Tomatoes are added if red wine is used instead of white wine.

CARIBBEAN SPICE BLENDS

Caribbean cooking is a beautiful collage of tropical ingredients with varied cultural influences. It blends indigenous ingredients grown by the Caribs and

Arawak Indians with ones imported by the different peoples who settled there, including Africans, Asian Indians, Arabs, Chinese, Spanish, French, Dutch and British. This cultural diversity has resulted in a popular fusion cooking with a wide range of tastes, textures and colors that differ from island to island. Some Caribbean products that are becoming more familiar in the U.S. are jerk seasoning, hot pepper sauces (Solomon Gundy, Jonkanoo hot pepper sauce, coconut run down sauce) and curry colombos (also called poudre de colombo).

REGIONAL CUISINES OF THE CARIBBEAN

The Caribbean islands are in close proximity, and they draw ingredients from spice, fruit and vegetable sources common to all of the islands, with certain ingredients and seasonings having mass appeal. At the same time, distinct taste preferences reign on each island. What makes each island's flavors distinct is the way food is seasoned and prepared. Cassava, habaneros and rice and beans are widely eaten in all of the regions, but diverse preparation techniques create variations on the theme. For example, rice and beans vary in flavor from Puerto Rico to Guadeloupe to Jamaica to Trinidad, depending on the type of beans, seasonings and added ingredients.

Through the years, authentic profiles have been modified to suit changing tastes. The Africans have influenced the cooking with groundnuts, cornmeal and oxtail; the Asian Indians have influenced it with curries, rotis and pigeon peas; and the Amer-Indians have influenced it with habaneros, cassava and annatto. Also, each island's flavors reflect the tastes of its European colonizers. For example, on the French-speaking islands of Guadeloupe and Martinique, bouquet garni, chives and cream-based sauces predominate; in the Spanish-speaking islands of Puerto Rico, the Dominican Republic and Cuba, cilantro, olives and garlic are commonly used; while in the British-speaking islands of Jamaica, Trinidad and Barbados, curries, chutneys and jerk foods are enjoyed; and in the Dutch islands of Curacao and St. Maarten, nasi goreng, satays, meat with Edam cheese fillings and ristaffels are popular.

The eastern Caribbean—Trinidad, Barbados and the French islands—generally eat hotter foods than the western Caribbean—Puerto Rico, Cuba and the Dominican Republic—except for Jamaica. Appearance is an important factor in Caribbean cooking. Colorful, sculptured fruits, chilies, flowers and vegetables sit on a backdrop of rice and beans, grilled chicken, fish and lobster and add visual pizazz to our plates.

Caribbean cooking is based on hot, fruity sauces, coconutty flavors, seasoned rices and marinated meats. Fruits, vegetables, seafoods and chile peppers are in abundance throughout the year and are easily available as fresh ingredients that add flavor, texture and color to foods.

Soups and stews such as sopito, conch chowder, bebele, pepperpot, quimbomba, caldo gallego, sancocho and pumpkin soup are staples of Caribbean diets. They are often made with meats and seafood (including leftovers) and are spiked with spices and hot peppers. Tubers or vegetables lend bulk, frequently making the soup or stew into an one-pot meal that is typically eaten with rice or breads.

Dasheen, pumpkin, breadfruit, yuca, plantains, callaloo, okra and chayote are added to create body and textures in stews and soups. Chayote, cho cho or christophene, a mild-flavored vegetable found in most regions, can be stuffed, steamed, sauced or added to soups and stews. Okra, also called gumbo or quimbombo, is traditionally used in dishes of African or Amer-Indian influence. Breadfruit has a starchy pulp that is boiled, grilled or roasted and added to soups and stews. Beans, such as pigeon peas (also called gandules or gungo peas), are curried and used in fillings for rotis or pastries, added to rice and peas or used in stews and soups. Tubers play a prominent role in Caribbean cooking, such as yuca (also called manioc or cassava), boniato, sweet potatoes and yams that are added to provide bulk and texture to soups and stews.

There is an abundance of fruits in the Caribbean. Tropical fruits such as mango, papaya, guava, soursop, citrus, cherimoya, passion fruit, tamarind, pineapple and ackee flavor nectars, preserves, cordials, relishes, hot sauces and stews. Ackee, which is often served with saltfish for breakfast in Jamaica, has the texture of scrambled eggs. Papaya, called paw paw, is used in stews, soups and salads when green. Coconut is a versatile ingredient in these islands' cooking. It is made into snacks, seasoned with allspice, clove, nutmeg and cinnamon, turned into a refreshing beverage or used as cream in curries, soups, stews, breads and desserts. Plantains (sometimes called platano macho), which are larger than their banana cousins, are popularly eaten as a side dish with rice and beans. When ripe and sweet, they are thinly sliced and fried or when green and mealy, cut thin for tostones. When ground, they thicken sauces and are used in breads and cakes. Plantain leaves are used to wrap meats, rice, cassava and desserts before cooking or serving.

Seafood, whether pickled, salted, sauced or fried is a common part of the diet in the Caribbean. Each island offers its own seafood specialty, curried, grilled, stewed, fried or served with hot condiments, such as conch fritters, fried flying fish, feroce de morue (salt cod with Scotch bonnet and lime juice), blaff du poisson (with bouquet garni, white wine, habaneros and allspice). Pork is a popular meat, fried, stewed, baked or processed into a sausage called boudin. Creole pork chops, Jug jug (a Barbadian dish with pigeon peas, thyme and habaneros) and a Caribbean take on Scottish haggis are popular dishes. Goat, especially curried, is frequently eaten in Trinidad and Jamaica. Beef is popular in the Spanish-speaking islands and is added to ropa vieja, beef al ajillo and picadillos. Chicken is made into poulet colombo, jerk chicken, calypso chicken, lemon chicken or Bajan chicken.

Cornmeal is made into many products such as funchi (dumplings with spices), coucou (with okra and spices) and tutu with black-eyed peas or is wrapped in banana leaves and boiled. Rice is popular on all of the islands, but flavors vary depending on the island. It is cooked with onions, salt cod, ginger, soy sauce, sofrito, adobo, curry seasoning, beans or peas to make Cuban Moros y Christianos (rice and black beans), Jamaican peas and rice, nasi goreng, Chinese fried rice or paella.

The spices commonly used in Caribbean cooking are allspice, thyme, black pepper, clove, nutmeg, cinnamon sticks, ground mustard, celery, turmeric, clove and garlic. Bay leaf, cilantro, recao leaf, lemongrass, chive and a variety of other fresh leafy herbs season foods and local drinks. Seasonings such as adobo, sofrito, mojo, curry, rouille and jerk create magic with fish, fried chicken, pork chops, stews, soups or sauces.

Indian curry powders and garam masalas found their way to most Caribbean islands. They are available in powder and paste forms and are mostly consumed in Trinidad, Jamaica and other islands where there is a greater Asian Indian population. Their flavors are generally mild and, depending on the islands' preference, come with added green mangoes, hot peppers, vinegar, vegetables or coconut milk. Curry seasonings often flavor dishes made with goat, lamb, fish, shrimp, conch, pork or chicken, and which are served with white rice or as fillings in rotis, the flat breads of the Caribbean. Curries called colombos in the French islands of Martinique and Guadeloupe are heavily laden with turmeric, ground mustard and vinegar that add a sweet and slightly sour taste.

Jerk barbecue, a classic Caribbean item that has captured the U.S. market is from Jamaica. Escovitch (or escabeche), a tart, hot and spicy marinade for fish, is derived from the Spanish and is an integral part of Puerto Rican and Jamaican cooking. Another popular sauce is Creole sauce, which comes from the French-speaking islands but is also popular elsewhere. It is added to steaks, hamburgers, grilled chicken and fried fish.

Hot pepper sauces are integral to Caribbean cooking. The first hot sauce was created by the Arawak and Carib Indians, the original settlers of these islands. Today, many varieties are becoming better known, with tangy, fruity, sweet, sour and fiery notes enhance barbecued steaks, grilled fish, chicken wings, stews, curries, soups and vegetables. Each island has its own preference for its hot sauce, using habanero peppers, bird peppers, cachuchas, rocotillos or pimientos as the principle ingredients. Hot sauces such as salsa ajillo, sauce chien, sauce Creole, rouille, mango chutneys, red pepper remoulade, sauce ti-malice, sauce picante or ajilimojili are some of the local popular sauces. Pickled and chopped Scotch bonnets with sliced onions and spices are important table condiments.

Below are some popular spice blends from the Caribbean.

POPULAR CARIBBEAN SPICE BLENDS

Jerk Seasoning

Description

Jerk seasoning is a hot, spicy marinade for pork, chicken and fish and is part of a Jamaican tradition of marinating and slowly grilling seasoned meat, pork or fish over a fire pit of pimiento wood. The technique of jerking came from the Maroons, the runaway Jamaican slaves who escaped the British in the 1600s. They hunted wild boar and pigs, which they preserved and cooked in deep pits lined with pimiento wood and leaves. The smoke drying was a way of preserving the meat. Jerking the meat from side to side was a method of texturizing it.

Ingredients

Jerk marinade consist of allspice, thyme, nutmeg, cloves, scallions and Scotch bonnet peppers.

Applications

Jerk seasoning is used as a marinade for meats, chicken or seafood or as a sauce to perk up boiled or grilled dishes. It is sold as a dry mix or a wet paste.

Escovitch/Escabeche Blend

Description

Escabeche is the Spanish word for a pickled product and has its origins in the Spanish-speaking islands. It is commonly used in the Spanish-speaking islands of Puerto Rico, Dominican Republic and Cuba and is also an integral part of Jamaican cooking. Escovitch is the Jamaican version of the blend that is tart, hot and spicy. This seasoning is generally used as a marinade for fish, chicken and pork.

Ingredients

Escabeche consists of pimientos, black pepper, onions, garlic, vinegar and Scotch bonnet peppers.

Applications

Fish, chicken, pork and other seafood or game meat are marinated with escovitch and are also simmered in this marinade. Sometimes, the whole fish is deep fried before it is simmered in the marinade. It also seasons tamale fillings and vegetables.

Creole/Criollo Sauce Blend

Description

Another popular Caribbean sauce is Creole or Criollo sauce, which is a cooked tomato-based sauce. It is popular in the English-, French- and Spanish-speaking Caribbean Islands and has flavor contributions from many ethnic groups, including Africans, Spanish, Chinese, French and North American Creoles. Where the islands have Spanish origins, this sauce is called Criollo sauce, and in the French-speaking or other islands, it is called Creole sauce. It has a spicy and slightly sour and sweet flavor. It is a spicy version of the Chinese sweet and sour sauce. Its heat varies with different regions.

Ingredients

Creole/Criollo sauce contains onions, tomatoes, hot pepper, vinegar and black pepper. Other ingredients such as bell peppers, tomato puree, ketchup, Worcestershire sauce, white wine, chicken stock, scallions, celery and lime juice are added, depending on the island preferences.

Applications

To make the sauce, onions are sliced and sauteed, then sliced tomatoes, hot peppers and spices are added and cooked until the sauce becomes slightly thick. Then, vinegar is added, and the whole mixture is simmered. The finished sauce is spooned over steaks, grilled chicken, pork, lobsters, crayfish, hamburgers and fried fish or is served as an accompanying condiment.

Hot Sauce Blends

Description

The Caribbean has a reputation for its tangy, fruity, sweet, sour and fiery hot sauces. Hot sauce to a Carib is like ketchup to a North American or sambal to a Southeast Asian. Hot peppers have been relished in the Caribbean throughout history and come in many varieties. Hot sauce can be aromatic with hints of apricot, peach, leafy spices and citrus or pungent with garlic and mustard.

The Carib and Arawak Indians, the original settlers of these islands, used hot pepper juice with cassava and spices to season their foods and developed the first Caribbean hot sauce.

Ingredients

The popular hot peppers in the Caribbean are the Scotch bonnets, bird peppers, rocotillos, pimientos and cachuchas. Scotch bonnets (also called Congo peppers, Bahama mama, Bonney peppers, Jamaican hots and pimient) form the principle ingredients of the hot sauces. They come in many colors and sizes, are aromatic and have hints of apricot, peach and citrus. Other ingredients typically used in the Caribbean hot sauces are mango, papaya, vinegar, chives or tomatoes with allspice, mustard, turmeric or black pepper.

Each island has developed its own preference for hot sauce. Barbados has its leafy spice and mustard-based sauces, Guadeloupe makes a sauce with its chive and garlic bases, while Trinidad is well known for its Scotch bonnet and papaya-based sauces.

Pickapeppa sauce, Matouk's calypso hot sauce, West Indian hot sauce, flambeau sauce and Busha Browne's hot sauce, each claiming to be hotter and tastier than the other, are popular condiments in the U.S. Pickapeppa is a hot steak sauce with tomatoes, onions, sugar, vinegar, peppers, mangoes and spices. It is a spicier version of A-1 sauce used in the U.S.

Applications

Caribbean hot sauces can enhance barbecued steaks, grilled fish, buffalo wings, stews, soups, pastas and vegetables. Pickled, chopped chile peppers, sliced onions and spices are important local table condiments. Pickapeppa sauce is used for basting, grilling, sautéing and stewing meats, seafood and vegetables. There are other numerous varieties of hot sauces on the market that include West Indian hot sauce, Matouks's, Solomon Gundy, Jonkanoo hot pepper sauce and calypso sauce, each one claiming to be hotter than the other.

Curry Blends/Colombo Blends/Garam Masala

Description

Asian Indians have had a significant influence in the Caribbean, especially in Trinidad, Jamaica, Guadeloupe and Martinique. When they emigrated to the Caribbean, they brought with them the curry blends of India, which were then modified using locally available ingredients. Garam masalas in the Caribbean are mild, do not usually contain turmeric or chile powder and are added with curry blends to curries. Colombos or poudre de colombo are fiery aromatic local curries in the French islands of Martinique and Guadeloupe.

Curries can be mild or hot and spicy, depending on the islands' preference. They include spices familiar to the islands, such as allspice, thyme, black pepper, onions, mangril, ginger, Scotch bonnets, green mangoes, hot peppers, vinegar, vegetables or coconut milk. A general curry powder contains cumin, coriander, turmeric, fenugreek, celery seed and fennel seed.

Poudre de colombos are heavily laden with mustard, turmeric and vinegar, yielding a sweet and slightly sour taste. They are generally sold in powder or paste forms.

Ingredients

A typical Caribbean curry blend contains coriander seeds, cumin seeds, poppy seeds, cloves, brown mustard seeds, anise seed, fenugreek, black peppercorns, ground turmeric and ground ginger. Scotch bonnets or Congo peppers are added for heat. Generally, the masalas do not have ground chilies. They are added to the curries toward the end of cooking or are served as a condiment. Spices can be in whole or ground form. Colombos have mustard seeds, toasted rice, hot peppers, coriander, turmeric or saffron, black pepper, garlic, onions, tamarind and/or vinegar. White wine, stock, coconut milk, eggplant, chayote, papaya, pumpkin, mangoes, pumpkins, okra, green tomatoes, taro, yam and lime juice are optional ingredients. Coconut milk is not added to fish or shellfish colombos.

Most whole spices are roasted and then ground. Chicken pieces are seasoned with curry blend and browned. Then, water, onions, tomatoes and other vegetables are added and simmered until they are cooked. Caribbeans usually add vegetables to curries.

Most Trinidadian curries are mild and contain ghee, turmeric and mangril but usually lack chile peppers. Those who want heat in their curries use hot sauces. In Jamaica, where curry goat is a specialty, curries are generally very hot and contain a good amount of black pepper, Scotch bonnets and allspice. Roucou (annatto oil), cilantro and vinegar mustard flour are some of the other added ingredients.

Applications

Caribbean islands are noted for their goat curries. Other ingredients cooked with curry are conch (lambi), fish, shellfish, chicken, lamb, pork, shrimp, beans and many vegetables. These dishes accompany boiled white rice or are used as fillings for the famous Caribbean flat breads, called "rotis," that are popular street foods. Curry blends are also added to barbecues, soups and stews. Curries are accompanied by mango chutney, grated coconut, salad and/or fried plantains.

AFRICAN SPICE BLENDS

Many groups, including Arabs, Europeans, Asian Indians and other Asians, through wars, colonization, slavery and exploration have influenced the indigenous cuisines of Africa. For purposes of discussion, this section has divided this vast land into four geographic sectors, North, East, West and South Africa, each with foods that have quite distinct flavors. Curry blends, egusi seeds, bitter leaf, pumpkin leaf, cassava, beans, okra, fenugreek, mace, cloves, groundnuts, rose petals, grains of paradise, cornmeal, cassava, palm oil and waterblommetjie are some typical ingredients used in African cuisine. Starch based meals and stews are the staples of most regions of Africa.

REGIONAL CUISINES OF AFRICA

North Africa

North African cuisine is influenced by the Berbers, Coptics, Greeks, Arabs, Jews, southern Europeans and Middle Easterners. This book classifies North African cooking under Mediterranean. Because North African cuisine follows a Mediterranean-style diet, olives, spicy sauces, fish, leafy spices, legumes and couscous are used abundantly. Tagines, stews, breads, couscous and condiments characterize the cuisine.

West Africa

West African cuisine is influenced by Africans, English, Arabs and French. Starch-based meals made from rice, breads, cassava (gari), potato, yams, millet and plantains predominate. Ground cornmeal (foofoo), grated roasted cassava porridge (gari), roasted corn flour (ablemanu) or moinmoin are served with spicy stews and sauces. Stews form the foundation of the meal. They are mostly made from beans and root or leafy vegetables, with little meat, poultry or fish. They are flavored with peanuts, red cayenne, peanut or palm oil, black pepper or cubebs. Except in South Africa, meats are used more as flavorings. Soups are thickened with okra and nuts and are also eaten with a starchy base. Sauces containing hot peppers, groundnuts, tomatoes or seafood are served with boiled roots, plantains and vegetables.

East Africa

East African cuisine is influenced by Africans, Asian Indians, Middle Easterners, Persians, Arabs, Dutch, English and Portuguese. The Indian influence is predominant in the local cuisine and comes mainly from Ismaili Muslims and Gujeratis, who are mostly vegetarians. Curries, kababs, pilaus, samosas, flat breads, stews, mashed root vegetables and rice are essential foods. Typical ingredients are lentils, ghee, mint, coconut milk, masalas and hot peppers. Ugali (cornmeal porridge), okra stew, groundnut stew, matoke or yams are typical foods of the indigenous population.

South Africa

South African cuisine is influenced by Europeans, Africans, Arabs, Indonesians, Malays from Indonesia and Asian Indians. A popular style of cooking is Cape Malay cuisine, a melting pot of flavors from Malays, Dutch, French Huguenots, indigenous Khoisan, Asian Indians and Portuguese. Cape Malay dishes include bobotie (sweet curried meat casserole), sosaties (sweet and pungent grilled lamb marinated with curry powder and coconut milk), bredies (vegetable/legume stews), beeriani (rice-lamb dish flavored with ginger, tomato, fennel seed, saffron, cumin and onion) and atjars (spicy, sweet pickled condiments). Curried dishes are served with chutneys and sambals. Cornmeal is made into pap or putu and served with seasoned stews. Common flavorings are curry powders, garam masalas, clove, allspice, black pepper, coriander, tomatoes, chile peppers and bay leaf.

POPULAR AFRICAN SPICE BLENDS

Curry Blends

Description

Curries in Africa are influenced by Muslim Indians, Malays and vegetarians from India. Different regions have their own curry blends influenced by different cultural groups. They can be hot, medium or mild and dry or soupy. Some of the different types of curry blends include frikkadel in South Africa, kimasin in East Africa and kefta tagine in North Africa. South African curries, influenced by Malays and Indians, are hot, spicy and sweet with dried fruits and apples. Berbere, Cape masalas, qalat dagga, ras-el hanout and Malawi curry powder are some examples of African curry blends.

Ingredients

There are many types of curry blends, especially in East and South Africa where they are popular. Usually, these curry blends contain cumin, coriander, clove, cinnamon, turmeric, ginger, fennel, cardamom, mustard, black peppercorn and fenugreek, with or without ground chile pepper. East African curries, such as kima, have chopped beef, red chile pepper, curry spices, eggplant and coconut. Mozambique's curries use piri-piri peppers and nuts to provide flavor and consistency. Ethiopians have their own versions of curries called w'ets with ajowan, pippali, mint, sweet basil paprika and tomato. In West Africa, groundnuts are added to give some crunchiness to their curries, and in South Africa, dried fruits, ginger and bird chilies flavor kormas, murghi kalya and atjars.

Applications

In Africa, curries accompany cooked rice, couscous or breads. They are used with lamb, poultry, fish, casseroles, vegetables, pastries, samosas, soups, stews or chutneys.

Berbere

Description

Berbere is a spicy, classic blend that resembles the curry blends of India. It flavors many Ethiopian and Eritrean mutton and lamb dishes, especially traditional stews called wats or w'ets. It is pungent and aromatic, with slightly sweet and peppery to very peppery notes.

Ingredients

Berbere has ajowan, cardamom, black cumin, fenugreek, cardamom, cinnamon stick, cloves, ground ginger, garlic, onion and ground dried chile powder. Other ingredients such as black pepper, long pepper, coriander leaf, tomato, mint or olive oil can be added. Basil, paprika, mustard, turmeric, nutmeg and allspice go well with it.

Applications

To make a berbere paste, the spices are first dry roasted for a few minutes until they become aromatic. Then, chile powder is added, and the mixture is dry roasted again for about a minute. Garlic and onions are blended with the other ingredients until smooth.

Berbere is an indispensable seasoning of Ethiopian dishes, especially the curry-like, stewed dishes called wats, which are the mainstay of their meals. Berbere is used to flavor lamb stews, chicken stew (doro wat) and vegetable or lentil stews (shiro wat). Muslims eat shiro wat during their fasting month of Ramadan.

Bebere primarily enhances fish and lentils. They are eaten with local sourdough flat bread called injera, which is made from a grain called teff. Berbere is also used in a condiment that accompanies grilled meats or poultry and is also added to soups. Berbere pairs well with salted fish, tomatoes, rice, potatoes, yams, butter, lemon juice and peanuts.

Piri-Piri/Pili-Pili/Peri-Peri/Peli-Peli

Description

Piri-piri is a slightly sweet, hot mixture that characterizes Mozambique's cuisines. There are several versions of it, depending on whether it is made in West Africa, Angola in South Africa or Mozambique. Piri-piri is also the name of a small, red, fiery capsicum pepper, which resembles bird pepper. Today, piri-piri refers to any hot spicy blend made from fresh or dried ground piri-piri pepper. It is used as a wet marinade, a dry rub, a table condiment or a hot dipping sauce.

Ingredients

Piri-piri is made from garlic, onions, ground red piri-piri peppers, hot paprika, bay leaf, parsley, black pepper, olive oil and lemon juice. Other versions have added cardamom, basil, ginger, butter and coconut milk. It pairs well with peanuts and shrimp paste. The West African version has tomatoes, onion, garlic, habanero pepper and prepared or grated horseradish.

Applications

Piri-piri is used as a marinade or rub for cooked, broiled, charcoal-grilled or barbecued chicken, fish, prawns and shrimp. To make a piri-piri marinade or rub, simply blend all of the ingredients. Piri-piri is also used to add extra flavor at the meal table. It gives added zest to stews, vegetables and cooking sauces. For use as a table condiment, blend all ingredients and cook gently for about five minutes. Coconut milk can be substituted for lemon juice.

In West Africa, piri-piri is used as a condiment and cocktail sauce and as a flavoring for grilled meats or sandwiches. Mayonnaise flavored with piri-piri sauce is used for cold chicken salad and coleslaw.

Atjar Blends

Description

The South African term "atjar" came from the Indonesian Malay word "acar" that originated from the Indian word "achar." Achar translates to spicy pickled vegetables or unripe fruits in oil or sauce. It is crisply textured, and its flavor varies depending on its ingredients. Generally, it is spicy and sweet. The South African favorite is a lime atjar, which is a sweet, spicy type that contains olives, mangoes and chile peppers. It is an essential flavoring in the

Cape Malay cuisine of South Africa. It is also used as a table condiment for the Indian-influenced cuisine of East Africa.

Ingredients

Atjar was originally made with fish oil and brine, but today, olive oil, peanut or other vegetable oils are added with brine. A general atjar marinade has anchovy fillets, garlic, ground fenugreek, turmeric, chile peppers, curry powder and vegetable oil. To prepare a lime atjar, limes, salt, mustard seeds, bay leaves and vinegar are combined. Other variations come with green mango, onions, cauliflower or green beans.

Application

Atjars are used to perk up entrees of fish, meats or rices. Vegetables such as green beans, okra, carrots, cucumber and brinjal can be atjared. They are served as accompaniments to rice, curries, fish, vegetable or meat dishes.

Dukkah (Also Called Za'Tar Or Do'Ah)

Description

Dukkah is a blend of spices, roasted chickpeas and nuts and is a staple flavoring in Egypt since ancient times. It has a spicy and nutty taste with a coarsely ground or crumbly texture. It usually accompanies breads on the mezze (or tapas-like) table. There are many versions of dukkah in the Middle East.

Ingredients

Dukkah is a dry mixture of coriander seeds, cumin seeds, lightly roasted sesame seeds, black peppercorns, mint leaves, unsalted roasted hazelnuts or peanuts and dried roasted chickpeas. Other ingredients added are thyme, savory, oregano, marjoram or savory to give varying tastes.

Application

To make dukkah, whole spices are roasted, cooled, mixed with the remaining ingredients and pounded in a mortar and pestle or blended in a food processor. It is eaten for breakfast with breads, as an appetizer or on the streets as a snack. It is sprinkled over many main entrees to add texture and flavor. In Egypt, a bread called aish is dipped into olive oil and then into this blend before it is consumed.

West African Hot Pepper Blends

Description

West African hot pepper blends are a fiery mix of peppers with other ingredients and are used commonly to add heat or spike West African fish, meat or vegetable dishes. They come as dry mixtures or as pastes.

Ingredients

A common hot sauce has chile peppers, onions, palm or peanut oil and dried shrimp. There are other variations based on applications and preference. A popular dry hot blend is made from black peppercorns, white peppercorns, green peppercorns, pink peppercorns, allspice berries, cubebs, ground ginger, grains of paradise, red cayenne pepper flakes and ground dried habaneros. Ata, a Nigerian hot sauce, is made from onions, red chile peppers, red bell peppers, tomatoes, dried shrimp and palm oil.

Applications

Hot pepper blends are used as rubs on meats or chicken before they are grilled. They are also used as a flavoring base for stews, soups and sauces and as condiments for boiled, stewed or fried root vegetables.

NORTH AMERICAN SPICE BLENDS

Prior to the arrival of Spanish, French and English explorers, Native Americans grew chile peppers, sassafras or allspice to season their wild rice, corn, venison, squash, beans, buffalo meat and other foods. Early European immigrants brought cinnamon, nutmeg, clove, mustard and molasses to the Americans. The African slaves brought groundnuts, benne and many West African ingredients. All of these influences created traditional American cooking. Poultry, old world, barbecue style, Chesapeake Bay, pumpkin spice and Old Bay seasonings are some traditional seasonings used for shrimp and crab boils, roast beef, steak, pork chops, pumpkin pie and pickled vegetables. The later immigrants from northern, central and southern Europe and East Asia contributed their own seasonings and foods. The Jews, Germans, Italians, Hungarians, Japanese, Chinese and Koreans gave olive oil, sesame, sweet basil, ginger, soy sauce, garlic, pizzas, pastas, bagels, coleslaw, dumplings, cheese, chow mein and fried rice. The latest migration of Latin Americans, eastern Europeans, South Asians, Southeast Asians, Caribbeans and Middle Easterners have brought cilantro, lemongrass, curry powders, sumac, habaneros, lentils, adobos and tropical fruits.

While North American cooking combined many different flavors, no uniform cooking style arose. Rather, many regional cuisines emerged that reflected the tastes of the local populations. In the Midwest, Scandinavians, Germans and Eastern Europeans contributed dairy and baked products, hearty soups, sauerkraut, frankfurters, schnitzel, sausages, bratwurst and cheeses. The Southern states had the most diverse flavors derived from Africans, Europeans, Native Americans and Mexicans, such as gravy, chili con carne, barbecued ribs and gumbos.

The cuisine of the Pacific West has been influenced by many ethnic groups—Chinese, Koreans, Vietnamese, Mexicans, Spanish, Japanese, Hawaiians and Native Americans. All of these influences are noted in the ingredients and preparation techniques found in this modern cuisine—stir-fries, olives, sushis, grills, pizzas, avocados, pork, figs, chile peppers, leafy greens, wine, taro, a variety of fruits and seafoods.

Today, Cajun, Creole, Southern, Southwest and other regional season ing blends of North America are evolving into new, popular, trendy flavorings. Overseas, these blends are characterizing North America to the world.

REGIONAL CUISINES OF NORTH AMERICA

Southern

Southern cooking is being revived as an important flavoring style today because many North Americans are going back to the basics. They are looking for foods they grew up with, the so-called comfort foods.

The South combines the home cooking of West Africans and Native Americans and the peasant cooking of southern Europe with influences from English and Spanish explorers. Of all of these various influences on Southern cooking, one of the greatest was from the West Africans who were brought here as slaves. Both before and after the Civil War, African-Americans contributed their unique soul food flavors that remain the heart of southern cooking. Soul food has also influenced other regional cuisines as African-American cooks migrated from the south.

Economics have also influenced Southern cooking. Through wars and depressions and poverty, Southern cooks improvised by using less expensive cuts of meat, such as fatback, to season beans, peas and vegetables. Pigs and chickens were less expensive to raise than beef, and foods were fried because lard was cheap and provided flavor.

Southern foods tend to be heavy and rich and generally not too spicy. Spices such as black pepper, cayenne pepper, onions and vinegar-based sauces are typically used with cream, sugar, lard, salt and butter. Tabasco sauce or other vinegar-based condiments provide heat. Mashed potatoes, macaroni and cheese, chicken-fried steak, roast turkey, corn muffins, biscuits, pickled pigs feet and ears, corn kernels, collard greens, black-eyed peas, relish, succotash and barbecued ribs are many of the typical comfort foods from the South.

There is tremendous variety in traditional southern cuisine. Virginia and Georgia are known for their seafood, ribs, corn dogs, biscuits and gravy, country ham, catfish stew, smothered pork chops and fried turkey. The Carolinas offer seafood, fried chicken, pork and rice. Northern Florida gives us seafood, sweet potatoes, citrus fruits and pecans. In Memphis, we enjoy its barbecues, breaded okra and mashed potatoes. The South also includes Louisiana, which is discussed separately because of its distinct spicy flavors.

Louisiana

Many ethnic groups have influenced the cooking of Louisiana, Arcadian French, Spanish, Native Americans, Haitians, Africans and Italians. Two major styles of Louisiana cooking have emerged: Cajun and Creole.

There are differences between these types of Louisiana cooking. Cajun (from the Native American pronunciation of Arcadian, French settlers from Nova Scotia) food is so-called "unsophisticated" country cooking that is influenced by African, Native American and Caribbean flavors. It is hot, very spicy and uses a dark roux and a lot of animal fat. Cajun food has many one-pot meals, including jambalaya and etouffee. Creole food is so-called "refined" city food (influenced more by Italians, Spanish and French) and uses a light roux with more cream, butter, garlic and tomato sauce.

These cuisines have given us "spiced up" dishes such as gumbos, blackened fish, dirty rice, shrimp Creole, Jezebal sauce, shrimp bisque, red rice and beans and crawfish etouffee. Roux is the basis of traditional Cajun-Creole cooking, whether a base for sauces, gumbos or etouffees. Other important ingredients of these cuisines are garlic, parsley, onions, bell peppers, sassafras, tomatoes, green onions, celery and fish flavorings.

Tex-Mex or Texan

Tex-Mex cooking incorporates many diverse flavors: Native American, Southern, Mexican and European. This diverse mix has given us fajitas, chilis, burritos, nachos, tacos, rice and beans, corn biscuits and many other familiar foods.

Texas is a large state, and its cuisine varies with neighboring influences. There are strong Mexican and Native American influences, especially in south and southwest Texas. Tomatillas, oregano, cumin, jalapenos, corn, black beans, avocados and beef are staple ingredients in this region. Cowboy cooking and the flavors of northern Mexico predominate in the west, resulting in barbecues and fajitas. In the East, there is a strong African influence, with okra, groundnuts, cornmeal and chile peppers. The Gulf Coast region is influenced by Louisiana, with rice, shellfish, oysters, jambalaya, Creole sauce and gumbos.

Southwest

The new Southwest cuisine spearheaded the modern concept of fusion cooking in North America. This cuisine focuses on visual appeal and taste and combines preparation techniques and ingredients from Europeans, Mexicans, Native Americans, and, recently, from Asians.

Native Americans such as Hopi, Pueblo, Navajos, Apaches and Zuni provided the corn, squash, berries, nuts, game and fish; Spanish and Mexicans brought their chile peppers, tomatoes, garlic, beans, chocolate, avocado and onions; while the European missionaries added their wheat, fruits and meats. Many varieties of chilies are used, including anchos, jalapenos, poblanos, chipotles and green New Mexican chilies. Juniper berries, black pepper, garlic, onions, paprika, oregano, Mexican cilantro, cumin, coriander and lime predominate their cooking. Recently, the Chinese and Japanese introduced newer seasonings, including wasabi, ginger, soy sauce and mustard. All of these spices and ingredients are added to stews, sauces, chilis, soups, grilled meats, chicken, fish and vegetables to create a Southwestern flair.

Today, the traditional regional North American cooking is evolving to new levels of visual appeal and creativity. Newer immigrants from Central and South America, the Caribbean, India and Southeast Asia are adding their touches and giving rise to newer versions of the regional cuisines. New, non-traditional regional foods are appearing, such as Floribbean, which combines habaneros, mojos, salsas, yuca, mango and marinated steaks from the Caribbean and Central America with key lime, fried chicken, tabasco pepper sauce and lima beans from the South.

Other regions of the U.S. are also developing new flavors that reflect their current populations. For example, in New York, new fusion foods such as Italian-Thai, Puerto Rican-Chinese or French-Indian will reflect its global diversity of Southeast Asians, Italians, Dominicans, Jamaicans and Asian Indians. Thus, we can expect many traditional American comfort foods to be revived in new regional flavors. Chili, mashed potatoes and Virginia ham will appear with many new twists such as chipotle chili, sofrito mashed potatoes or curried ham.

Below is a description of several of the more popular U.S. regional blends.

POPULAR NORTH AMERICAN REGIONAL SPICE BLENDS

Cajun Blend

Description

Cajun blend is a classic Louisiana spice blend used with barbecued meats, fish and rice and beans. There are many variations, depending upon its end usage—whether to flavor rice, as a dry rub to "blacken" fish and chicken or as a cooking sauce. It is spicy, hot and pungent. With blackened foods, it leaves a slightly bitter aftertaste.

Ingredients

Cajun blend combines sweet paprika, ground celery, garlic, black pepper, onion, cayenne red pepper, caraway, Greek oregano, cumin, dillseed, mace, cardamom, turmeric, basil, marjoram, thyme rosemary and bay leaf.

Applications

To prepare Cajun blend, all of the spices are ground, except for the leafy spices—bay, thyme, marjoram, oregano and basil, which are crushed and added at the end.

Cajun blend is used as a dry rub on chicken, steak or fish to create "blackened" finished products. The leafy spice particulates get burnt in the process, whereby adding the authentic blackened flavor and crunchy texture to the product. Cajun blend gives spicy, aromatic and nutty notes to the finished applications. It is used with red beans, corn and red and green bell peppers in "dirty rice," which is a popular flavored rice in the south.

Creole Sauce Blend

Description

Creole blend is a smooth, spicy, slightly sweet seasoning that is used with Burgundy wine, tomatoes, bell peppers, celery and shrimps. It can be considered the spicy ketchup of Louisiana.

Ingredients

Cajun sauce blend is made from onions, garlic, seafood stock, bay leaf, thyme, parsley, ground black pepper, ground red cayenne pepper and green onions. Other flavorings used are oregano and lemon juice.

Applications

To make Cajun sauce blend, all spices are mixed well, except the leafy spices. Then bay, thyme, parsley and green onions are added and blended gently. The sauce is simmered with chicken, fish or shrimp and topped over white rice. This sauce can be used as a spicy accompaniment to grilled meats and fish or as a rub for barbecues. It can also be used as a dip for chips or fresh cut vegetables.

Green Chile Blend

Description

Green chile blend is a Southwestern-style seasoning with a mix of European, Mexican, Native American and, more recently, Asian spices.

Ingredients

The typical chilies used are anchos, poblanos, chipotles and green New Mexican chilies. They are mixed with juniper berries, black pepper, garlic, onions, paprika, oregano, Mexican cilantro, cumin, coriander and lime flavor. Wasabi, ginger and mustard can also be added.

Applications

All of the spices are ground or pureed to make a smooth or coarsely textured sauce. Green chile blends enhance stews, chilis, soups, grilled meats, chicken, fish or vegetables that require a Southwestern flair.

Chili Powder

Description

Chili powder was created by the poor Mexicans of Texas who added some spices and chile peppers to meat to make a hearty stew. Mexican cooks served this to the local cowboys who called this stew, chili con carne because of the chile peppers that predominated the flavor. The seasoning blend that went into this stew finally came to be called chili powder.

Today, chili powder is a mildly pungent seasoning that is used commonly in chilis, Tex-Mex foods, Texan barbecues and chili con carnes. Mexicans use this spice blend for burritos enchiladas and tamale fillings. It has a red to burgundy color and a deep, rich flavor with very little heat. There are medium to hot chili powders for those who desire varying heat levels.

Ingredients

Chili powder is a blend of ground chile pepper with ground cumin, garlic, Mexican oregano and salt. The ground chile pepper used is usually ancho chile pepper, which is a mild chile pepper. About 80% of the blend is ground chile pepper. The hotter versions have the ground red cayenne pepper.

Applications

To prepare chili powder, all of the ground ingredients are well mixed. This chili blend is then mixed with lime juice and water and used as a rub to prepare poultry, beef or pork for grilling, broiling or barbecuing. For chili con carne, chili powder is added to tomato, onion, coriander, red and green bell peppers, black pepper, allspice, cloves and cilantro. Then, this mixture is added to ground beef and cooked to a thick consistency. Red kidney beans, lamb or mutton can be substituted for beef. Corn and other beans such as pinta, black and white can be used to make a vegetarian chili with chili powder.

Chili blend can also be applied to curries, sauces and condiments to create varying notes, to add zest or to enhance existing flavors.

Barbecue Sauce Blends

Description

Barbecue is meat, poultry and seafood, whole or pieces, that are slowly cooked in a pit or on a spit using hot coals or flavored wood chips or hardwood. They are typically marinated or covered with barbecue sauces that are dark reddish brown with spicy, smoky, sweet and/or mildly tart flavors. There are many types of North American barbecue sauces. Texas, South Carolina, Memphis and other regions all have their own distinctive, traditional-style barbecue sauces. Newer versions and concepts of barbecue sauces are emerging that incorporate many Latin, Asian and Caribbean concepts, such as tandoori, hoisin or chipotle flavors.

Ingredients

The basic barbecue sauce is made from tomatoes, mustard, onions, vinegar, garlic and brown sugar with smoky or grilled notes. Beer, wine, molasses and other spices are optional ingredients. Texas sauces are thick, spicy and hot and consist of paprika, chili powder, black pepper, dark brown sugar and vinegar, Worcestershire sauce, ketchup, dry mustard and jalapenos. The Carolina sauce is tart and thin with cider vinegar, ketchup, crushed red pepper and brown sugar. The Memphis sauce is spicy and sweet, with paprika, garlic, brown sugar, onion and black pepper. Optional ingredients are olive oil, bell peppers, honey, corn syrup, orange juice, chipotle, serranos, ancho, thyme, cilantro soy sauce, tomato juice and lemongrass.

Applications

Chicken, ribs, beef, pork or seafood is basted with these sauces and slowly barbecued over charcoal, hickory wood smoke chips or other flavored chips until the finished product develops a charred or dark brown color and an aromatic flavor.

GLOBAL SPICE BLENDS

Seasoning or spice blends are generally unique to particular regions, yet there are some spice blends that are found throughout the world. These have a common theme, but contain different spices because of their availability and regional and cultural preferences. Two primary examples are the green spice blends and chile blends. Every country has its own green blend and chile blend that are used as seasonings, whether in soups, sauces, stews, rices, pastas, salads or as marinades.

POPULAR REGIONAL ETHNIC GREEN BLENDS

Green Blends

Description

Green blends are savory mixtures of green leafy spices and or green chile peppers, with other spices and ingredients. To North Americans, Italian pesto is the most familiar green blend, but there are other green blends, such as zhug, salsa verde, achari or chat masala. Almost every ethnic cuisine has its own version of a green blend, with varying flavors and textures based upon the ingredients, preparation styles, applications and how it is presented in a dish. A green blend's color can range from a bright to dull green, depending upon the type and amount of green ingredients and spices and whether it is served fresh or cooked.

Ingredients

Essential to any green blend are the ingredients that provide the characteristic color—green leafy spices or green chile peppers. Other ingredients such as nuts, garlic, ginger, tomatoes or lemon are added to give the blend added character or flavor. The most commonly used green leafy spices are basil, cilantro, parsley and mint. They are popular because their flavors combine well with a wide variety of ingredients—onions, cucumbers, coconut milk, green mango, lemongrass, coriander and cheeses. Some green leafy spices, such as chives and parsley, provide the greenness but with little flavor, while basil, cilantro, sage or mint provide the green blend's dominant flavor. Vegetables such as cucumber, celery or bell peppers are sometimes chopped and added for a crunchy texture. Some of the well-known global green spice blends and their typical ingredients are listed in Table 36.

TABLE 36. Global Green Spice Blends.

Green Blend	Ingredients
Thai green curry paste	Thai basil, lemongrass, green cayenne chilies, cilantro, shallot, galangal, coriander seed, black pepper, shrimp paste, coconut milk, kaffir, lime leaf
Italian pesto	Sweet basil, pine nut, parmesan and romano cheese, garlic, olive oil
Mexican salsa verde	Cilantro, green tomato, green jalapeno, onion, green bell pepper
Puerto Rican recaito	Cilantro, culantro, black pepper, garlic
Middle Eastern tabbouleh-style dressing	Parsley, mint, bulgar wheat, lemon juice, tomato
Malaysian kurma	Cilantro, spearmint, green chile, coriander seed, star anise, cardamom, coconut milk
Indian green masala	Green cayenne chile, cilantro, coriander seed, ginger, dried mango
Argentinean chimmichurri	Parsley, cilantro, garlic, lime juice, black pepper
Kenyan achari	Green cayenne chile pepper, unripe mango, cilantro and shredded coconut.

Applications

Green blends can be added as a major seasoning in a sauce or curry or to provide zest to a stew, salad or soup. They are added fresh or cooked to flavor salsas, chutneys, mustards, dips or salad dressings. Commercially, green blends come in dry or paste forms. They can be smooth, coarsely textured or in chopped forms, depending on the end usage. Whether one desires a green blend's domineering notes or its enhancing profile, the result is a wonderful balance of flavors in the finished product.

Each ethnic green blend has its own favorite leafy spice that gives its distinctive flavor profile. For example, sweet basil characterizes Italian pesto, while the more pungent Thai basil is favored in Thai green curry, and cilantro is preferred in Mexican green salsas. Green cayenne, jalapeno, serrano, Cubanelle or New Mexican chile peppers provide color, as well as heat, flavor and texture to the blend. Other spices or flavorings such as lemongrass, culantro, sorrel, lime, avocado or curry leaf are added in smaller amounts to enhance its characteristic ethnic profile.

With the increasing availability of new ethnic ingredients and the growing demand for variety and stronger flavors, adventurous consumers will seek other "new" pestos. Concepts for new pestos can be derived from the other green blends around the world. Lemongrass, sweet basil, cilantro, pine nuts, parmesan cheese and green chilies can be combined to create a new version of the traditional Italian pesto, such as a spicier Thai-Italian pesto or Mexican-Italian pesto.

Chile Blends

Description

First cultivated in the Andes of South America, the chile pepper has traveled globally to become the dominant component of many spice blends. Chile blends now provide flavor and color to foods in countries in all corners of the world—India, Sri Lanka, Southeast Asia, North Africa, the Caribbean and Latin America. Each of these regions has its own favorite chile peppers—Thais and Malaysians prefer cayennes and bird peppers; Caribbeans enjoy habaneros; Mexicans choose poblanos, jalapenos and serranos; and Peruvians like ajis, rocotos and pequins.

Chile blends are prized not only for their heat but also for their delicate flavors. Chile blends incorporate one or more types of chile peppers, as well as other ingredients, thus creating many different tastes. Chile blends can be a simple mixture of chilies, salt, vinegar and sugar, such as cili boh or sambal olek, or a complex mixture with other added ingredients.

Chile peppers can be added fresh or dried, pureed, chopped or ground to create spice blends. The particular form used depends on the end product and how it is used. Chile peppers can be roasted to modify their flavors and used seeded or deseeded to modify their heat.

Ingredients

Many ingredients are mixed with chile peppers to create unique flavor releases, to tone down their heat or even to add color to the finished product. Garlic, shallots, lemongrass, coriander, cilantro, tomato, green mango, tamarind, brown sugar and lime juice lend well to chile blends. Other spices, onions, nuts, seeds, chocolate, vegetables, fruits, fermented shrimp or fish, and coconut can also be used. The choices depend on the particular application, ethnic preferences and availability. Chile blends lend well to pickling or preserving in brine or vinegar and with other seasonings and, if desired, fruits or vegetables. Chile blends come in dry, liquid and paste forms.

Some global chile blends and their typical ingredients are shown in Table 37.

Applications

The number of applications that use chile blends is almost endless, and they are not limited to foods. Every ethnic group has its own chile blends that contain locally available and typically preferred ingredients—Mexicans with their

TABLE 37. Popular Global Chile Blends.

Chile Blend	Ingredients
Tunisian harissa	Cayenne/New Mexican, garlic, cumin, cinnamon, coriander seed, caraway, olive oil
Malaysian sambal belacan	Cayenne, garlic, shallot, shrimp paste, lime juice
Bajan hot sauce	Habanero, mustard, papaya, sugar, chives, black pepper, vinegar
Kenyan bajia	Cayenne, tomato, coconut milk and fermented liquor of coconut
Arentinian chimichurri	Jalapeno, onion, parsley, oregano, black pepper, lemon juice, vinegar
South African blatjang	Cayenne/bird pepper, onion, almond paste, dried apricot, apricot puree, garlic, bay leaf, lime juice
North American hot sauce/ tabasco sauce	Cayenne/Tabasco pepper, vinegar, salt

moles and ceviches, Asian Indians with their curries and achars, Thais with their salads and kaengs, Malaysians with their sambals or cili boh, Mediterraneans with their rouille and harissa, Caribbeans with jerk blends or Indipep sauce and Africans with their piri-piri blatjang and berbere.

Even certain ethnic drinks, such as atole, tepache or chilote contain chile blends. Atole, which is a corn-, chile- and chocolate-based drink enjoyed by the Aztecs, continues to be a favorite Mexican breakfast item. Whether to provide heat, flavor or simply to perk up a dish, chile blends can be added during cooking to provide zest to sauces, soups or stews, used as marinades or served as a dip or freshly made condiment/relish to a meal.

Chile blends go well with grilled, steamed or roasted meats, chicken or fish. Their flavors are complemented with certain spices and flavorings. They are given a creamy sweet mouthfeel with coconut milk; sweet and sour notes with tomatoes and palm sugar; sharp and sour tastes with lime juice and tamarind; rich roasted brown and bitter notes with coffee or chocolate; and savory fermented flavors with shrimp paste and spices.

Certain profiles are added to chile blends to appeal specifically to different ethnic groups: a tartness for North Americans (Louisiana hot sauce); a fishy, pungent note for the Vietnamese (nuoc cham); a sweet and sour note for Chinese; a sweet garlicky note for the Thais (Sriracha sauce); a pungent and slightly bitter taste for the Oaxacans (moles); a sour and spicy green note for the Indians (achars); and a fruity, mustardy note for the Caribbeans.

Bibliography

Achaya, K. T., 1994, *Indian Food: A Historical Companion*. Oxford University Press, Delhi, New York.

Andrews, Jean S., 1984, *Peppers, Domesticated Capsicums*. University of Texas Press, Austin, TX.

Ashurst, P. R., 1995, *Food Flavorings*, 2nd edition. Blackie Academic and Professional, an imprint of Chapman and Hall, New York.

ASTA Cleanliness Specifications for Spices Seeds and Herbs, Revised April 28, 1999, ASTA, Inc., Engelwood Cliffs, NJ.

Bremness, Lesley, 1991, *The Complete Book of Herbs*. Color Library Books Library Ltd., England.

Burdock, George A., 1994, *Fenarolis Handbook of Flavor Ingredients*, 3rd edition. CRC Press Inc., FL.

Chopra, Deepak, 1991, *Perfect Health*. Harmony Books, a div. of Crown Publishers, Inc., New York.

Corn, Charles, 1998, *The Scents of Eden—A Narrative of the Spice Trade*. Kodansha International, Japan.

Czarnecki, Jack, 1995, *A Cook's Book of Mushrooms*. Artisan, a div. of Workman Publishing Company, Inc., New York.

Desai, Urmila, 1995, *The Ayurvedic Cookbook*. Lotus Press, Twin Lakes, WI.

DeWitt, Dave and Gerlach, Nancy, 1990, *The Whole Chile Pepper Book*. Little Brown & Company (Canada).

DeWitt, Dave, 1999, *The Chile Pepper Encyclopedia*. William Morrow & Company, Inc.

Disney, A. R., 1978, *Twilight of the Pepper Empire*. Harvard University Press, Cambridge, MA, London, England.

Farrell, Kenneth T., 1985, *Spices, Condiments and Seasonings*. The Avi Publishing Company, Inc., U.S.

Hadady, Letha, 1996, *Asian Health Secrets*. Crown Publishers Inc., New York.

Heinerman, John, 1994, *Encyclopedia of Healing Juices*. Prentice Hall, NJ.

Johari, Harish, 1994, *The Healing Cuisine*. Healing Arts Press, Rochester, VT.

Katzer, Gernot, 1999, *Spice Pages.*
www-ang.kfunigrar.Ac.At./~katzer/engl/index.html.

Kenji, Hirasa and Mitsuo, Takemasa, 1998, *Spice Science and Technology,* pp. 163–198. Marcel Dekker Inc., New York, Basel, Hong Kong.

Khan, M. T., *Spices in Indian Economy*. Academic Foundation, Delhi, India.

Lawless, Julia, 1995, *The Illustrated Encyclopedia of Essential Oils*. Barnes and Noble, Inc., London.

Lust, John B., Oct. 1974, *The Herb Book,* 2nd printing. Benedict Lust Publications, Simi Valley, CA.

Mahindru, S. N., 1982, *Spices in Indian Life,* First edition. Sultan Chand & Sons (Pvt) Ltd., New Delhi, India.

Mazza, G. and Oomah, B. D., eds. 2000. *Herbs, Botanicals and Teas.* Technomic Publishing Company, Inc., Lancaster, PA.

Mindell, Earl, 1994, *Food as Medicine.* Simon & Schuster, New York.

Naj, Amal, 1993, *Peppers—A Story of Hot Pursuits.* Vintage Books, a div. of Random House, Inc., New York.

Norman, Jill, 1991, *The Complete Book of Spices.* Viking Studio Books, a div. of Penguin Books, New York.

Onstad, Dianne, 1996, *Whole Foods Companion.* Chelsea Green Publishing Company, VT.

Ortiz, Elisabeth L., 1994, *Encyclopedia of Herbs, Spices and Flavorings.* Carroll and Brown Ltd., London.

Parry, John W., 1953, *Story of Spices.* Chemical Publishing Co., Inc., New York.

Pennington, Jean A. T., 1998, *Food Values of Portions Commonly Used.* Lippincott-Raven Publishers, Philadelphia, PA.

Pruthi, J. S., 1976. *Spices and Condiments.* National Book Trust, New Delhi, India.

Purseglove, J. W., Brown, E. G., Green, C. L. and Robbins, S. R. J., 1981, *Spices, Volume 1.* Longman Group UK Ltd., Harlow, England.

Purseglove, J. W., Brown, E. G., Green, C. L. and Robbins, S. R. J., 1987–88, *Spices, Volume 2.* Longman Group UK Ltd., Harlow, England.

Reineccius, Gary, 1994. *Source Book of Flavors,* pp. 186–199, 255-360. Chapman & Hall, New York, London.

Ridley, Henry N., 1912, *Spices.* Macmillan and Co., London.

Rosengarten Jr., Frederic, 1979, *The Book of Spices.* Pyramid Communications, Inc., New York.

Schivelbusch, Wolfgang, 1993, *Taste of Paradise,* pp. 3–14. Vintage Books, a div. of Random House, Inc., New York.

Stobart, Tom, 1982. *Herbs, Spices and Flavorings.* Milton Peter Corp. in association with The Overlook Press, Woodstock, NY.

Stuckey, Maggie, 1997, *The Complete Spice Book.* St. Martin's Press, New York.

Tiwari, Maya Ayurveda, 1995, *A Life of Balance.* Healing Arts Press, VT.

Toussaint-Samat, Maguelonne, Reprinted 1993, *A History of Foods.* Blackwell Publishers, Cambridge, MA.

Uhl, Susheela R., 1995, Authentic Ethnic Foods. Food Product Design (*FPD*), Weeks Publishing Inc., Chicago, IL.

Uhl, Susheela R., 1996, Designing for Hispanic Marketplace. *FPD,* Weeks Publishing Inc., Chicago, IL.

Uhl, Susheela R., 1996, *Ingredients—The Building Blocks of Ethnic Foods, Food Technology*. IFT, Chicago, IL.

Uhl, Susheela R., 1996, Legumes. *FPD,* Weeks Publishing Inc., Chicago, IL.

Uhl, Susheela R., 1997, Ethnic Side Dishes. *FPD,* Weeks Publishing Inc., Chicago, IL.

Uhl, Susheela R., 1997, Spices & Spice Extractives. *FPD,* Weeks Publishing Inc., Chicago, IL.

Uhl, Susheela R., 1998, 4 Steps to Create Authentic Ethnic Foods. Special Edition, *FPD,* Weeks Publishing Inc., Chicago, IL.

Uhl, Susheela R., 1998, Asian Ingredients in American Cooking. *Food Forum* (IACP publication), Louisville, KY.

Uhl, Susheela R., 1998, Chile Blends. World Fare, *FPD,* Weeks Publishing Inc., Chicago, IL.

Uhl, Susheela R., 1998, Cilantro—Flavorings, Fine Cooking. The Taunton Press, Newton, CT.

Uhl, Susheela R., 1998. Developing New and Ethnic Foods. *FPD,* Weeks Publishing Inc., Chicago, IL.

Uhl, Susheela R., 1998, Ethnic Entrees. *FPD,* Weeks Publishing Inc., Chicago, IL.

Uhl, Susheela R., 1998, Green Blends. World Fare, *FPD,* Weeks Publishing Inc., Chicago, IL.

Uhl, Susheela R., 1998, Hot & Spicy Foods. *FPD,* Weeks Publishing Inc., Chicago, IL.

Uhl, Susheela R., 1998, Mole Blends. World Fare, *FPD,* Weeks Publishing Inc., Chicago, IL.

Uhl, Susheela R., 1998, New Ethnic Entrees. *FPD,* Weeks Publishing Inc., Chicago, IL.

Uhl, Susheela R., 1998, Putting the Home Back in Meal Solutions. *Refrigerated and Frozen Foods,* Stagnito Communications Inc., div. of Medical Communications Inc.

Uhl, Susheela R., 1999, Creatively Carribean. *FPD,* Weeks Publishing Inc., Chicago, IL.

Uhl, Susheela R., 1999, Exploring Asian Ethnic. *FPD,* Weeks Publishing Inc., Chicago, IL.

Uhl, Susheela R., 1999, Mediterranean Medley. *FPD,* Weeks Publishing Inc., Chicago, IL.

Uhl, Susheela R., 1999, How to Spice Up HMR. *Refrigerated and Frozen Foods,* Stagnito Communications Inc., div. of Medical World Communications, Inc.

Uhl, Susheela, R., 2000, Ethnic Extravaganza. *Refrigerated and Frozen Foods,* Stagnito Communications Inc., div. of Medical World Communications, Inc.

Uhl, Susheela, R., 2000, Spice Translator. www.SusheelaConsulting.com.

COOKBOOKS

AFRICAN

Cusick, Heidi Haughy, 1995, *Soul and Spice*. Chronicle Books, San Francisco, CA.

Hachten, Harva, 1998, *Best of Regional African Cooking*. Hippocrene Books Inc., New York.

Hafner, Dorinda, 1993, *A Taste of Africa*. Ten Speed Press, Berkley, CA.

Harris, Jessica B., 1998, *The Africa Cookbook—Taste of a Continent*. Simon & Schuster, New York.

ASIAN

Alejandro, Reynaldo G., 1998, *The Food of Philippines*. Periplus Edns (HK) Ltd.

Bhumichitr, Vatcharin, 1994, *Vatch's Thai Cookbook*. Pavilion Books, Ltd., Great Britain.

Brennen, Jennifer, 1984, *Encyclopedia of Chinese and Oriental Cooking*. St. Martins Press, New York.

Choi, Trieu Thi and Isaak, Marcel, 1997, *The Food of Vietnam*. Periplus Edns (HK) Ltd.

DeWitt, Dave and Pais, Arthur 1994, *A World Of Curries*. Little, Brown & Company, Boston, MA.

Holzen, Heinz von and Arsana, Lother, 1995, *The Food of Indonesia*. Periplus Edns (HK) Ltd.

Kosaki, Takawaki and Wagner, Walter, 1996, *The Food of Japan*. Periplus Edns (HK) Ltd.

Krauss, Sven, Ganguillet, Laurent and Sanguanwong, 1996, *The Food of Thailand*. Periplus Edns (HK) Ltd.

Madhur, Jaffrey, 1995, *Flavors of India*. BBC Books, London.

Millon, Marc and Kim, 1991, *Flavors of Korea*. Andre Deutsch Limited, London, Great Britain.

Pham, Mai, 1996, *The Best of Vietnamese and Thai Cooking*. Prima Publishing, Roclin, CA.

Sahni, J., 1996, *Savoring Spices and Herbs*. William Morrow and Company, Inc., New York.

Sri, Owen, 1994, *Indonesian Regional Cooking*. Doubleday Publishing, Great Britain.

Tettoni, Luca Invernizzi and Hutton, Wendy, 1995, *The Food of Malaysia*. Perplus Edns (HK) Ltd.

Tettoni, Luca Invernizzi and Hutton, Wendy, 1996, *The Food of China*. Periplus Edns (HK) Ltd.

Yan, Martin, 1995, *Culinary Journey through China*. Publishers Group West, San Francisco, CA.

CARIBBEAN

Dedeaux, Devra, 1991, *The Sugar Reef Caribbean Cookbook*. Dell Publishing, a div. of Bantam Doubleday Dell Publishing Inc., New York.

Hafner, Dorinda, 1996, *Taste of the Caribbean*. Ten Speed Press, Berkley, CA.

EUROPEAN

Child, Julia, 1998, *The French Chef Cookbook*. Ballantine Publishing Group, New York.

Ducasse, Alain, 1998, *The Flavors of France*. Artisan, a div. of Workman Publishing Inc., New York.

Root, Waverly, 1992, *The Food of France*. Vintage Books, a div. of Random House Inc., New York.

LATIN AMERICAN

Botafogo, Dolores, 1993, *The Art of Brazilian Cooking*. Hippocrene Books Inc., New York.

Green, Linette, 1994, *A Taste of Cuba*. Penguin Books USA, New York.

Karoff, Barbara, 1990, *South American Cooking*. Aris Books, Editorial Offices and Test Kitchen, Berkley, CA.

Ortiz, Elisabeth Lambert, 1994, *The Book of Latin American Cooking*. The Ecco Press, Hopewell, NJ.

Ortiz, Yvonne, 1997, *A Taste Of Puerto Rico*. Penguin Books USA Inc., New York.

Quintana, Patricia, 1993, *The Taste of Mexico*. Stewart, Tabori and Chang, New York.

Romano, Dora, 1993, *Rice and Beans and Tasty Things—A Puerto Rican Cookbook*. Ramallo Bros. Printing, Inc., Puerto Rico.

Zarela, Martínez, 1992, *Foods from My Heart*. MacMillan Publishing Company, New York.

Zaslavsky, Nancy, 1995, *A Cooks Tour of Mexico*. St. Martin's Press, New York.

MEDITERRANEAN

Chapman, Pat, 1989, *Homestyle Middle Eastern Cooking*. The Crossing Inc., Freedom, CA.

Hazan, Giuliano, 1993, *The Classic Pasta Cookbook*. Dorling Kindersley, Inc., New York.

Hazan, Marcella, 1984, *More Classic Italian Cooking*. Ballantine Books, a div. of Random House, New York.

Hazan, Marcella, 1997, *Marcella Cucina*. Harper Collins Publisher's Inc., New York.

Hess, Reinhardt and Sabine, Salzer, 1999, *Regional Italian Cuisine*. Barron's Educational Series Inc., Hauppauge, NY.

Roden, Claudia, 1974, *A Book of Middle-Eastern Food*. Random House Inc., New York.

Roden, Claudia, 1987, *Mediterranean Cookbook*. Alfred A. Knopf Inc., New York.

Root, Waverly, 1992, *The Food of Italy*. Vintage Books, a div. of Random House Inc., New York.

Wolfert, Paula, 1994, *Mediterranean Cooking*. Harper Collins Publishers Inc., New York.

NORTH AMERICAN

Pyles, Stephan, 1993, *The New Texas Cuisine*. Doubleday, a div. of Bantam Doubleday Dell Publishing Group Inc., New York.

Raichlen, Steven, 1993, *Miami Spice—The New Florida Cuisine*. Workman Publishing Company Inc., New York.

Stromquist, Joan, 1993, *Sante Fe—Hot and Spicy Recipe*. Tierra Publications, Sante Fe, NM.

Index